Performance Evaluation and Benchmarking of Intelligent Systems

R. Madhavan · E. Tunstel · E. Messina
Editors

Performance Evaluation and Benchmarking of Intelligent Systems

 Springer

Editors
Raj Madhavan
Computational Sciences
  and Engineering Division
Oak Ridge National Laboratory
1 Bethel Valley Road
Oak Ridge
TN 37831
USA
and
Intelligent Systems Division
National Institute of Standards
  and Technology (NIST)
Gaithersburg
MD 20899
USA
raj.madhavan@ieee.org

Elena Messina
National Institute of Standards &
  Technology (NIST)
Intelligent Systems Division
100 Bureau Drive
Gaithersburg MD 20899
Mail Stop 8230
USA
elena.messina@nist.gov

Edward Tunstel
Johns Hopkins University
Applied Physics Laboratory
11100 John Hopkins Road
Laurel MD 20723
USA
edward.tunstel@jhuapl.edu

ISBN 978-1-4899-8300-8   ISBN 978-1-4419-0492-8 (eBook)
DOI 10.1007/978-1-4419-0492-8
Springer New York Dordrecht Heidelberg London

© Springer Science+Business Media, LLC 2009
Softcover re-print of the Hardcover 1st edition 2009

Printed on acid-free paper

Springer is part of Springer Science+Business Media (www.springer.com)

To Priti, Priyanka and Kalpana – RM
To my wife and children whose well-being is my
performance benchmark – ET
In loving memory of my parents, Margherita and
Nicola – EM

Preface

To design and develop capable, dependable, and affordable intelligent systems, their performance must be measurable. Scientific methodologies for standardization and benchmarking are crucial for quantitatively evaluating the performance of emerging robotic and intelligent systems' technologies. There is currently no accepted standard for quantitatively measuring the performance of these systems against user-defined requirements; and furthermore, there is no consensus on what objective evaluation procedures need to be followed to understand the performance of these systems. The lack of reproducible and repeatable test methods has precluded researchers working towards a common goal from exchanging and communicating results, inter-comparing system performance, and leveraging previous work that could otherwise avoid duplication and expedite technology transfer. Currently, this lack of cohesion in the community hinders progress in many domains, such as manufacturing, service, healthcare, and security. By providing the research community with access to standardized tools, reference data sets, and open source libraries of solutions, researchers and consumers will be able to evaluate the cost and benefits associated with intelligent systems and associated technologies. In this vein, the edited book volume addresses performance evaluation and metrics for intelligent systems, in general, while emphasizing the need and solutions for standardized methods.

To the knowledge of the editors, there is not a single book on the market that is solely dedicated to the subject of performance evaluation and benchmarking of intelligent systems. Even books that address this topic do so only marginally or are out of date. The research work presented in this volume fills this void by drawing from the experiences and insights of experts gained both through theoretical development and practical implementation of intelligent systems in a variety of diverse application domains. The book presents a detailed and coherent picture of state-of-the-art, recent developments, and further research areas in intelligent systems.

This edited book volume is a collection of expanded and revised papers presented at the 2008 Performance Metrics for Intelligent Systems (PerMIS'08: *http://www.isd.mel.nist.gov/PerMIS_2008/*) workshop held at the National Institute of Standards and Technology (NIST) from August 19–21, 2008. PerMIS is the only workshop of its kind dedicated to defining measures and methodologies of evaluating performance of intelligent systems. The Intelligent Systems Division of NIST, under the leadership of Dr. John Evans as Division Chief, initiated this series in 2000 to address the lack of measurements and basic understanding of the performance of intelligent systems. Prof. Alexander Meystel of Drexel University who had a long-standing collaboration with NIST's Intelligent Systems Division was a prime mover behind PerMIS from the very beginning. Dr. Meystel was instrumental in shaping the PerMIS workshops and their content for several years, with an emphasis on the theoretical foundations of the field of intelligent systems. Dr. James Albus, Senior NIST Fellow, provided valuable guidance throughout the workshop series, through his deep insights into intelligence. Over the years, the workshops have increased their focus on applications of performance measures to practical problems in commercial, industrial, homeland security, military, and space applications, while still retaining elements of theoretical examination. It has proved to be an excellent forum for discussions and partnerships, dissemination of ideas, and future collaborations between researchers, graduate students, and practitioners from industry, academia, and government agencies. Financial sponsorship has been primarily by NIST and Defense Advanced Research Projects Agency (DARPA). Additional support throughout the years in logistical terms has come, at various times, from the IEEE (originally known as the Institute of Electrical and Electronic Engineers, Inc.), IEEE Control Systems Society, IEEE Neural Net Council (which became the IEEE Computational Intelligence Society), IEEE Systems, Man, and Cybernetics Society, IEEE Robotics and Automation Society, the National Aeronautics and Space Administration (NASA), and the Association for Computing Machinery (ACM).

In the years since its inception in 2000, the PerMIS workshop series has brought to light several key elements that are necessary for the development of the science and engineering of performance measurement for intelligent systems. The endeavor of measuring the performance of intelligent systems requires development of a framework that spans the theoretical foundations for performance measures to the pragmatic support for applied measurements. The framework would serve to guide development of specific performance metrics and benchmarks for individual domains and projects. The chapters in this book provide a broad overview of several of the different, yet necessary perspectives that are needed to attain a discipline for performance measurement of intelligent systems. If the field of intelligent systems is to be a true engineering or scientific endeavor, it cannot exist without quantitative measurements.

There exists a need to balance the desire for overarching theories and generality of measures and the pragmatic specificity required for applied evaluations. Similarly, there is a need for the measurement of performance of components and subsystems and for the overall integrated system. The papers that were selected from PerMIS'08 illustrate many of the dimensions of performance evaluation. Biologically-inspired measures are an example of foundational, overarching principles for the discipline. Areas covered in these selected papers include a broad range of applications, such as assistive robotics, planetary surveying, urban search and rescue, and line tracking for automotive assembly. Subsystems or components described in this book include human-robot interaction, multi-robot coordination, communications, perception, and mapping. In this emerging and challenging field of performance evaluation, supporting tools are essential to making progress. Chapters devoted to simulation support and open source software for cognitive platforms provide examples of the type of enabling underlying technologies that can help intelligent systems to propagate and increase in capabilities.

The edited book volume is primarily intended to be a collection of chapters written by experts in the field of intelligent systems. We envisage the book to serve as a professional reference for researchers and practitioners in the field and also for advanced courses for graduate level students and robotics professionals in a wide range of engineering and related disciplines including computer science, automotive, healthcare, manufacturing, and service robotics. The book is organized into 13 chapters. As noted earlier, these chapters are significantly expanded and revised versions of papers presented at PerMIS'08. Out of 58 papers presented at the 2008 workshop, these papers were selected based on feedback and input from the reviewers during the workshop paper acceptance process. A summary of the chapters follows:

> Multiagent systems (MAS) are becoming increasingly popular as a paradigm for constructing large-scale intelligent distributed systems. In "Metrics for Multiagent Systems", Robert Lass, Evan Sultanik, and William Regli provide an overview of MAS and of currently-used metrics for their evaluation. Two main classes of metrics are defined: Measures of Effectiveness (MoE), which quantify the system's ability to complete a task within a specific environment, and Measures of Performance (MoP), which measure quantitatively performance characteristics, such as resource usage, time to complete a task, or other relevant quantities. Metrics can be further classified by type of mathematical scale, namely, whether these quantities are defined as nominal, ordinal, intervals, or ratios. The authors classify a set of existing agent metrics according to whether they are MoE, MoP, type of mathematical formulation, their community of origin, and which layer they apply to within a reference model for agent systems. Using the defined metrics, the authors present a framework for determining how to select metrics for a new project, how to collect them, and ultimately analyze a system based on the chosen metrics and demonstrate this process through a case study of distributed constraint optimization algorithms.
>
> A list of evaluation criteria is developed in the chapter by Birsen Donmez, Patricia Pena and Mary Cummings entitled "Evaluation Criteria for Human-Automation Performance Metrics" to assess the quality of human-automation performance to facilitate the selection of metrics for designing evaluation experiments in many domains such as human-robot interaction and medicine. The evaluation criteria for assessing the quality of a metric are based on five categories: experimental constraints, comprehensive understanding, construct validity, statistical efficiency, and measurement technique efficiency. Combining the criteria with a list of resulting metric costs and benefits, the authors provide substantiating examples for evaluating different measures.

Several areas of assistive robotic technologies are surveyed in the chapter
"Performance Evaluation Methods for Assistive Robotic Technology" by
Katherine Tsui, David Feil-Seifer, Maja Matarić, and Holly Yanco to derive
and demonstrate domain-specific means for evaluating the performance of
such systems. Two cases studies are included to detail the development of
performance metrics for an assistive robotic arm and for socially assistive
robots. The authors provide guidelines on how to select performance measures for end-user evaluations of such assistive systems. It is their conclusion
that end-user evaluations should focus equally on human performance measures as on system performance measures.

Given that biological systems set the standard for intelligent systems, it is useful to examine how bio-inspired approaches can actually assist researchers
in understanding which aspects derived from biology can enable desirable
functionality, such as cognition. Gary Berg-Cross and Alexei Samsonovich
discuss several schools of thought as applied to intelligent systems and how
these approaches can aid in understanding what elemental properties enable
adaptability, agility, and cognitive growth. High-level, "Theory-of-Mind"
approaches are described, wherein a pre-existing cognitive framework exists.
But what are the essential elements of cognition (what the authors term
"critical mass")? "Cognitive decathlons" are appealing methods of trying to
elucidate exactly what this critical mass is. An alternative view to the high-
level one is that of developmental robotics, which posits that cognitive processes emerge through learning and self-organization of many interacting
sub-systems within an embodied entity. Developmental robotics offers the
potential advantage of side-stepping the rigid engineering that may be present
in top-down approaches. The authors conclude that there are yet pitfalls in
the developmental approaches, including biases in the systems' design. They
point the way forward by suggesting that developmental approaches should
be guided by principles of how intelligence develops, but should not slavishly
follow bio-inspiration on all fronts.

In the chapter "Evaluating Situation Awareness of Autonomous Systems", Jan
Gehrke argues that higher-level situation analysis and assessment are crucial
for autonomous systems. The chapter provides a survey of situation awareness for autonomous systems by analyzing features and limitations of existing approaches and proposes a set of criteria to be satisfied by situation-aware
agents. An included example ties together these ideas by providing initial
results for evaluating such situation-aware systems.

The role of simulation in predicting robot performance is addressed by Stephen
Balakirsky, Stefano Carpin, George Dimitoglou, and Benjamin Balaguer
in the chapter "From Simulation to Real Robots with Predictable Results:
Methods and Examples". The authors reinforce the theoretical argument
that algorithm development in simulation accelerates the motion planning
development cycle for real robots. With a focus on model deficiencies as
sources of simulation system brittleness, the chapter presents a methodology for simulation-based development of algorithms to be implemented on

real robots including comparison of the simulated and physical robot performance. Examples using simulated robot sensor and mobility models are used to demonstrate the approach and associated techniques for validating simulation models and algorithms.

The chapter "Cognitive Systems Platforms using Open Source" by Patrick Courtney, Olivier Michel, Angelo Cangelosi, Vadim Tikhanoff, Giorgio Metta, Lorenzo Natale, Francesco Nori and Serge Kernbach reports on various open source platforms that are being developed under the European Union Cognitive Systems program. In particular, significant research efforts in cognitive robotics with respect to benchmarking are described in detail.

The chapter "Assessing Coordination Demand in Coordinating Robots" by Michael Lewis and Jijun Wang presents a means to characterize performance of the dual task of coordinating and operating multiple robots using a difficulty measure referred to as coordination demand. Several approaches are presented for measuring and evaluating coordination demand in applications where an operator coordinates multiple robots to perform dependent tasks. Simulation experiments involving tasks related to construction and search and rescue reveal the utility of the measure for identifying abnormal control behaviors and facilitating operator performance diagnosis as well as aspects of multi-robot tasks that might benefit from automation.

Intelligent systems such as mobile robots have to operate in demanding and complex environments. Receiving transmissions from human operators, other devices or robots and sending images or other data back are essential capabilities required by many robots that operate remotely. In their chapter "Measurements to Support Performance Evaluation of Wireless Communications in Tunnels for Urban Search and Rescue Robots", Kate Remley, George Hough, Galen Koepke, and Dennis Camell describe the complexities that confront the wireless communications systems for a robot that is used to explore unknown and difficult environments. They define the types of measures that can be taken to characterize the environment in which a robot must operate. The data can be used to help define reproducible test methods for the communications sub-systems and can enable modeling performance so as to allow a more advanced robot to compensate for degradation in its network by changing parameters or deploying repeaters. A detailed description of an experiment conducted in a subterranean tunnel illustrates the environment characterization and modeling process.

In order for intelligent mobile robots to operate effectively and safely in the world, they need to be able to determine where they are with respect to their surroundings and to form a map of their environment, either for their own navigation purposes or, if exploring environments for humans, to transmit to their operators. Chris Scrapper, Raj Madhavan, Rolf Lakaemper, Andrea Censi, Afzal Godil, Asim Wagan, and Adam Jacoff discuss various approaches aimed at defining quantitative measures for localization and mapping in their chapter "Quantitative Assessment of Robot-Generated

Maps". First off, physical test environments that abstract challenges that robots may encounter in real-world applications are described. Some theoretical approaches that leverage statistical analysis techniques to compare experimental results are discussed. Force Field Simulation is put forward as a means of evaluating the consistency of maps. Finally, quantitative measures of quality, based on features extracted by three algorithms—Harris corner detector, the Hough transform, and the Scale Invariant Feature Transform—are presented.

Performance metrics for evaluating and prescribing approaches to surveying land areas on other planets using mobile robots are presented in the chapter by Edward Tunstel, John Dolan, Terrence Fong, and Debra Schreckenghost entitled "Mobile Robotic Surveying Performance for Planetary Surface Site Characterization". The authors apply a geometry-based area coverage metric to several different surveying approaches and assess trends in relative performance when applied to surveys of a common area. Limitations of the metric are highlighted to motivate the need for richer metrics that account for more than geometric attributes of the task. Examples of metrics that further account for system, environment, or mission attributes that impact performance are presented in the context of NASA research on human-supervised robotic surveying.

The chapter "Performance Evaluation and Metrics for Perception in Intelligent Manufacturing" by Roger Eastman, Tsai Hong, Jane Shi, Tobias Hanning, Bala Muralikrishnan, Susan Young, and Tommy Chang summarizes contributions and results of a workshop special session. Various aspects of the general problem of benchmarking complex perception tasks in the context of intelligent manufacturing applications are covered ranging from camera calibration to pose estimation of objects moving in an environment with uncontrolled lighting and background. The authors discuss the interrelationships of the underlying approaches and bring them together to form a three-step framework for benchmarking perception algorithms and sensor systems.

The final chapter provides a detailed description of how the results of performance evaluations can be used to understand where advanced technologies can be applied in industry. In "Quantification of Line Tracking Solutions for Automotive Applications", Jane Shi, Rick Rourke, Dave Groll, and Peter Tavora describe in depth experiments that their team conducted to quantify performance of line tracking for automotive robotic assembly applications. They selected as performance metrics the range of relative positional tracking error between the robot and the vehicle body and the repeatability of the relative positional tracking error as measured in three standard deviations. These metrics provided insight in inter-comparing results from three line tracking solutions. Their analysis of system performance can lead to conclusions about where current technology is applicable and where there are technology gaps that must be addressed before robotic line assembly tracking is possible within the necessary tolerances.

It would be remiss on our part if we did not acknowledge the countless number of people who variously contributed to this effort. Firstly, we would like to express our sincere thanks to the authors of the chapters for reporting their thoughts and experiences related to their research and also for patiently addressing reviewers' comments and diligently adhering to the hectic deadlines to have the book sent to the publisher in a timely manner. We are indebted to the reviewers for providing insightful and thoughtful comments on the chapters which tremendously improved the quality of the expanded workshop papers to be included in this book. Even though not directly involved with the production of this book, we acknowledge the entire PerMIS'08 crew including program and organizing committee members, reviewers, and our local support staff who tirelessly ensure the success of PerMIS making it a pleasant and memorable experience for everyone involved. Our thanks are due to Springer for publishing this book and for accommodating us at various stages of the publication process. Lastly, but certainly not in the order of importance, we are grateful to our immediate family members who paid the "full price of the book" many times over!

We believe that this book is an important contribution to the community in assembling research work on performance evaluation and benchmarking of intelligent systems from various domains. It is our sincere hope that many more will join us in this time-critical endeavor. Happy reading!

Oak Ridge, TN   Raj Madhavan
Laurel, MD   Edward Tunstel
Gaithersburg, MD   Elena Messina
May 2009

Contents

Contributors

B. Balaguer University of California, Merced, CA, USA

S. Balakirsky National Institute of Standards and Technology, Gaithersburg, MD, USA, stephen.balakirsky@nist.gov

Gary Berg-Cross Engineering, Management and Integration, 13 Atwell Ct., Potomac, MD 20854, USA, gbergcross@gmail.com

Dennis Camell NIST Electromagnetics Division, Boulder, CO, USA

Angelo Cangelosi Adaptive Behaviour & Cognition Group, University of Plymouth, Plymouth, UK

S. Carpin University of California, Merced, CA, USA, scarpin@ucmerced.edu

A. Censi Control & Dynamical Systems, California Institute of Technology, Pasadena, CA 91125, USA, andrea@cds.caltech.edu

Tommy Chang National Institute of Standards and Technology, Gaithersburg, MD, USA

Patrick Courtney PerkinElmer, Beaconsfield, UK, patrick.courtney@perkinelmer.com

M. L. Cummings Department of Aeronautics and Astronautics, Massachusetts Institute of Technology, Cambridge, MA 02139, USA, bdonmez@mit.edu

G. Dimitoglou Hood College, Frederick, MD, USA

John M. Dolan Robotics Institute, Carnegie Mellon University, 5000 Forbes Ave., Pittsburgh, PA 15213, USA, jmd@cs.cmu.edu

Birsen Donmez Department of Aeronautics and Astronautics, Massachusetts Institute of Technology, Cambridge, MA 02139, USA, bdonmez@mit.edu

Roger Eastman Loyola University Maryland, Baltimore, USA, reastman@loyola.edu

David Feil-Seifer University of Southern California, Department of Computer Science, Los Angeles, CA, USA, dfseifer@usc.edu

Terrence Fong Intelligent Robotics Group, NASA Ames Research Center, Moffett Field, CA 94035, USA, terry.fong@nasa.gov

Jan D. Gehrke Center for Computing and Communication Technologies – TZI, Universität Bremen, 28359 Bremen, Germany, jgehrke@tzi.de

A. Godil Information Technology Laboratory, National Institute of Standards and Technology (NIST), Gaithersburg, MD 20899, USA, afzal.godil@nist.gov

Dave Groll General Motors Corporation, Warren, MI, USA

Tobias Hanning University of Passau, Passau, Germany

Tsai Hong National Institute of Standards and Technology, Gaithersburg, MD, USA

George Hough New York City Fire Department, New York City, NY, USA

A. Jacoff Intelligent Systems Division, National Institute of Standards and Technology (NIST), Gaithersburg, MD 20899, USA, adam.jacoff@nist.gov

Serge Kernbach University of Stuttgart, Stuttgart, Germany

Galen Koepke NIST Electromagnetics Division, Boulder, CO, USA

R. Lakaemper Department of Computer and Information Sciences, Temple University, Philadelphia, PA 19122, USA, lakamper@temple.edu

Robert N. Lass Department of Computer Science, College of Engineering, Drexel University, Philadelphia, PA 19104, USA, urlass@cs.drexel.edu

Michael Lewis School of Information Sciences, University of Pittsburgh, Pittsburgh, PA 15260, USA, ml@sis.pitt.edu

Raj Madhavan Computational Sciences and Engineering Division, Oak Ridge National Laboratory, 1 Bethel Valley Road, Oak Ridge, TN 37831, USA and Intelligent Systems Division, National Institute of Standards and Technology (NIST), Gaithersburg, MD 20899, USA raj.madhavan@ieee.org

Maja Matarić University of Southern California, Department of Computer Science, Los Angeles, CA, USA, mataric@usc.edu

Giorgio Metta Italian Institute of Technology, University of Genoa, Genoa, Italy

Olivier Michel Cyberbotics Ltd., Lausanne, Switzerland

Bala Muralikrishnan National Institute of Standards and Technology, Gaithersburg, MD, USA

Lorenzo Natale Italian Institute of Technology, Genoa, Italy

Francesco Nori Italian Institute of Technology, Genoa, Italy

Patricia E. Pina Department of Aeronautics and Astronautics, Massachusetts Institute of Technology, Cambridge, MA 02139, USA

William C. Regli Department of Computer Science, College of Engineering, Drexel University, Philadelphia, PA 19104, USA, regli@cs.drexel.edu

Kate A. Remley NIST Electromagnetics Division, Boulder, CO, USA, kate.remley@nist.gov

Rick F. Rourke General Motors Corporation, Warren, MI, USA

Alexei V. Samsonovich Krasnow Institute for Advanced Study, George Mason University, Fairfax, VA 22030-4444, USA, samsonovich@cox.net

Debra Schreckenghost TRACLabs Inc., Houston, TX 77058, USA, schreck@traclabs.com

C. Scrapper The MITRE Corporation, McLean, VA 22102, USA, cscrapper@mitre.org

Jane Shi General Motors Company, Warren, MI, USA, jane.shi@gm.com

Evan A. Sultanik Department of Computer Science, College of Engineering, Drexel University, Philadelphia, PA 19104, USA, eas28@cs.drexel.edu

Peter W. Tavora General Motors Corporation, Warren, MI, USA

Vadim Tikhanoff Adaptive Behaviour & Cognition Group, University of Plymouth, Plymouth, UK

Katherine Tsui University of Massachusetts Lowell, Department of Computer Science, One University Avenue, Lowell, MA, USA, ktsui@cs.uml.edu

Edward Tunstel Space Department, Johns Hopkins University, Applied Physics Laboratory, Laurel, MD 20723, USA, edward.tunstel@jhuapl.edu

A. Wagan Information Technology Laboratory, National Institute of Standards and Technology (NIST), Gaithersburg, MD 20899, USA, asim.wagan@nist.gov

Jijun Wang Quantum Leap Innovations Inc., Newark, DE 19711, USA, jijunwang.cn@gmail.com

Holly Yanco University of Massachusetts Lowell, Department of Computer Science, One University Avenue, Lowell, MA, USA, holly@cs.uml.edu

S. Susan Young Army Research Laboratory, Adelphi, MD, USA

Chapter 1
Metrics for Multiagent Systems

Robert N. Lass, Evan A. Sultanik, and William C. Regli

Abstract A Multiagent System (MAS) is a software paradigm for building large scale intelligent distributed systems. Increasingly these systems are being deployed on handheld computing devices that rely on non-traditional communications mediums such as mobile ad hoc networks and satellite links. These systems present new challenges for computer scientists in describing system performance and analyzing competing systems. This chapter surveys existing metrics that can be used to describe MASs and related components. A framework for analyzing MASs is provided and an example of how this framework might be employed is given for the domain of distributed constraint reasoning.

1.1 Introduction

An agent is a situated computational process with one or more of the following properties: autonomy, proactivity and interactivity. A multiagent system (MAS) is a system with one or more agents. MASs are a popular software paradigm for building large scale intelligent distributed systems. Increasingly, these systems are being deployed on handheld computing devices, or on non-traditional networks such as mobile ad hoc networks (MANETs) and satellite links. These systems present new challenges for computer scientists in describing system performance and analyzing competing systems.

Much of the research in MASs is entirely theoretical, insofar as no examples of large scale systems of this type exist. As a result, most research utilize simulation or metrics not validated against real-world results. Furthermore, there is a lack of standard terminology or even an agreed upon set of functions that

R.N. Lass (✉)
Department of Computer Science, College of Engineering, Drexel University, Philadelphia,
PA 19104, USA
e-mail: urlass@cs.drexel.edu

R. Madhavan et al. (eds.), *Performance Evaluation and Benchmarking
of Intelligent Systems*, DOI 10.1007/978-1-4419-0492-8_1,

a MAS must provide. The recently published Agent Systems Reference Model [1], however, does attempt to provide such a universal language for discourse to the MAS community, much in the same way that the Open Systems Interconnection reference model has standardized terminology for the field of computer networking.

This is not to say that simulators or metrics have no value. So many variables exist that comparing a fielded MAS against another fielded system is not a straightforward task. In some cases, researchers may have neither access to adequate hardware nor sufficient practical experience to successfully run experiments with live systems [2].

Problems with current methods of evaluating decentralized systems are well known. Haeberlen, for example, argues that current practices have a tendency to be inappropriately generalized, to use technically inappropriate but "standard" evaluation techniques, to focus too heavily on feasible systems and to ossify research [3]. Generalization is caused by only evaluating the performance of the system in a small portion of the environmental and workload space. Standard evaluation techniques bias research toward systems that perform well with regard to those techniques. The difficulty of establishing new methods may cause systems to be evaluated at operating points that are not commensurate with their intended use. As a result of these three practices, research may become ossified, making new discoveries difficult. Finally, robustness suffers: focusing on only a few evaluation points may not uncover behavior that might occur in a more dynamic environment.

The novelty of this chapter is in its bringing together of a number of seemingly disjoint areas of study and is organized as follows: First an anatomy of MASs is presented. Next, a taxonomy of types of metrics is given. Next, an overview of a framework for quantitative decision making, based on work by Basili, is given. This framework allows for choosing which metrics to apply to a system for quantitative decision making. This is followed by a survey of metrics commonly used to describe multiagent systems, primarily from the networking and intelligent systems communities, along with the abstract components in a MAS to which they can be applied. Finally, as an example of how these metrics and techniques might be applied to a real MAS, an analysis of DCOPolis [4]—a framework for distributed constraint optimization—is presented.

1.2 Background on Multiagent Systems and Metrics

Models describe relationships between components of a system to facilitate reasoning about the system. Many abstract models of MASs have been proposed. This chapter adheres to the terminology and concepts given in the Agent Systems Reference Model (ASRM) [1] and classifies metrics based on the layer of the ASRM to which they are applied. We believe that this model is more relevant for applied researchers because the various components of the ASRM were informed by reverse engineering of existing multiagent systems, rather than theories about how multiagent systems *ought* to operate. In addition, the ASRM was not written by a single research group, but a collection of scientists and engineers from industry, government and academia.

1.2.1 Anatomy of a Multiagent System

An abstract representation of a MAS, as presented in the ASRM, is given in Fig. 1.1. At the top of the diagram are *agents*, represented as triangles. Conceptually, an agent is a process with a *sensor* interface that determines the state of the world. It gives information about the world to a *controller*, which performs some computation, and may result in the *effector* taking some action to modify the world. A thermostat could be considered a simple example of an agent. The sensor consists of a

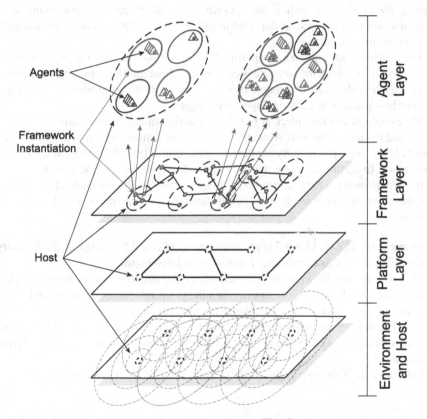

Fig. 1.1 An abstract representation of a multiagent system. This diagram shows all of the different agents, frameworks, platforms (along with their non-agent software), and hosts needed to deliver the functionality of the system. Different shades of circle represent different types of frameworks. The framework layer links show the logical connections that exist between hosts, the platform layer shows the communications connections that exist between hosts, and the environment layer shows the state of the physical medium (in this case, radio signals) used for communications between hosts

thermometer, the controller decides whether or not to turn the air conditioner or heater on, and the effector is the air conditioner or heater interface. There are many different paradigms available for modeling the overall goals of MASs, and for translating these into descriptions of how the individual agents should behave [5].

The agent is supported by an *agent framework*. An agent framework is the set of software components that facilitate agent execution. In some agent systems the agent framework may be trivial, for example, if the agents run natively on the platform (as opposed to in a virtual machine or some other local execution environment). Most agent systems, however, are based on a framework that supports key functionality that agents commonly employ, such as services for migration, messaging, and matchmaking. Examples of such systems include JADE [6], Cougaar [7], and A-Globe [8].

Under the framework is the *platform*. The platform consists of all of the non-agent software present, such as the operating system, databases, networking software or window managers. As depicted in Fig. 1.1, each platform may concurrently support multiple frameworks.

The platform executes on a computing device, or a *host*. This is the physical computing device on which the software is executed. A host may have multiple platforms executing on it. Hosts are distributed in and interact with the *environment*—the physical world—which is the bottom layer in the figure.

Measurement can take place at four layers in the model: agent, framework, platform, and environment/host. Although there is not a functional mapping between ASRM layers and the Open System Interconnection (OSI) [9] layers, there is some correlation. In addition, *system* measurements that cover the whole system (i.e. all of the components functioning together) can be taken. Within each of these layers, there are different levels and classifications of components which could be measured.

- *Framework*: The OSI layer 7 application protocol could be analyzed, the memory footprint, CPU usage, and other framework related metrics.
- *Platform*: Except for in the trivial case where the agents run directly on the platform, the OSI [9] layers 2–6 occur within the platform. Measurement could occur at any of these levels. This means the performance of 802.11, Internet Protocol (IP), Transmission Control Protocol (TCP), Session Initiation Protocol (SIP), and Secure Sockets Layer (SSL) may all be measured, each of which is at a different OSI layer.
- *Environment*: This layer is primarily composed of the OSI layer one.

1.2.2 Types of Metrics

There are a number of types of metrics that can be applied at each layer of the diagram. Metrics may be classified along two axes: their effectiveness (or performance) and the types of data they represent. There are other ways to organize metrics, but these classifications are chosen in particular for two reasons. First, the role a metric plays lies primarily in its use to plot and compare data. The former is virtually always a plot of effectiveness versus performance and making correct comparisons with the latter requires knowing the classification of the data. Second, these two methods of classification provide total coverage; virtually any metric can be classified using each of these schemes.

1.2.2.1 Effectiveness vs Performance

Measures of Effectiveness (MoE) quantify the system's ability to complete its task in a given environment. In some cases this could be a binary value (e.g., "the system succeeds or it fails") while in other cases it could be a range of values (e.g., "the system correctly identified 87% of the targets").

Measures of Performance (MoP) are quantitative measures of some secondary performance characteristic, usually being resource consumption, such as bandwidth required, power consumed, communications range or time to perform a certain task. MoP metrics do not describe the quality of the solution but the quality of obtaining the solution.

A common use for metrics is to allow for visualization of some aspect of the system. Such visualizations will almost always consist of a MoE plotted against a MoP.

1.2.2.2 Data Classification

In Stevens' seminal paper on scales of measurement, metrics are divided into one of four different categories [10]: *ratio*, *interval*, *ordinal* or *nominal*. The membership of a metric in a category informs the type of empirical determinations, mathematical transformations and statistical operations that are permitted. They are summarized as follows.

Nominal measurements are labels that are assigned to data. This type of measure can determine equality of objects, but nothing else. The only permissible transformations are substitutions. Suitable statistical operations are the number of cases, the mode, and the contingency correlation. An example of this is labeling test subjects as being either male or female.

Ordinal measurements are rankings or orderings of data that allow for determination of a "greater than" or "less than" property between data. Any order preserving transformation is admissible. The median and percentiles, along with the nominal statistical operations are allowed. An example of this measurement is a runner's "place" in a marathon.

Interval measurements are like ordinal measurements, but the differences between any two measures has meaning. Any linear transformation is permitted. In addition to the ordinal statistical operations, the mean, standard deviation, rank-order correlation and product-moment correlation are acceptable. An example of interval measurements are Fahrenheit or Centigrade temperature scales.

Ratio measurements are the same as interval measurements except that there is an absolute (not arbitrarily chosen) zero point. Multiplying by a constant is the only allowed transformation. Any statistical operation is allowed. An example of ratio measurements is the Kelvin temperature scale.

1.3 Metrics

Given the last section's discussion of types of metrics and how they relate to MASs, we are now able to introduce specific metrics. This section contains a survey of metrics that describe aspects of MASs about which scientists are often concerned. These metrics are organized by both the layer in the ASRM model (Section 2.1) to which they are most likely to be relevant and the community from which they originate. Some of these metrics could be useful at multiple layers; they are therefore presented in Table 1.1 visualizing this taxonomy.

Table 1.1 Classification of all of the metrics mentioned in this paper in terms of the community the come from, the level at which they are taken, and their scale. Note that some papers do not focus on a single metric, and therefore are not classified by scale

	Agent/Framework	Platform	Environment	System
Distributed systems	[11, 12]	[11, 12]		
Intelligent systems	[13][b] [14–17][d]	[13][b] [14–17][d]	[13][b] [18, 19]	[13, 20–29] [30–32][c] [33–35][d]
Networking			[36, 37]	

[b]ordinal
[c]interval
[d]ratio

1.3.1 Agents and Frameworks

An agent is a situated computational process. For sake of this discussion, agents are the components that achieve the desired high-level functionality of a system. The agent framework is a part of the system that provides functionality to the agents. Because of the way that agents and frameworks are differentiated, metrics that are relevant for the agents in one system may be relevant to the framework in another system, and vice versa. This is because different frameworks provide varying amounts of built-in functionality for the agents. In the latter case, the "intelligence" is part of the agents, in the former, it is part of the framework. Therefore, both agents and frameworks are categorized together in this section.

Quantifiable measures for autonomy are described by Barber and Martin in [14]. The autonomy metric is a number between 0 and 1, and is always defined with respect to a goal and an agent. The number is calculated by determining the percentage of decisions that are made by the agent to reach the goal. For example, consider an agent a_1 working to achieve goal g_1. If one out of every four decisions used to reach g_1 were made by a_1, the autonomy metric with respect to a_1 and g_1

is 0.25. If the agents vote on decisions, then the metric is the weight that a given agent's vote has on the final decision. For example, if five agents each cast votes of equal weight for each decision, each agent's autonomy metric is 0.2.

For conflict resolution and coordination, metrics such as those used by the distributed constraint reasoning community may be used. Cycle-based runtime (CBR) [15], NCCC [16] and ENCCC [17] are three popular metrics. Despite their origin, these metrics could be applied to many more general distributed asynchronous algorithms with dependencies between processes.

1.3.2 Platform

The platform is the software on top of which the agent framework runs. This includes the operating system, compilers, and hardware drivers.

1.3.2.1 Distributed Systems

Many MASs are distributed systems, and it follows that techniques for analyzing distributed systems can be applied to MAS analysis. Hollingsworth summarized metrics [12] and techniques [11] for evaluating the performance of distributed systems. The reader is referred to these two publications for a survey of the area.

1.3.2.2 Networking

The majority of the literature on measuring communication with respect to the asymptotic complexity of distributed algorithms consist of defining metrics that project the cost of communication axis onto the cost of computation [38–40, 16, 15, 41, 17]. Doing so, however, does not capture the multivariate nature of communication. For example, a message sent over the Internet from one side of the world to the other will always[1] have at least 66 ms of latency.[2] A sensor network may have very low latency but may be constrained by bandwidth/power. Therefore, despite transmitting an equivalent number of messages, a single distributed algorithm that is deployed on the Internet may have very different asymptotic properties than the same algorithm deployed on a sensor network; properties that cannot be solely captured in the computation of the algorithm (e.g. power usage from bandwidth). Specialized metrics for measuring the network performance and communication cost are therefore vital for measuring MASs.

Connectivity refers to the state of the communications links between computers in a network. This is often represented as a graph with the nodes representing hosts and the edges representing communications links. If one host can communicate with another, there is a communications link between them.

---

[1] Assuming correctness of the Special Theory of Relativity, correctness of the Principle of Causality, and lack of subterranean networks.
[2] Taking the diameter of the Earth as 40,008 km and the speed of light as 299,792,458 m/s.

The volatility of these links depends on the type of network. On traditional wired networks, the communications links between hosts are relatively static. On a MANET, however, the links between hosts change as the hosts move spatially.

The performance of the network at the third layer of the OSI model—specifically that of the routing protocol—is also critical. In [37], Jackson et al., test three different ad hoc routing protocols under three different scenarios. In these experiments, the scenarios differed in terms of the network load.

There are several ways of measuring cross-layer network performance. By "cross-layer" we mean the aggregate system performance; a system may be deployed with a very efficient link layer protocol but exhibit overall poor behavior due to an inefficient routing protocol running at a higher layer. By using a program such as `netperf` [42], throughput can be measured for a given network topology and configuration.

Various metrics for evaluating MANETs are described by RFC 2501 [36], mainly at the four lowest layers of the OSI model. These metrics include end-to-end throughput & delay, route acquisition time, percentage of out-of-order delivery and efficiency. The latter is a general term describing the overhead involved in sending data. Three example efficiency ratios are given: bits transmitted to bits delivered, control bits transmitted to data bits delivered and control & data packets transmitted to data packets delivered. The RFC describes different *contexts* under which a MANET may operate. A context is a description of the environment under which the MANET operates, and may include network size, connectivity, topological rate of change, link capacity, fraction of unidirectional links, traffic patterns, mobility and the fraction and frequency of sleeping nodes. When evaluating the performance of a MANET system, it is important to note the context, or the state of the environment, under which the researcher performs the evaluation.

1.3.3 Environment/Host Metrics

These metrics describe some aspect of the environment in which the system was tested. In the case of a robot, this might be the physical world. In the case of a software agent, this may be the services, users, and other agents with which it interacts.

In [18], Anderson proposes the quantification of the complexity of a test environment. The intent is to allow one to measure the task performance of an agent with respect to the tests' complexities. A more specific example of the complexity of a test environment is given in [19], which describes three metrics for describing the "traverseability" of terrain by a robot. Two such metrics describe roughness and one is for "crossability."

Since humans can be part of the environment in which agents operate, it may be useful to describe the extent to which humans interact with them. A classification system for autonomous systems is proposed by Huang in [13], based on the complexity of the mission, independence from humans and difficulty of the operational environment. Terms and other metrics for autonomous systems are also defined in this work.

1.3.4 System

System metrics measure macroscopic aspects of the system as a whole. When comparing different systems' performance at a specific task, an evaluator often desires a brief summary they can present to others on the overall performance or effectiveness of the system, making these metrics some of the most relevant with respect to describing a system's overall behavior.

There has been a lot of work in the robotics community on evaluating the effectiveness of robotic systems, which have many similarities with—and are often modeled as—MASs. The robots' performance usually cannot be measured in terms of optimality, as they exist in an inherently stochastic environment, rendering objectivity difficult. This often means that robots are given a number of tasks that are taken to be representative and evaluated based on how well they are able to complete the tasks. Software agents in a messy environment may be able to derive metrics based on these robotics metrics. For example, Balakirsky et al., give a disaster scenario relating to autonomous robots. The authors propose metrics that award points based on how well the robots are able to map their environment and find disaster victims. In a hybrid human-robot system, one method of analysis is to measure the effect of the robot on the human user's effectiveness at a task [23]. Also concerned with human-robot interaction is [24], which presents a framework for evaluating performance and testing procedures for human-robot teams.

Another human-machine hybrid system is the integrated automobile crash warning system, presented in [34]. Three metrics describing effectiveness are given based on the warnings produced by the system: percent true, percent false and percent missed. A measure of performance is also used to describe how far before an area of danger a driver will be able to stop.

An approach to evaluating performance in surveillance systems is presented by in [20], along with domain specific metrics.

A discussion of endurance testing for robots and a recommendation for all safety, security and rescue robots to undergo endurance testing is given in [25]. Using WEKA [43], statistical analysis was performed on the failure data collected to determine the causes of fault.

The results of a long term experimental use of robots in a joint effort between Swedish academic and military organizations is described in [32]. It includes a qualitative analysis of the users' attitude toward the robots before, during and after the study.

When the optimal or actual solution (i.e. ground truth) is known, one way to evaluate effectiveness is to compare the optimal or actual solution to the solution produced by the MAS. In [26], Foedisch, Schlenoff and Madhavan present a new approach to analysis of road recognition algorithms. In the approach, the feature extraction results are compared to actual features from a National Institute of Standards and Technology database. The feature search trees are also used to describe the computational complexity of the search.

When it is not clear what the optimal solution is, or when there are many ways to describe the effectiveness of a system's solution, several metrics may be needed. In

[27], Stanton, Antonishek and Scholtz set up a slalom course and sought to measure the performance of a hybrid human-robot system at navigating the course. There is no single metric that describes how well the system performed, so a number of performance measures were recorded such as time to navigate the course, gates passed through, symbols seen, and the results of human users' survey. The latter covered efficiency, effectiveness and user satisfaction.

Some work is less system specific. For example, in [33] Kannan and Parker propose a general effectiveness metric $P = A - B$, where A is the success metric, B is the failure metric and P is the combined performance metric. This type of metric works for a wide variety of systems that have some notion of success and failure. Similarly, in [35] Chellappa and Roy-Chowdhury propose an information theoretic metric for evaluating the quality of the amount of information processed by an intelligent system.

If a single performance metric is desired, a group of metrics can be combined. In [28], the authors propose a number of performance metrics for the Mars Rover and a formula for generating a composite performance score. The scores are combined using a technique inspired by information theory, described in [30].

A definition of performance, scalability and stability in terms of MASs, and an example of analyzing a MAS for these factors is presented in [29]. In general, performance is computational cost & throughput (i.e. computational complexity and message complexity), scalability is the rate at which the overhead increases as the agent population increases, and stability is a measure of whether or not there is an equilibrium point that the system will return to after perturbations. An example of analyzing these factors is given for a MAS that solves a standard contract net [44] problem.

In many cases, a number of different metrics are needed to get a sense of the performance of the system. An example of this is presented in [21], in which Commuri et al., investigate benchmarks for unmanned ground vehicles in terms of reconfiguration, communications and adaptation & learning. Another is [22], in which Brendle uses four metrics (two MoE & two MoP) to evaluate an algorithm for transforming disparate data sets to a common coordinate system: convergence (MoP), speed (MoP), translation error (MoE) and rotation error (MoE).

1.4 Analysis Framework for Multiagent Systems

Now that a number of metrics have been surveyed, this section presents a framework for applying such metrics to a decision making task. There are three main components: *selection, collection*, and *application*. First the evaluator decides which metrics to use, which must be grounded in some overall goal for the system. Next, the metrics are collected by performing experiments. Finally, the metrics are applied to the original goals to determine if the system meets the goal or perhaps if one system performs better than another.

1.4.1 Selection

There are an infinite number of metrics that could be applied to a system, and an infinite number of ways to apply them. How does a researcher go about deciding which metrics need to be measured for their system?

Many times when analyzing a MAS, evaluators will pick a few or even one metric and show how the system performs. This will focus the analysis on a single layer of the system, and, absent of some theoretical analysis, this may not provide adequate evidence that the system will scale to larger problems. Collecting metrics broadly from different layers of the framework will provide more insight into how the system will perform under different conditions.

The Goal, Question, Metric (GQM)[45] approach for evaluation was developed during a series of projects at the NASA Software Engineering Lab. This technique is intended to provide a focus to investigation and evaluation of systems.

In the GQM approach, the evaluator first chooses goals for different products, processes, and/or resources. There are four parts to a goal: the purpose, the issue, the object and the viewpoint. The example given in [45] is "Improve (*purpose*) the timeliness of (*issue*) change request processing (*object*) from the managers viewpoint (*viewpoint*)."

Next, the evaluator selects questions—usually with quantifiable answers—that must to be answered to understand whether or not the system meets the goal. Each goal may necessitate multiple questions.

Finally, the metric is a set of data associated with the questions that can be subjective (depending on the point of view, such as ease of use of a user interface) or objective (which is independent of the point of view, such as program size). These data are used to answer the *questions*, which in turn informs the evaluator about the *goals*.

As an example scenario to illustrate how the GQM approach works, consider evaluating two systems for solving a Distributed Constraint Reasoning (DCR) problem [46] on a MANET. A DCR problem consists of a number of agents that must assign values to variables. These variables may be constrained with variables owned by other agents. The goal is to minimize the global cost of the value assignments. A more detailed description is given below in Section 5. The computation and communication may be over a MANET, which, as described above, is a network with more limitations and dynamism than a traditional wired network.

The first step in applying the GQM metric to DCR is to decide on a goal: "Select (*purpose*) the system (*object*) providing the lowest average runtime in a bandwidth constrained environment (*issue*) from the point of view of the last agent to converge on a solution (*viewpoint*)." There is still some ambiguity to the statement, but the scope is narrower. For example, this goal is not concerned with the networking cost of a system, the amount of information an algorithm leaks to other agents (privacy), or memory utilization. There are usually multiple goals in a real evaluation, but for the rest of this example we will only consider this single goal.

The next step is to select questions that allow to characterize the object with respect to the goal, such as "How much time is required for the system to converge given test input data A?" "How much time is required for dataset B?" Since we are concerned with bandwidth constrained environments, one might also ask, for each algorithm, "How much bandwidth does this algorithm require?"

Alternatively, there may be metrics or tests that are commonly used in the domain in which the system operates. For example, a number of metrics specific to DCR have been proposed by the distributed artificial intelligence community, as described in Section 3.1.

1.4.2 Collection

After the questions are chosen, as described in the previous section, there is a need to select a set of metrics on which to collect data that will allow the questions to be answered. The example questions were only concerned with time to completion; therefore, continuing the example, runtime information for the system needs to be collected. This could entail a sophisticated solution, such as instrumenting the code to record timing information, or it could be something more informal such as having an experimenter time the system with a stopwatch.

It is worth noting the context (state of the environment) in which the data are collected. In some cases this may be very simple, as little variance is expected in the environment in which the system will operate. In other cases, there may be many variables in the environment that could affect the outcome of the experiments, and must be recorded.

Publications describing practices for conducting research, methods for analyzing data are classified here as "empirical methods." Some of the literature are general, such as that of Balakirsky and Kramer [47] in which some "rules of thumb" for comparing algorithms are given, in addition to an example of the application of each of such rules. One widely cited resource for empirical methods for MASs is a text of Cohen [48].

1.4.3 Application

In the example scenario, we asked the questions, "How long does the system take to converge with test data A," and "How long does the system take to converge with the test data B," to help us meet the goal "Select the system providing the lowest average runtime from the point of view of the last agent to converge on a solution." All that is left is to compare the runtime of each system to determine which is the lowest. If the runtime using both sets of test data is lower for one system, clearly that is the system to select. If one system has a lower runtime for test data A and another has a lower runtime for test data B, then one will have to either decide which data set is most similar to what the system will encounter when deployed, or one needs to create new goals and reassess the system.

1.5 Case Study: DCOP Algorithms and DCOPolis

As an example case study, the following section gives an analysis of a set of Distributed Constraint Optimization (DCOP) algorithms using the framework described above.

A large class of multiagent coordination and distributed resource allocation problems can be modeled as Distributed Constraint Reasoning problems. DCR has generated a lot of interest in the constraint programming community and a number of algorithms have been developed to solve DCR problems [49–51, 41]. A formal treatment of DCOP is outside of the scope of this chapter; the reader is referred to [46] for an introduction to the topic.

Informally, DCR is a method for agents to collaboratively solve constraint reasoning problems distributedly with only local information. The four main components of a DCR problem are variables, domains, agents and constraints. Each *agent* has a set of *variables*, to which it must assign *values*. Each variable has an associated *domain*, which is the set of all possible value assignments to the variable. *Constraints* are a set of functions that specify the cost of any set of partial variable assignments. Finally, each agent is assigned one or more variables for which it is responsible for value assignment. DCOP algorithms work by exchanging messages between agents, who give each other just enough information to allow each agent to make a globally optimal variable assignment.

DCOPolis [4] is a framework for both *comparing* and *deploying* DCR software in heterogeneous environments. DCOPolis has three key components. Different communications *platforms*, DCR *algorithms* and *problems* can be plugged in for a truly comprehensive analysis of algorithmic performance. As advocated in [3], the framework separates the implementation of the algorithms being studied from the platform (simulator, real 802.11 network, etc) to allow code to be written once and then tested in the simulator or run as part of a real system. This contributes to comparative analysis of DCR algorithms by allowing different state-of-the-art algorithms to run in the same simulator under the same conditions or to be deployed on "real" hardware in "real" scenarios. The latter also allows the simulation data, as well as algorithm metrics to be verified. Finally, DCOPolis introduces a new form of distributed algorithm simulation that shows promise of accurate prediction of real-world runtime.

DCOPolis has three primary abstract components: *problems*, *algorithms*, and *platforms*. The main function of DCOPolis is to provide an interface through which the three components can interact. By writing a new instance of any of these components that properly adheres to DCOPolis' API, any algorithm can solve any instance of any problem while running on any platform—even without prior knowledge of such. This makes implementation and testing of new algorithms and platforms trivial.

In keeping with the example given in Section 4 ("Select the system providing the lowest average runtime in a bandwidth constrained environment from the point of view of the last agent to converge on a solution."), the framework is applied to DCOPolis running two different algorithms. We are only using one goal in this example, but we will describe other metrics that could be taken at each layer.

- *Agent*: DCOPolis agents are instantiated with a local view of the problem and then assign values to their variables, send messages to other agents, and change the assigned values based on the messages they received from other agents. Here, we need to record the time each agent takes to converge upon a solution.
- *Framework*: In this case, the framework is what the ASRM refers to as a "null framework." The functionality is contained within the agents, which interact directly with the platform through the Java Virtual Machine. There is nothing to be measured here.
- *Platform*: Any of the metrics in Section 1.2.2 can measure the performance of the network at the platform layer. Given such metrics, an estimation can be made of the system's performance in situations in which the bandwidth is below that of the test environment. Also at this level—if one were interested in improving as well as comparing the systems—the metric in [52] could also be employed to determine which bottlenecks could be optimized.
- *Environment*: The goal stated that we must be concerned with a bandwidth constrained environment. The primary environmental metric to measure, then, is the bandwidth used by each of the algorithms.

The problem in which we are interested is graph coloring. Formally, given a graph $G = \langle N, E \rangle$ and a set of colors C, assign each vertex, $n \in N$, a color, $c \in C$, such that the number of adjacent vertices with the same color is minimized. Graph coloring is a commonly-cited problem used for evaluating DCOP algorithms [49, 50], and can model many real world problems.

This problem is encoded as a DCOP as follows: for each vertex $n_i \in N$, create a variable in the DCOP $v_i \in V$ with domain $D_i = C$. For each pair of adjacent vertices $\langle n_i, n_j \rangle \in E$, create a constraint of cost 1 if both of the associated variables are assigned the same color: $(\forall c \in C : f(\langle v_i, c \rangle, \langle v_j, c \rangle) \mapsto 1)$. A and α cannot be generically defined for graph coloring; they will depend on the application. Most publicly-available benchmark problem sets create one agent per variable [53], which is done here as well.

1.5.1 Experimental Setup

First we describe in detail the software used to run the experiments, including how it generates pseudotrees. Then we describe the computing devices on which it runs and the datasets it solves.

As previously mentioned, DCOPolis is chosen as the experimental testbed because it was originally designed as framework for comparing and deploying distributed decision processes in heterogeneous environments. At the time the experiments were performed, DCOPolis had three DCOP algorithms implemented: Adopt, DPOP and a naïve algorithm called Distributed Hill Climbing. Only Adopt and DPOP were used for the experiments.

A similarity between Adopt and DPOP is that they both assume the existence of a tree ordering over all of the variables in the problem. The pseudotree has an invariant that for each pair of variables $\langle \upsilon_i, \upsilon_j \rangle$ that are neighboring in the constraint graph it must be the case that υ_i is either an ancestor or descendent of υ_j in the pseudotree. The pseudotree also contains a backedge between all pairs of neighbors in the constraint graph that do not have a parent/child relationship in the pseudotree. For each $\upsilon \in V$, $\alpha(v)$ must know the relative tree position (i.e., ancestor, descendent, or parent) of each constraint graph neighbor of υ. The authors of both Adopt and DPOP assume that the agents would simply elect one agent to create this ordering which is then broadcast to the rest of the group. Since the runtime of both algorithms is highly dependent on the structure of the pseudotree, it is ensured for sake of experimentation that for each problem instance the algorithms are given identical pseudotrees.

Five HP-TC4200 tablet PCs with 1.73 Ghz Intel Pentium M processors and 512 M of RAM were connected via Ethernet to a Netgear FS108 switch. No machines were connected to the switch other than the ones taking part in the experiment and the switch was not connected to the Internet or any other network. All the machines were running Ubuntu 7.04 Linux with a 2.6.22 kernel.

The USC Teamcore project has a variety of sample problem data files in their DCOP repository [53] which are used in the analysis. The graph coloring problems are from the "Graph coloring dataset" and range from 8 to 30 variables. The dataset is composed of 3-coloring problems (i.e., problems in which one of three colors must be assigned to each vertex). The problems are encoded as above (i.e., there is one variable for each vertex and the variables' domains are the set of possible colors). In the interest of experimentally analyzing the effect of domain size, the dataset are augmented by also solving 4- and 5-coloring problems.

1.5.2 Results and Analysis

The results of the graph coloring experiments on a wired network can be seen in Fig. 1.2. For the problems in the experiment, this implementation of DPOP seemed to use less bandwidth than Adopt, at the expense of a greater runtime. Although the answers to the questions are not totally in favor of a single algorithm, the trend is clear. This leads to the refinement of the question regarding bandwidth: "Does the system use more bandwidth than we expect to be available?" If neither does, then one should use Adopt. If only Adopt does, one should use DPOP. If both do, then one needs to reconsider his or her approach to solving this problem.

1.6 Summary

MASs are complicated systems comprising a number of interconnected components. Measuring these systems presents new challenges, especially when they are deployed in dynamic environments such as mobile ad hoc networks. This chapter

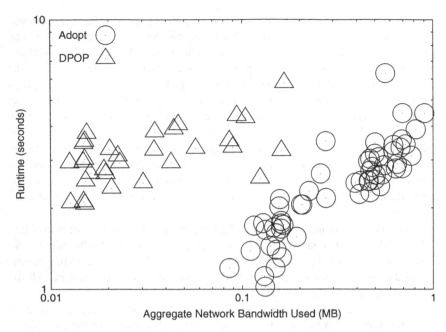

Fig. 1.2 A plot of the bandwidth used and time to solution (labeled "runtime") for two different DCOP algorithms—DPOP (×) and Adopt (Δ)—on a set of graph coloring problems

provided a survey of metrics that can be applied to such systems. A taxonomy of the metrics was produced based on the terminology and organization of the Agent Systems Reference Model. Given these metrics, a general framework for applying the metrics to MASs was proposed, based on Basili's GQM approach. Finally, as an example, the proposed framework was used to compare the performance of some distributed constraint reasoning algorithms.

References

1. Mayk, I., Regli, W.C., eds.: Agent Systems Reference Model. Intelligent Agents Integrated Product Sub-Team, Networking Integrated Product Team, Command and Control Directorate, Headquarters, US Army Research, Development, and Engineering Command, Communications-Electronics Research, Development, and Engineering Center, Department of the Army (November 2006) http://gicl.cs.drexel.edu/people/regli/reference_model-v1a.pdf.
2. Sirer, E.G.: Sextant Deployment (accessed 10/31/2007) http://www.cs.cornell.edu/People/egs/sextant/deployment.php.
3. Haeberlen, A., Mislove, A., Post, A., Druschel, P.: Fallacies in Evaluating Decentralized Systems. In: Proceedings of the 5th International Workshop on Peer-to-Peer Systems (February 2006)
4. Sultanik, E.A., Lass, R.N., Regli, W.C.: eDCOPolis: A Framework for Simulating and Deploying Distributed Constraint Optimization Algorithms. In: Proceedings of the Ninth Annual Workshop on Distributed Constraint Reasoning (September 2007)

5. Wooldridge, M.: Introduction to Multiagent Systems. John Wiley & Sons, Inc. New York, NY, USA (2001)
6. Italia, T.: Java Agent DEvelopment Framework (JADE) http://jade.tilab.com/.
7. Corporation, B.: Cognitive Agent Architecture (Cougaar) http://www.cougaar.org/.
8. Group, A.T.: A-globe http://agents.felk.cvut.cz/aglobe/.
9. Zimmerman, H.: OSI Reference Model—the ISO Model of Architecture for Open System Interconnection. IEEE Transactions on Communications **28**(4) (April 1980) 425–432
10. Stevens, S.S.: On the Theory of Scales of Measurement. Science (1946)
11. Hollingsworth, J.K., Lumpp, J., Miller, B.P.: Techniques for Performance Measurement of Parellel Programs. In: Parallel Computers: Theory and Practice. IEEE Press, Hoboken, NJ, USA (1994)
12. Hollingsworth, J.K., Miller, B.P.: Parallel Program Performance Metrics: A Comparison and Validation. In: Proceedings of the 1992 ACM/IEEE Conference on Supercomputing. IEEE Computer Society Press, Los Alamitos, CA, USA (1992) 4–13
13. Huang, H.M.: The Autonomy Levels for Unmanned Systems ALFUS Framework. In: Proceedings of the Performance Metrics for Intelligent Systems Workshop, NIST Special Publication Gaithersburg, MD, USA (2006) 47–51
14. Barber, K.S., Martin, C.E.: Agent Autonomy: Specification, Measurement, and Dynamic Adjustment. In: Proceedings of the Autonomy Control Software Workshop. (1999) 8–15
15. Davin, J., Modi, P.J.: Impact of problem centralization in distributed constraint optimization algorithms. In: Proceedings of the Fourth International Joint Conference on Autonomous Agents and Multiagent Systems. ACM Press, New York, NY, USA (2005) 1057–1063
16. Meisels, A., Kaplansky, E., Razgon, I., Zivan, R.: Comparing Performance of Distributed Constraints Processing Algorithms. In: Proceedings of the Third International Workshop on Distributed Constraint Reasoning, Bologna, Italy (July 2002)
17. Silaghi, M., Lass, R.N., Sultanik, E.A., Regli, W.C., Matsui, T., Yokoo, M.: Constant Cost of the Computation-Unit in Efficiency Graphs for DCOPs. In: Proceedings of the International Conference on Intelligent Agent Technology (December 2008)
18. Anderson, M.L.: A Flexible Approach to Quantifying Various Dimensions of Environmental Complexity. In: Proceedings of the Performance Metrics for Intelligent Systems Workshop (2004)
19. Molino, V., Madhavan, R., Messina, E., Downs, T., Jacoff, A., Balakirsky, S.: Treversability Metrics for Urban Search and Rescue Robots on Rough Terrain. In: Proceedings of the Performance Metrics for Intelligent Systems Workshop, NIST Special Publication Gaithersburg, MD, USA (2006)
20. Freed, M., Harris, R., Shafto, M.: Measuring Autonomous UAV Surveillance. In: Proceedings of the Performance Metrics for Intelligent Systems Workshop. (2004)
21. Commuri, S., Li, Y., Hougen, D., Fierro, R.: Evaluating Intelligence in Unmanned Ground Vehicle Teams. In: Proceedings of the Performance Metrics for Intelligent Systems Workshop (2004)
22. Brendle, B.: 3D Data Registrnation Based on Human Perception. In: Proceedings of the Performance Metrics for Intelligent Systems Workshop, NIST Special Publication Gaithersburg, MD, USA (2006)
23. Burke, J.L., Murphy, R.R., Riddle, D.R., Fincannon, T.: Task Performance Metrics in Human-Robot Interaction: Taking a Systems Approach. In: Proceedings of the Performance Metrics for Intelligent Systems Workshop (2004)
24. Freedy, A., McDonough, J., Jacobs, R., Freedy, E., Thayer, S., Weltman, G.: A Mixed Initiative Human-Robots Team Performance Assessment System for Use in Operational and Training Environments. In: Proceedings of the Performance Metrics for Intelligent Systems Workshop (2004)
25. Kramer, J.A., Murphy, R.R.: Endurance Testing for Safety, Security and Rescue Robots. In: Proceedings of the Performance Metrics for Intelligent Systems Workshop, NIST Special Publication Gaithersburg, MD, USA (2006) 247–254

26. Foedisch, M., Schlenoff, C., Madhavan, R.: Performance Analysis of Symbolic Road Recognition for On-road Driving. In: Proceedings of the Performance Metrics for Intelligent Systems Workshop, NIST Special Publication Gaithersburg, MD, USA (2006)

27. Stanton, B., Antonishek, B., Scholtz, J.: Development of an Evaluation Method for Acceptable Usability. In: Proceedings of the Performance Metrics for Intelligent Systems Workshop, NIST Special Publication Gaithersburg, MD, USA (2006) 263–267

28. Tunstel, E.: Performance Metrics for Operational Mars Rovers. In: Proceedings of the Performance Metrics for Intelligent Systems Workshop, NIST Special Publication Gaithersburg, MD, USA (2006) 69–76

29. Lee, L., Nwana, H., Ndumu, D., Wilde, P.D.: The Stability, Scalability and Performance of Multi-agent Systems. BT Technology Journal **16**(3) (1998) 94–103

30. Rodriguez, G., Weisbin, C.R.: A New Method to Evaluate Human-Robot System Performance. Autonomous Robots **14**(2) (2003) 165–178

31. Balakirsky, S., Scrapper, C., Carpin, S., Lewis, M.: USARSim: Providing a Framework for Multi-Robot Performance Evaluation. In: Proceedings of the Performance Metrics for Intelligent Systems Workshop, NIST Special Publication Gaithersburg, MD, USA (2006)

32. Lundberg, C., Christensen, H., Reinhold, R.: Long Term Study of a Portable Field Robot in Urban Terrain. In: Proceedings of the Performance Metrics for Intelligent Systems Workshop, NIST Special Publication Gaithersburg, MD, USA (2006)

33. Kannan, B., Parker, L.E.: Fault Tolerance Based Metrics for Evaluating System Performance in Multi-Robot Teams. In: Proceedings of the Performance Metrics for Intelligent Systems Workshop, NIST Special Publication Gaithersburg, MD, USA (2006)

34. Ference, J.J., Szabo, S., Najm, W.G.: Performance Evaluation of Integrated Vehicle-Based Safety Systems. In: Proceedings of the Performance Metrics for Intelligent Systems Workshop, NIST Special Publication Gaithersburg, MD, USA (2006) 85–89

35. Chellappa, R., Roy-Chowdhury, A.K.: An Information Theoretic Evaluation Criterion for 3D Reconstruction Algorithms. In: Proceedings of the Performance Metrics for Intelligent Systems Workshop. (2004)

36. Corson, S., Macker, J.: Mobile Ad hoc Networking (MANET): Routing Protocol Performance Issues and Evaluation Considerations. RFC 2501 (January 1999)

37. Johansson, P., Larsson, T., Hedman, N., Mielczarek, B., Degermark, M.: Scenario-Based Performance Analysis of Routing Protocols for Mobile Ad-Hoc Networks. In: Proceedings of the 5th Annual ACM/IEEE International Conference on Mobile Computing and Networking. ACM Press, New York, NY, USA (1999) 195–206

38. Lamport, L.: Time, Clocks and the Ordering of Events in a Distributed System. Communications of the ACM **21**(7) (July 1978) 558–565

39. Zhang, Y., Mackworth, A.K.: Parallel and Distributed Algorithms for Finite Constraint Satisfaction Problems. In: Proceedings of the Third IEEE Symposium on Parallel and Distributed Processing (1991) 394–397

40. Hamadi, Y., Bessière, C., Quinqueton, J.: Backtracking in Distributed Constraint Networks. In: Proceedings of the European Conference on Artificial Intelligence (1998) 219–223

41. Chechetka, A., Sycara, K.: No-Commitment Branch and Bound Search for Distributed Constraint Optimization. In: Proceedings of the Fifth International Joint Conference on Autonomous Agents and Multiagent Systems. ACM Press, New York, NY, USA (2006) 1427–1429

42. Jones, R.: Netperf http://www.netperf.org/netperf/NetperfPage.html.

43. Witten, I.H., Frank, E.: Data Mining: Practical Machine Learning Tools and Techniques. Morgan Kaufmann San Fransisco, CA, USA (2005)

44. Smith, R.G.: The Contract Net Protocol: High Level Communication and Control in a Distributed Problem Solver. IEEE Transactions on Computers **12**(C-29) (December 1980) 1004–1113

45. Basili, V., Caldiera, G., Rombach, H.: The goal question metric approach. Encyclopedia of Software Engineering **1** (1994) 528–532

46. Yokoo, M., Durfee, E., Ishida, T., Kuwabara, K.: The Distributed Constraint Satisfaction Problem: Formalization and Algorithms. IEEE Transactions on Knowledge and Data Engineering **10**(5) (1998) 673–685
47. Balakirsky, S., Kramer, T.R.: Comparing Algorithms: Rules of Thumb and an Example. In: Proceedings of the Performance Metrics for Intelligent Systems Workshop (2004)
48. Cohen, P.R.: Empirical Methods for Artificial Intelligence. MIT Press, Cambridge, MA, USA (1995)
49. Modi, P.J., Shen, W.M., Tambe, M., Yokoo, M.: An Asynchronous Complete Method for Distributed Constraint Optimization. In: Proceedings of the Second International Joint Conference on Autonomous Agents and Multiagent Systems. ACM Press, New York, NY, USA (2003) 161–168
50. Mailler, R., Lesser, V.: Solving Distributed Constraint Optimization Problems Using Cooperative Mediation. In: Proceedings of the Third International Joint Conference on Autonomous Agents and Multiagent Systems. IEEE Computer Society, Washington, DC, USA (2004) 438–445
51. Petcu, A., Faltings, B.: A Distributed, Complete Method for Multi-Agent Constraint Optimization. In: Proceedings of the Fifth International Workshop on Distributed Constraint Reasoning, Toronto, Canada (September 2004)
52. Anderson, T.E., Lazowska, E.D.: Quartz: a Tool for Tuning Parallel Program Performance. In: Proceedings of the ACM SIGMETRICS Conference on Measurement and Modeling of Computer Systems. ACM Press, New York, NY, USA (1990) 115–125
53. Pearce, J.P.: University of Southern California DCOP Repository (2007) http://teamcore.usc.edu/dcop/.

Chapter 2
Evaluation Criteria for Human-Automation Performance Metrics

Birsen Donmez, Patricia E. Pina, and M.L. Cummings

Abstract Previous research has identified broad metric classes for human-automation performance in order to facilitate metric selection, as well as understanding and comparing research results. However, there is still a lack of a systematic method for selecting the most efficient set of metrics when designing evaluation experiments. This chapter identifies and presents a list of evaluation criteria that can help determine the quality of a metric in terms of experimental constraints, comprehensive understanding, construct validity, statistical efficiency, and measurement technique efficiency. Based on the evaluation criteria, a comprehensive list of potential metric costs and benefits is generated. The evaluation criteria, along with the list of metric costs and benefits, and the existing generic metric classes provide a foundation for the development of a cost-benefit analysis approach that can be used for metric selection.

2.1 Introduction

Human-automation teams are common in many domains, such as command and control operations, human-robot interaction, process control, and medicine. With intelligent automation, these teams operate under a supervisory control paradigm. Supervisory control occurs when one or more human operators intermittently program and receive information from a computer that then closes an autonomous control loop through actuators and sensors of a controlled process or task environment [1]. Example applications include robotics for surgery and geologic rock sampling, and military surveillance with unmanned vehicles.

A popular metric used to evaluate human-automation performance in supervisory control is mission effectiveness [2, 3]. Mission effectiveness focuses on performance

B. Donmez (✉)
Massachusetts Institute of Technology, Department of Aeronautics and Astronautics, Cambridge, MA 02139, USA
e-mail: bdonmez@mit.edu

R. Madhavan et al. (eds.), *Performance Evaluation and Benchmarking of Intelligent Systems*, DOI 10.1007/978-1-4419-0492-8_2, © Springer Science+Business Media, LLC 2009

as it relates to the final output produced by the human-automation team. However, this metric fails to provide insights into the process that leads to the final mission-related output. A suboptimal process can lead to a successful completion of a mission, e.g., when humans adapt to compensate for design deficiencies. Hence, focusing on just mission effectiveness makes it difficult to extract information to detect design flaws and to design systems that can consistently support successful mission completion.

Measuring multiple human-computer system aspects such as workload and situation awareness can be valuable in diagnosing performance successes and failures, and in identifying effective training and design interventions. However, choosing an efficient set of metrics for a given experiment still remains a challenge. Many researchers select their metrics based on their past experience. Another approach to metric selection is to collect as many measures as possible to supposedly gain a comprehensive understanding of the human-automation team performance. These methods can lead to insufficient metrics, expensive experimentation and analysis, and the possibility of inflated type I errors. There appears to be a lack of a principled approach to evaluate and select the most efficient set of metrics among the large number of available metrics.

Different frameworks of metric classes are found in the literature in terms of human-autonomous vehicle interaction [4–7]. These frameworks define metric taxonomies and categorize existing metrics into high-level metric classes that assess different aspects of the human-automation team performance and are generalizable across different missions. Such frameworks can help experimenters identify system aspects that are relevant to measure. However, these frameworks do not include evaluation criteria to select specific metrics from different classes. Each metric set has advantages, limitations, and costs, thus the added value of different sets for a given context needs to be assessed to select an efficient set that maximizes value and minimizes cost.

This chapter presents a brief overview of existing generalizable metric frameworks for human-autonomous vehicle interaction and then suggests a set of evaluation criteria for metric selection. These criteria and the generic metric classes constitute the basis for the future development of a cost-benefit methodology to select supervisory control metrics.

2.2 Generalizable Metric Classes

For human-autonomous vehicle interaction, different frameworks of metric classes have been developed by researchers to facilitate metric selection, and understanding and comparison of research results. Olsen and Goodrich proposed four metric classes to measure the effectiveness of robots: task efficiency, neglect tolerance, robot attention demand, and interaction effort [4]. This set of metrics measures the individual performance of a robot, but fails to measure human performance explicitly.

Human cognitive limitations often constitute a primary bottleneck for human-automation team performance [8]. Therefore, a metric framework that can be generalized across different missions conducted by human-automation teams should include cognitive metrics to understand what drives human behavior and cognition.

In line with the idea of integrating human and automation performance metrics, Steinfeld et al. [7] suggested identifying common metrics in terms of three aspects: human, robot, and the system. Regarding human performance, the authors discussed three main metric categories: situation awareness, workload, and accuracy of mental models of device operations. This work constitutes an important effort towards developing a metric toolkit; however, this framework suffers from a lack of metrics to evaluate collaboration effectiveness among humans and among robots.

Pina et al. [5] defined a comprehensive framework for human-automation team performance based on a high-level conceptual model of human supervisory control. Figure 2.1 represents this conceptual model for a team of two humans collaborating, with each controlling an autonomous platform. The platforms also collaborate autonomously, depicted by arrows between each collaborating unit. The operators receive feedback about automation and mission performance, and adjust automation behavior through controls if required. The automation interacts with the real world through actuators and collects feedback about mission performance through sensors.

Based on this model, Pina et al. [5] defined five generalizable metric classes: mission effectiveness, automation behavior efficiency, human behavior efficiency,

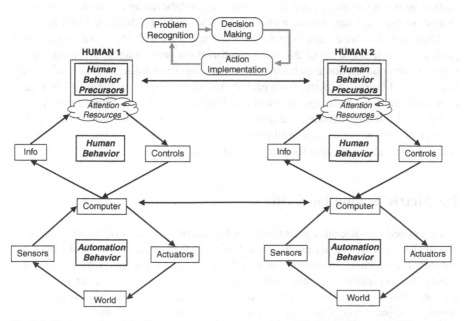

Fig. 2.1 Conceptual model of human-supervisory control (modified from Pina et al. [5])

Table 2.1 Human supervisory control metric classes [9]

Metric classes
Mission effectiveness (e.g., key mission performance parameters)
Automation behavior efficiency (e.g., usability, adequacy, autonomy, reliability)
Human Behavior Efficiency – Attention allocation efficiency (e.g., scan patterns, prioritization) – Information processing efficiency (e.g., decision making)
Human behavior precursors – Cognitive precursors (e.g., situational awareness, mental workload) – Physiological precursors (e.g., physical comfort, fatigue)
Collaborative metrics – Human/automation collaboration (e.g., trust, mental models) – Human/human collaboration (e.g., coordination efficiency, team mental model) – Automation/automation collaboration (e.g., platform's reaction time to situational events that require autonomous collaboration)

human behavior precursors, and collaborative metrics (Table 2.1). Mission effectiveness includes the previously discussed popular metrics and measures concerning how well the mission goals are achieved. Automation and human behavior efficiency measure the actions and decisions made by the individual components of the team. Human behavior precursors measure a human's internal state, including attitudes and cognitive constructs that can be the cause of and influence a given behavior. Collaborative metrics address three different aspects of team collaboration: collaboration between the human and the automation, collaboration between the humans that are in the team, and autonomous collaboration between different platforms.

These metric classes can help researchers select metrics that result in a comprehensive understanding of the human-automation performance, covering issues ranging from automation capabilities to human cognitive abilities. A rule of thumb is to select at least one metric from each metric class. However, there still is a lack of a systematic methodology to select a collection of metrics across these classes that most efficiently measures the performance of human-automation systems. The following section presents a preliminary list of evaluation criteria that can help researchers evaluate the quality of a set of metrics.

2.3 Metric Evaluation Criteria

The proposed metric evaluation criteria for human supervisory control systems consist of five general categories, listed in Table 2.2. These categories focus both on the metrics, which are constructs, and on the associated measures, which are mechanisms for expressing construct sizes. There can be multiple ways of measuring a metric. For example, situational awareness, which is a metric, can be measured based on objective or subjective measures [10]. Different measures for the same metric can generate different benefits and costs. Therefore, the criteria presented in this section evaluate a metric set by considering the metrics (e.g., situational

Table 2.2 Metric evaluation criteria

Evaluation criteria	Example
Experimental constraints	Time required to analyze a metric
Comprehensive understanding	Causal relations with other metrics
Construct validity	Power to discriminate between similar constructs
Statistical efficiency	Effect size
Measurement technique efficiency	Intrusiveness to subjects

awareness), the associated measures (e.g., subjective responses), and the measuring techniques (e.g., questionnaires given at the end of experimentation).

The costs and benefits of different research techniques in human engineering have been previously discussed in the literature [11, 12]. The list of evaluation criteria presented in this chapter is specific to the evaluation of human-automation performance and was identified through a comprehensive literature review of different metrics, measures, and measuring techniques utilized to assess human-automation interaction [9]. Advantages and disadvantages of these methods, which are discussed in detail in Pina et al. [9], fell into five general categories that constitute the proposed evaluation criteria.

These proposed criteria target human supervisory control systems, with influence from the fields of systems engineering, statistics, human factors, and psychology. These fields have their own flavors of experimental metric selection including formal design of experiment approaches such as response surface methods and factor analyses, but often which metric to select and how many are left to heuristics developed through experience.

2.3.1 Experimental Constraints

Time and monetary costs associated with measuring and analyzing a specific metric constitute the main practical considerations for metric selection. Time allocated for gathering and analyzing a metric also comes with a monetary cost due to man-hours, such as time allocated for test bed configurations. Availability of temporal and monetary resources depends on the individual project; however, resources will always be a limiting factor in all projects.

The stage of system development and the testing environment are additional factors that can guide metric selection. Early phases of system development require more controlled experimentation in order to evaluate theoretical concepts that can guide system design. Later phases of system development require a less controlled evaluation of the system in actual operation. For example, research in early phases of development can assess human behavior for different proposed automation levels, whereas research in later phases can assess the human behavior in actual operation in response to the implemented automation level.

The type of testing environment depends on available resources, safety considerations, and the stage of research development. For example, simulation environments give researchers high experimental control, which allows them the ability to manipulate and evaluate different system design concepts accordingly. In simulation environments, researchers can create off-nominal situations and measure operator responses to such situations without exposing them to risk. However, simulation creates an artificial setting and field testing is required to assess system performance in actual use. Thus, the types of measures that can be collected are constrained by the testing environment. For example, responses to rare events are more applicable for research conducted in simulated environments, whereas observational measures can provide better value in field testing.

2.3.2 Comprehensive Understanding

It is important to maximize the understanding gained from a research study. However, due to the limited resources available, it is often not possible to collect all required metrics. Therefore, each metric should be evaluated based on how much it explains the phenomenon of interest. For example, continuous measures of workload over time (e.g., pupil dilation) can provide a more comprehensive dynamic understanding of the system compared to static, aggregate workload measures collected at the end of an experiment (e.g., subjective responses).

The most important aspect of a study is finding an answer to the primary research question. The proximity of a metric to answer the primary research question defines the importance of that metric. For example, a workload measure may not tell much without a metric to assess mission effectiveness, which is what the system designers are generally most interested in understanding. However, this does not mean that the workload measure fails to provide additional insights into the human-automation performance. Another characteristic of a metric that is important to consider is the amount of additional understanding gained using a specific metric when a set of metrics are collected. For example, rather than having two metrics from one metric class (e.g., mission effectiveness), having one metric from two different metric classes (e.g., mission effectiveness and human behavior) can provide a better understanding of human-automation performance.

In addition to providing additional understanding, another desired metric quality is its causal relations with other metrics. A better understanding can be gained if a metric can help explain other metrics' outcomes. For example, operator response to an event, hence human behavior, will often be dependent on the conditions and/or the operator's internal state when the event occurs. The response to an event can be described in terms of three set of variables [13]: a pre-event phase that defines how the operator adapts to the environment; an event-response phase that describes the operator's behavior in accommodating the event; and an outcome phase that describes the outcome of the response process. The underlying reasons for the operator's behavior and the final outcome of an event can be better understood if the initial conditions and operator's state when the event occurs are also measured. When used as covariates in statistical analysis, the initial conditions of the environment and the operator can help explain the variability in other metrics of interest. Thus, in addition to human behavior, experimenters are encouraged to measure human behavior precursors in order to assess the operator state and environmental conditions, which may influence human behavior.

High correlation between different measures, even if they are intended to assess different metrics, is another limiting factor in metric/measure selection. A high correlation can be indicative of the fact that multiple measures can assess the same metric or the same phenomenon. Hence, including multiple measures that are highly correlated with each other can result in wasted resources and also bring into question construct validity, which is discussed next.

2.3.3 Construct Validity

Construct validity refers to how well the associated measure captures the metric or construct of interest. For example, subjective measures of situational awareness ask subjects to rate the amount of situational awareness they had on a given scenario or task. These measures are proposed to help in understanding subjects' situational awareness [10, 14]. However, self-ratings assess meta-comprehension rather than comprehension of the situation: it is unclear whether operators are aware of their lack of situational awareness. Therefore, subjective responses on situational awareness are not valid to assess actual situational awareness, but rather the awareness of lack of situational awareness.

Good construct validity requires a measure to have high sensitivity to changes in the targeted construct. That is, the measure should reflect the change as the construct moves from low to high levels [15]. For example, primary task performance generally starts to break down when the workload reaches higher levels [15, 16]. Therefore, primary task performance measures are not sensitive to changes in the workload at lower workload levels, since with sufficient spare processing capacity, operators are able to compensate for the increase in workload.

A measure with high construct validity should also be able to discriminate between similar constructs. The power to discriminate between similar constructs is especially important for abstract constructs that are hard to measure and difficult to define, such as situational awareness or attentiveness. An example measure that fails to discriminate two related metrics is galvanic skin response. Galvanic skin response is the change in electrical conductance of the skin attributable to the stimulation of the sympathetic nervous system and the production of sweat. Perspiration causes an increase in skin conductance, thus galvanic skin response has been proposed and used to measure workload and stress levels (e.g., [17]). However, even if workload and stress are related, they still are two separate metrics. Therefore, galvanic skin response alone cannot suggest a change in workload.

Good construct validity also requires the selected measure to have high inter- and intra-subject reliability. Inter-subject reliability requires the measure to assess the same construct for every subject, whereas intra-subject reliability requires the measure to assess the same construct if the measure was repeatedly collected from the same subject under identical conditions.

Intra- and inter-subject reliabilities are especially of concern for subjective measures. For example, self-ratings are widely utilized for mental workload assessment [18, 19]. This technique requires operators to rate the workload or effort experienced while performing a task or a mission. Self-ratings are easy to administer, non-intrusive, and inexpensive. However, different individuals may have different interpretations of workload, leading to decreased inter-subject reliability. For example, some participants may not be able to separate mental workload from physical workload [20], and some participants may report their peak workload, whereas others may report their average workload. Another example of low inter-subject reliability is for subjective measures of situational awareness. Vidulich and Hughes [10] found that about half of their participants rated situational awareness by gauging the amount of information to which they attended; while the other half of the participants rated their SA by gauging the amount of information they thought they had overlooked. Participants may also have recall problems if the subjective ratings are collected at the end of a test period, raising concerns on the intra-subject reliability of subjective measures.

2.3.4 Statistical Efficiency

There are three metric qualities that should be considered to ensure statistical efficiency: total number of measures collected, frequency of observations, and effect size.

Analyzing multiple measures inflates type I error. That is, as more dependent variables are analyzed, finding a significant effect when there is none becomes more likely. The inflation of type I error due to multiple dependent variables can be handled with multivariate analysis techniques, such as Multivariate Analysis of Variance (MANOVA) [21]. However, it should be noted that multivariate analyses are harder to conduct, as researchers are more prone to include irrelevant variables in multivariate analyses, possibly hiding the few significant differences among many insignificant ones. The best way to avoid failure to identify significant differences is to design an effective experiment with the most parsimonious metric/measure set that specifically addresses the research question.

Another metric characteristic that needs to be considered is the frequency of observations required for statistical analysis. Supervisory control applications require humans to be monitors of automated systems, with intermittent interaction. Because humans are poor monitors by nature [22], human monitoring efficiency is an important metric to measure in many applications. The problem with assessing monitoring efficiency is that, in most domains, errors or critical signals are rare, and operators can have an entire career without encountering them. For that reason, in order to have a realistic experiment, such rare events cannot be included in a study with sufficient frequency. Therefore, if a metric requires response to rare events, the associated number of observations may not enable the researchers to extract meaningful information from this metric. Moreover, observed events with a low frequency of occurrence cannot be statistically analyzed unless data is obtained from a very large number of subjects, such as in medical studies on rare diseases. Conducting such large scale supervisory control experiments is generally cost-prohibitive.

The number of subjects that can be recruited for a study is especially limited when participants are domain experts such as pilots. The power to identify a significant difference, when there is one, depends on the differences in the means of factor levels and the standard errors of these means, which constitute the effect size. Standard errors of the means are determined by the number of subjects. One way to compensate for limited number of subjects in a study is to use more sensitive measures that will provide a large separation between different conditions, that is, a high effect size. Experimental power can also be increased by reducing error variance by collecting repeated measures on subjects, focusing on sub-populations (e.g., experienced pilots), and/or increasing the magnitude of manipulation for independent variables (low and high intensity rather than low and medium intensity). However, it should also be noted that increased experimental control, such as using sub-populations, can lead to less generalizable results, and there is a tradeoff between the two.

2.3.5 Measurement Technique Efficiency

The data collection technique associated with a specific metric should not be intrusive to the subjects or to the nature of the task. For example, eye trackers are used for capturing operators' visual attention [23, 24]. However, head-mounted eye trackers can be uncomfortable for the subjects, and hence influence their responses. Wearing an eye-tracker can also lead to an unrealistic situation that is not representative of the task performed in the real world.

Eye trackers are an example of how a measurement instrument can interfere with the nature of the task. The measuring technique itself can also interfere with the realism of the study. For example, off-line query methods are used to measure operators' situational awareness [25]. These methods are based on briefly halting the experiment at randomly selected intervals, blanking the displays, and administering a battery of queries to the operators. This situational awareness measure assesses global situational awareness by calculating the accuracy of an operator's responses. The collection of the measure requires the interruption of the task in a way that is unrepresentative of real operating conditions. The interruption may also interfere with other metrics such as operator's performance and workload, as well as other temporal-based metrics.

2.4 Metric Costs vs. Benefits

The evaluation criteria discussed previously can be translated into potential cost-benefit parameters as seen in Table 2.3, which can be ultimately used to define cost and benefit functions of a metric set for a given experiment. The breakdown in

Table 2.3 Representative cost-benefit parameters for metric selection

Costs		
Data gathering	Preparation	Time to setup
		Expertise required
	Data collection	Equipment
		Time
		Measurement error likelihood
	Subject recruitment	Compensation
		IRB preparation and submission
		Time spent recruiting subjects
Data analysis	Data storage/transfer	Equipment
		Time
	Data reduction	Time
		Expertise required
		Software
	Statistical analysis	Error proneness given the required expertise
		Time
		Software
		Expertise

Benefits	
Comprehensive understanding	Proximity to primary research question
	Coverage – Additional understanding given other metrics
	Causal relations to other metrics
Construct validity	Sensitivity
	Power to discriminate between similar constructs
	Inter-subject reliability
	Intra-subject reliability
Statistical efficiency	Effect size — Difference in means — Error variance
	Frequency of observations
	Total number of measures collected
Measurement technique efficiency	Non-intrusiveness to subjects
	Non-intrusiveness to task nature
Appropriateness for system development phase/testing environment	

Table 2.3 is based on the ability to assign a monetary cost to an item. Parameters listed as cost items can be assigned a monetary cost, whereas the parameters listed as benefit items cannot be assigned a monetary cost but nonetheless can be expressed in some kind of a utility function. However, some of the parameters listed under benefits can also be considered as potential costs in non-monetary terms, leading to a negative benefit.

It should be noted that the entries in Table 2.3 are not independent of each other, and tradeoffs exist. For example, recruiting experienced subjects can enhance construct validity and statistical efficiency, however, this may be more time consuming. Figure 2.2 presents results of an experiment conducted to evaluate an automated navigation path planning algorithm in comparison to manual path planning using paper charts in terms of time to generate a plan [26]. Two groups of subjects were recruited for this experiment: civilian and military. The variability of responses of the military group was less than the civilian group, resulting in smaller error variance and larger effect size. However, recruiting military participants requires more effort as these participants are more specialized. Such tradeoffs need to be evaluated by individual researchers based on their specific research objectives and available resources.

Fig. 2.2 Data variability for different subject populations

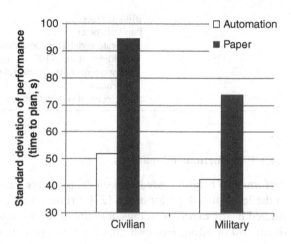

In order to demonstrate how metrics, measures, and measurement techniques can be evaluated using Table 2.3 as a guideline, the following sections present two human behavior metrics, i.e., mental workload and attention allocation efficiency, as examples for evaluating different measures.

2.4.1 Example 1: Mental Workload Measures

Workload is a result of the demands a task imposes on the operator's limited resources. Thus, workload is not only task-specific, but also person-specific. The measurement of mental workload enables, for example, identification of bottlenecks in the system or the mission in which performance can be negatively impacted. Mental workload measures can be classified into three main categories: performance, subjective, and physiological (Table 2.4). This section presents the limitations and advantages associated with each measure guided by Table 2.3. The discussions are summarized in Table 2.5.

Table 2.4 Example measures of mental workload

Measures		Techniques
Performance	Speed or accuracy for the primary task	Primary task
	Time to respond to messages through an embedded chat interface	Secondary task
Subjective (self-ratings)	Modified Cooper-Harper Scale for workload	Unidimensional questionnaires
	NASA TLX	Multidimensional questionnaires
Physiological	Blink frequency	Eye tracking
	Pupil diameter	Eye tracking
	Heart rate variability coefficient	Electrocardiogram
	Amplitudes of the N100 and P300 components of the event-related potential	Electroencephalogram
	Skin electrical conductance	Galvanic skin response

2.4.1.1 Performance Measures

Performance measures are based on the principle that workload is inversely related to the level of task performance [27]. Primary task performance should always be studied in any experiment, thus, utilizing it to assess workload comes with no additional cost or effort. However, this measure presents severe limitations as a mental workload metric, especially in terms of construct validity. Primary task performance is only sensitive in the "overload" region, when the task demands more resources from the operator than are available. Thus, it does not discriminate between two primary tasks in the "underload" region (i.e., the operator has sufficient reserve capacity to reach perfect performance). In addition, primary task performance is not only affected by workload levels, but also by other factors such as correctness of the decisions made by the operator.

Secondary task performance as a workload measure can help researchers assess the amount of residual attention an operator would have in case of an unexpected system failure or event requiring operator intervention [28]. Therefore, it provides additional coverage for understanding human-automation performance. Secondary task measures are also sensitive to differences in primary task demands that may not be reflected in primary task performance, so have better construct validity. However, in order to achieve good construct validity, a secondary task should be selected with specific attention to the types of resources it requires. Humans have different types of resources (e.g., perceptual resources for visual signals vs. perceptual resources for auditory signals) [20]. Therefore, workload resulting from the primary task can be greatly underestimated if the resource demands of the secondary task do not match those of the primary task.

Table 2.5 Evaluation of workload measures

Measures	Advantages	Limitations
Primary task performance	*Cost:* – Can require major cost/effort. However, no additional cost/effort required if already collected to assess mission effectiveness. *Comprehensive Understanding:* – High proximity to primary research question	*Construct Validity:* – Insensitive in the "underload" region – Affected by other factors
Secondary task performance	*Comprehensive Understanding:* – Coverage (assesses the residual attention an operator has) *Construct Validity:* – Sensitivity	*Cost:* – Some level of additional cost/effort *Measurement Technique Efficiency:* – Intrusive to task nature (if not representative of the real task)
Subjective measures	*Cost:* – Cheap equipment, easy to administer *Measurement Technique Efficiency:* – Not intrusive to subjects or the task	*Cost:* – More expertise required for data analysis – More subjects required to achieve adequate power *Construct Validity:* – Inter-subject reliability – Intra-subject reliability – Power to discriminate between similar constructs *Statistical Efficiency:* – Large number of observations required
Physiological measures	*Comprehensive Understanding:* – Continuous, real-time measure	*Cost:* – High level of equipment cost and expertise required – Data analysis is time consuming and requires expertise – Measurement error likelihood *Construct Validity:* – Power to discriminate between similar constructs *Measurement Technique Efficiency:* – Intrusive to subjects and task nature *Appropriateness for system development phase:* – Typically appropriate only for laboratory settings

Some of the secondary tasks that have been proposed and employed include producing finger or foot taps at a constant rate, generating random numbers, or reacting to a secondary-task stimulus [27]. Secondary tasks that are not representative of operator's real tasks may interfere with and disrupt performance of the primary task. However, problems with intrusiveness can be mitigated if embedded secondary tasks are used. In those cases, the secondary task is part of operators' responsibilities but has lower priority in the task hierarchy than the primary task. For example, Cummings and Guerlain used a chat interface as an embedded secondary task measurement tool [29]. Creating an embedded secondary task resolves the issues related to intrusiveness, however, it also requires a larger developmental cost and effort.

2.4.1.2 Subjective Measures

Subjective measures require operators to rate the workload or effort experienced while performing a task or a mission. Unidimensional scale techniques involve asking the participant for a rating of overall workload for a given task condition or at a given point in time [18, 30]. Multidimensional scale techniques require the operator to rate various characteristics of perceived workload [19, 31], and generally possess better diagnostic abilities than the unidimensional scale techniques. Self-ratings have been widely utilized for workload assessment, most likely due to their ease of use. Additional advantages are their non-intrusive nature and low cost. Disadvantages include recall problems, and the variability of workload interpretations between different individuals. In addition, it is unclear whether subjects' reported workload correlates with peak or average workload level. Another potential problem is the difficulty that humans can have when introspectively diagnosing a multidimensional construct, and in particular, separating workload elements [20]. Moreover, self-ratings measure perceived workload rather than actual workload. However, understanding how workload is perceived can be sometimes as important as measuring actual workload.

Self-ratings are generally assessed using a Likert scale that generates ordinal data. The statistical analysis appropriate for such data (e.g., logistic regression, nonparametric methods) requires more expertise than simply conducting analysis of variance (ANOVA). Moreover, the number of subjects needed to reach adequate statistical power for this type of analysis is much higher than it is for ANOVA. Thus, even if subjective measures are low cost during the experimental preparation phase, they may impose substantial costs later by requiring additional expertise for data analysis as well as additional data collection.

2.4.1.3 Physiological Measures

Physiological measures such as heart rate variability, eye movement activity, and galvanic skin response are indicative of operators' level of effort and engagement, and have also been used to assess operator workload. Findings indicate that blink rate, blink duration, and saccade duration all decrease with increased workload, while pupil diameter, number of saccades, and the frequency of long fixations all increase [32]. Heart rate variability is generally found to decrease as workload increases [33]. The electroencephalogram (EEG) has been shown to reflect subtle shifts in workload. However, it also reflects subtle shifts in alertness and attention, which are related to workload, but can reflect different effects. In addition, significant correlations between EEG indices of cognitive state changes and performance have been reported [34–36]. As discussed previously, galvanic skin response (GSR) can be indicative of workload, as well as stress levels [17].

It is important to note that none of these physiological measures directly assess workload. These measures are sensitive to changes in stress, alertness, or attention, and it is almost impossible to discriminate whether the physiological parameters vary as a consequence of mental workload or due to other factors. Thus, the construct validity of physiological measures to assess workload is questionable.

An advantage of physiological measures is the potential for a continuous, real-time measure of ongoing operator states. Such a comprehensive understanding of operator workload can enable researchers to optimize operator workload, using times of inactivity to schedule less critical tasks or deliver non-critical messages so that they do not accumulate during peak periods [37]. Moreover, this type of knowledge could be used to adapt automation, with automation taking on more responsibilities during high operator workload [38].

Some additional problems associated with physiological measures are sensor noise (i.e., high levels of measurement error likelihood), high equipment cost, intrusiveness to task nature and subjects, and the level of expertise as well as additional time required to setup the experiment, collect data, and analyze data. Moreover, due to the significant effort that goes into setting up and calibrating the equipment, physiological measures are very difficult to use outside of laboratory settings.

2.4.2 Example 2: Attention Allocation Efficiency Measures

In supervisory control applications, operators supervise and divide their attentiveness across a series of dynamic processes, sampling information from different channels and looking for critical events. Evaluating attention allocation efficiency involves not only assessing if operators know where to find the information or the functionality they need, but also if they know when to look for a given piece of information or when to execute a given function [39]. Attention allocation measures aid in the understanding of whether and how a particular element on the display is effectively used by the operators. In addition, attention allocation efficiency measures also assess operators' strategies and priorities. It should be noted that some researchers are interested in comparing actual attention allocation strategies with optimal strategies; however, optimal strategies might ultimately be impossible to know. In some cases, it might be possible to approximate optimal strategies via dynamic programming or some other optimization technique [40]. Otherwise, the expert operators' strategy or the best performer's strategy can be used for comparison.

As shown in Table 2.6, there are three main approaches to study attention allocation: eye movements, hand movements, and verbal protocols. Table 2.7 presents the limitations and advantages associated with different measures in terms of the cost-benefit parameters identified in Table 2.3.

Table 2.6 Example attention allocation efficiency measures

Measures	Techniques
Proportion of time that the visual gaze is within each "area of interest" of an interface	Eye tracking
Average number of visits per min to each "area of interest" of an interface	Human interface-inputs
Switching time for multiple tasks	Human interface-inputs
Information used	Human interface-inputs
Operators' task and event priority hierarchies	Verbal protocols

Extensive research has been conducted with eye trackers and video cameras to infer operators' attention allocation strategies based on the assumption that the length and the frequency of eye fixations on a specific display element indicate the level of attention on the element [39, 41]. Attention allocation metrics based on eye movement activity can be dwell time (or glance duration) and glance frequency spent within each "area of interest" of the interface. While visual resources are not the only human resources available, as information acquisition typically occurs through vision in supervisory control settings, visual attention can be used to infer operators' strategies and the employment of cognitive resources. Eye tracking to assess attention allocation efficiency comes with similar limitations to physiological measures used for workload assessment, which have been discussed in Section 2.4.1.

The human interface-inputs reflect operators' physical actions, which are the result of the operators' cognitive processes. Thus operators' mouse clicking can be used to measure operators' actions, determine what information was used, and to infer operators' cognitive strategies [23, 42]. A general limitation with capturing human interface-inputs is that directing attention does not necessarily result in an immediate action, so inferring attention allocation in this manner could be subject to missing states.

Verbal protocols require operators to verbally describe their thoughts, strategies, and decisions, and can be employed simultaneously while operators perform a task, or retrospectively after a task is completed. Verbal protocols are usually videotaped so that researchers can compare what subjects say, while simultaneously observing the system state through the interface the subjects used. This technique provides insights into operators' priorities and decision making strategies, but it can be time consuming and is highly dependent on operators' verbal skills and memory. Moreover, if the operator is interrupted while performing a task, verbal protocols can be intrusive to the task.

2.5 Discussion

Supervisory control of automation is a complex phenomenon with high levels of uncertainty, time-pressure, and a dynamic environment. The performance of human-automation teams depends on multiple components such as human behavior,

Table 2.7 Evaluation of different attention allocation efficiency measures

Measures	Advantages	Limitations
Eye movements (eye tracking)	*Comprehensive Understanding:* – Continuous measure of visual attention allocation	*Cost:* – High level of equipment cost and expertise required – Data analysis is time consuming and requires expertise – Measurement error likelihood *Construct Validity:* – Limited correlation between gaze and thinking *Measurement Technique Efficiency:* – Intrusive to subjects and task nature *Appropriateness for System Development Phase:* – Appropriate for laboratory settings
Interface clicks (human interface-inputs)	*Comprehensive Understanding:* – Continuous measure of subjects' actions	*Cost:* – Time consuming during data analysis *Construct Validity:* – Directing attention does not always result in an immediate interface action
Subjective measures (verbal protocols)	*Comprehensive Understanding:* – Insight into operators' priorities and decision making strategies	*Cost:* – Time intensive *Construct Validity:* – Inter-subject reliability (dependent on operator's verbal skills) – Intra-subject reliability (recall problems with retrospective protocols) *Measurement Technique Efficiency:* – Intrusive to task nature (interference problems with real-time protocols) *Appropriateness for System Development Phase:* – Appropriate for laboratory settings

automation behavior, human cognitive and physical capabilities, team interactions, etc. Because of the complex nature of supervisory control, there are many different metrics that can be utilized to assess performance. However, it is not feasible to collect all possible metrics. Moreover, collecting multiple metrics that are correlated can lead to statistical problems such as inflated type I errors.

This chapter presented a list of evaluation criteria and cost-benefit parameters based on the criteria for determining a set of metrics for a given supervisory control research question. Thus, a limitation of this list of evaluation criteria is that it is not comprehensive enough to address all issues relevant to assessing human-technology interactions. The most prominent issues for assessing human-automation interaction were identified through a comprehensive literature review [9] and were populated under five major categories: experimental constraints, comprehensive understanding, construct validity, statistical efficiency, and measurement technique efficiency. It should be noted that there are interactions between these major categories. For example, the intrusiveness of a given measuring technique can affect the construct validity for a different metric. In one such case, if situational awareness is measured by halting the experiment and querying the operator, then the construct validity for the mission effectiveness or human behavior metrics become questionable. Therefore, the evaluation criteria presented in this chapter should be applied to a collection of metrics rather than each individual metric, taking the interactions between different metrics into consideration.

The list of evaluation criteria and the relevant cost-benefit parameters presented in this chapter are guidelines for metric selection. It should be noted that there is not a single set of metrics that are the most efficient across all applications. The specific research aspects such as available resources and the questions of interest will ultimately determine the relative metric quality. Moreover, depending on the specific research objectives and limitations, the cost-benefit parameters presented in Table 2.3 can have different levels of importance. Thus, these parameters can receive a range of weights in cost-benefit functions created for different applications. Identifying the most appropriate technique for helping researchers to assign their subjective weights is under investigation as part of an ongoing research effort. Thus, future research will further develop this cost-benefit analysis approach, which will systematically identify an efficient set of metrics for classifications of research studies.

Acknowledgments This research was funded by the US Army Aberdeen Test Center. The authors would like to thank Dr. Heecheon You for reviewing the manuscript.

References

1. T. B. Sheridan, *Telerobotics, Automation, and Human Supervisory Control*. Cambridge, MA: The MIT Press, 1992.
2. J. Scholtz, J. Young, J. L. Drury, and H. A. Yanco, "Evaluation of human-robot interaction awareness in search and rescue," in *Proceedings of the IEEE International Conference on Robotics and Automation (ICRA)*. New Orleans, 2004.
3. N. J. Cooke, E. Salas, P. A. Kiekel, and B. Bell, "Advances in measuring team cognition," in *Team Cognition: Understanding the Factors that Drive Process and Performance*, E. Salas and S. M. Fiore, Eds. Washington, D.C.: American Psychological Association, 2004, pp. 83–106.
4. R. O. Olsen and M. A. Goodrich, "Metrics for evaluating human-robot interactions," in *Proceedings of NIST Performance Metrics for Intelligent Systems Workshop*, 2003.

5. P. E. Pina, M. L. Cummings, J. W. Crandall, and M. Della Penna, "Identifying generalizable metric classes to evaluate human-robot teams," in *Proceedings of Metrics for Human-Robot Interaction Workshop at the 3rd Annual Conference on Human-Robot Interaction*. Amsterdam, The Netherlands, 2008.

6. J. W. Crandall and M. L. Cummings, "Identifying predictive metrics for supervisory control of multiple robots," *IEEE Transactions on Robotics – Special Issue on Human-Robot Interaction*, vol. 23, pp. 942-951, 2007.

7. A. Steinfeld, T. Fong, D. Kaber, M. Lewis, J. Scholtz, A. Schultz, and M. A. Goodrich, "Common metrics for human-robot interaction," in *Proceedings of the 1st Annual IEEE/ACM Conference on Human Robot Interaction (Salt Lake City, Utah)*. New York, NY: ACM Press, 2006.

8. C. D. Wickens, J. D. Lee, Y. Liu, and S. G. Becker, *An Introduction to Human Factors Engineering*, 2nd ed. Upper Saddle River, New Jersey: Pearson Education, Inc., 2004.

9. P. E. Pina, B. Donmez, and M. L. Cummings, *Selecting Metrics to Evaluate Human Supervisory Control Applications*, MIT Humans and Automation Laboratory, Cambridge, MA HAL2008-04, 2008.

10. M. A. Vidulich and E. R. Hughes, "Testing a subjective metric of situation awareness," in *Proceedings of the Human Factors Society 35th Annual Meeting*. Santa Monica, CA: The Human Factors and Ergonomics Society, 1991, pp. 1307–1311.

11. A. Chapanis, *Research Techniques in Human Engineering*. Baltimore: The Johns Hopkins Press, 1965.

12. M. S. Sanders and E. J. McCormick, *Human Factors in Engineering and Design*. New York: McGraw-Hill, 1993.

13. B. Donmez, L. Boyle, and J. D. Lee, "The impact of distraction mitigation strategies on driving performance," *Human Factors*, vol. 48, pp. 785–804, 2006.

14. R. M. Taylor, "Situational awareness rating technique (SART): the development of a tool for aircrew systems design," in *Proceedings of the NATO Advisory Group for Aerospace Research and Development (AGARD) Situational Awareness in Aerospace Operations Symposium (AGARD-CP-478)*, 1989, p. 17.

15. F. T. Eggemeier, C. A. Shingledecker, and M. S. Crabtree, "Workload measurement in system design and evalution," in *Proceeding of the Human Factors Society 29th Annual Meeting*. Baltimore, MD, 1985, pp. 215–219.

16. F. T. Eggemeier, M. S. Crabtree, and P. A. LaPoint, "The effect of delayed report on subjective ratings of mental workload," in *Proceedings of the Human Factors Society 27th Annual Meeting*. Norfolk, VA, 1983, pp. 139–143.

17. S. Levin, D. J. France, R. Hemphill, I. Jones, K. Y. Chen, D. Ricard, R. Makowski, and D. Aronsky, "Tracking workload in the emergency department," *Human Factors*, vol. 48, pp. 526–539, 2006.

18. W. W. Wierwille and J. G. Casali, "A validated rating scale for global mental workload measurement applications," in *Proceedings of the Human Factors Society 27th Annual Meeting*. Santa Monica, CA, 1983, pp. 129–133.

19. S. G. Hart and L. E. Staveland, "Development of NASA-TLX (Task Load Index): results of empirical and theoretical research," in *Human Mental Workload*, P. Hancock and N. Meshkati, Eds. Amsterdam, The Netherlands: North Holland B.V., 1988, pp. 139–183.

20. R. D. O'Donnell and F. T. Eggemeier, "Workload assessment methodology," in *Handbook of Perception and Human Performance: Vol. II. Cognitive Processes and Performance*, K. R. Boff, L. Kaufmann, and J. P. Thomas, Eds. New York: Wiley Interscience, 1986, pp. 42-1–42-49.

21. R. A. Johnson and D. W. Wichern, *Applied Multivariate Statistical Analysis*, 5th ed. NJ: Pearson Education, 2002.

22. T. B. Sheridan, *Humans and Automation: System Design and Research Issues*. New York, NY: John Wiley & Sons Inc., 2002.

23. M. E. Janzen and K. J. Vicente, "Attention allocation within the abstraction hierarchy," *International Journal of Human-Computer Studies*, vol. 48, pp. 521–545, 1998.

24. B. Donmez, L. Boyle, and J. D. Lee, "Safety implications of providing real-time feedback to distracted drivers," *Accident Analysis & Prevention*, vol. 39, pp. 581–590, 2007.

25. M. R. Endsley, B. Bolte, and D. G. Jones, *Designing for Situation Awareness: An Approach to User-Centered Design*. Boca Raton, FL: CRC Press, Taylor & Francis Group, 2003.

26. M. Buchin, "Assessing the impact of automated path planning aids in the maritime community," in *Electrical Engineering and Computer Science M.Eng*. Cambridge, MA: Massachusetts Institute of Technology, 2009.

27. C. D. Wickens and J. G. Hollands, *Engineering Psychology and Human Performance*, 3rd ed. New Jersey: Prentice Hall, 1999.

28. G. D. Ogden, J. M. Levine, and E. J. Eisner, "Measurement of workload by secondary tasks," *Human Factors*, vol. 21, pp. 529–548, 1979.

29. M. L. Cummings and S. Guerlain, "Using a chat interface as an embedded secondary tasking tool," in *Proceedings of the 2nd Annual Human Performance, Situation Awareness, and Automation Technology Conference*. Daytona Beach, FL, 2004.

30. A. H. Roscoe and G. A. Ellis, *A Subjective Rating Scale for Assessing Pilot Workload in Flight: A Decade of Practical Use*, Royal Aeronautical Establishment, Farnborough, England TR90019, 1990.

31. G. B. Reid and T. E. Nygren, "The subjective workload assessment technique: a scaling procedure for measuring mental workload," in *Human Mental Workload*, P. Hancock and N. Meshkati, Eds. Amsterdam, The Netherlands: North Holland, 1988, pp. 185–218.

32. U. Ahlstrom and F. Friedman-Berg, *Subjective Workload Ratings and Eye Movement Activity Measures*, US Department of Transportation, Federal Aviation Administration DOT/FAA/ACT-05/32, 2005.

33. A. J. Tattersall and G. R. J. Hockey, "Level of operator control and changes in heart rate variability during simulated flight maintenance," *Human Factors*, vol. 37, pp. 682–698, 1995.

34. C. Berka, D. J. Levendowski, M. Cventovic, M. M. Petrovic, G. F. Davis, M. N. Lumicao, M. V. Popovic, V. T. Zivkovic, R. E. Olmstead, and P. Westbrook, "Real-time analysis of EEG indices of alertness, cognition, and memory acquired with a wireless EEG headset," *International Journal of Human Computer Interaction*, vol. 17, pp. 151–170, 2004.

35. J. B. Brookings, G. F. Wilson, and C. R. Swain, "Psychophysiological responses to changes in workload during simulated air-traffic control," *Biological Psychology*, vol. 42, pp. 361–377, 1996.

36. K. A. Brookhuis and D. De Waard, "The use of psychophysiology to assess driver status," *Ergonomics*, vol. 36, 1993.

37. S. T. Iqbal, P. D. Adamczyk, S. Zheng, and B. P. Bailey, "Towards an index of opportunity: understanding changes in mental workload during task execution," in *Proceedings of the ACM Conference on Human Factors in Computing Systems*. Portland, Oregon, 2005, pp. 311–320.

38. R. Parasuraman and P. A. Hancock, "Adaptive control of mental workload," in *Stress, Workload, and Fatigue*, P. A. Hancock and P. A. Desmond, Eds. Mahwah, New Jersey: Lawrence Erlbaum Associates, Publishers, 2001, pp. 305–320.

39. D. A. Talluer and C. D. Wickens, "The effect of pilot visual scanning strategies on traffic detection accuracy and aircraft control," in *Proceedings of the 12th International Symposium on Aviation Psychology*. Dayton, OH, 2003.

40. M. Puterman, *Markov Decision Processes: Discrete Stochastic Dynamic Programming*. New Jersey: Wiley, 2005.

41. C. D. Wickens, J. Helleberg, J. Goh, Xu, X., and W. J. Horrey, *Pilot Task Management: Testing and Attentional Expected Value Model of Visual Scanning*, NASA Ames Research Center, Moffett Field, CA ARL-01-14/NASA-01-7, 2001.

42. S. Bruni, J. Marquez, A. S. Brzezinski, and M. L. Cummings, "Visualizing operators' cognitive strategies in multivariate optimization," in *Proceedings of the Human Factors and Ergonomics Society's 50th Annual Meeting*. San Francisco, CA, 2006.

Chapter 3
Performance Evaluation Methods for Assistive Robotic Technology

Katherine M. Tsui, David J. Feil-Seifer, Maja J. Matarić, and Holly A. Yanco

Abstract Robots have been developed for several assistive technology domains, including intervention for Autism Spectrum Disorders, eldercare, and post-stroke rehabilitation. Assistive robots have also been used to promote independent living through the use of devices such as intelligent wheelchairs, assistive robotic arms, and external limb prostheses. Work in the broad field of assistive robotic technology can be divided into two major research phases: *technology development*, in which new devices, software, and interfaces are created; and *clinical*, in which assistive technology is applied to a given end-user population. Moving from technology development towards clinical applications is a significant challenge. Developing performance metrics for assistive robots poses a related set of challenges. In this paper, we survey several areas of assistive robotic technology in order to derive and demonstrate domain-specific means for evaluating the performance of such systems. We also present two case studies of applied performance measures and a discussion regarding the ubiquity of functional performance measures across the sampled domains. Finally, we present guidelines for incorporating human performance metrics into end-user evaluations of assistive robotic technologies.

3.1 Introduction

Assistive robots have the potential to provide therapeutic benefits in health care domains ranging from intervention for Autism Spectrum Disorders to post-stroke rehabilitation to eldercare. However, it is invariably challenging to transition an assistive device developed in the lab to the target domain. This problem can occur even when the device was designed with a specific end-user in mind. Römer et al. provided guidelines for compiling a technical file for an assistive device for transfer

K.M. Tsui (✉)
University of Massachusetts Lowell, Department of Computer Science,
One University Avenue, Lowell, MA, USA
e-mail: ktsui@cs.uml.edu

R. Madhavan et al. (eds.), *Performance Evaluation and Benchmarking of Intelligent Systems*, DOI 10.1007/978-1-4419-0492-8_3,
© Springer Science+Business Media, LLC 2009

from academic development to manufacturing [80]. Their guidelines state that documentation of an assistive device must include its "intended use, design specifications, design considerations, design methods, design calculations, risk analysis, verification of the specifications, validation information of performance of its intended use, and compliance to application standards" [80]. Academic and industrial research labs are the piloting grounds for new concepts. Due to the institutional separation between the research environment and end-users, special care must be taken so that a technology, developed in the lab, properly addresses the needs of end-users in the real world. It is thus imperative for the development of assistive robotic technologies to involve the end-user in the design and evaluations [44]. These end-user evaluations, with the proper performance measures, can provide the basis for performance validation needed to begin the transition from research pilot to end product.

Does there exist a ubiquitous set of performance measures for the evaluation of assistive robotic technologies? Time to task completion and time on task are common measures. Römer et al. propose an absolute measure for time to task completion, in which the time is compared to that of an able-bodied person's performance [80]. Task completion time fits many robotic applications, such as retrieving an object with a robotic manipulator. However, it may not suit other applications, such as a range of motion exercise in the context of rehabilitation of an upper limb. Römer et al. also acknowledge other factors in determining performance measures, namely "user friendliness, ease of operation, (and) effectiveness of input device" [80].

Aside from the very general metrics described above, should we even seek a ubiquitous set of performance metrics? The lack of ubiquitous performance metrics is a result of necessarily domain-specific performance needs. Most metrics do not translate well between domains or even sub-domains. Thus, the field of assistive robotic technology has used a wide variety of performance measures specific to the domains for end-user evaluations. However, there are observable similarities between various employed metrics and how they are devised. In order to evaluate an assistive robotic technology within a particular domain, clinical performance measures are needed to lend validity to the device.

Clinical evaluation is the mechanism used to determine the clinical, biological, or psychological effects of an intervention. Clinical evaluations use The Good Clinical Practice Protocol, which requires clearly stated objectives, checkpoints, and types and frequency of measurement [100]. Well-established domains have developed generally agreed-upon performance measures over time. For example, the Fugl-Meyer motor assessment, created in 1975, is commonly used in evaluating upper limb rehabilitation for patients post-stroke recovery [40]. On the other hand, FIM (formerly known as the Functional Independence Measure) is popular for measuring a person's functional independence with respect to activities of daily living (ADLs) [65]. The two evaluations have little, if any, relation to each other, because they emerged from different domains. However, they are both used broadly, albeit for different user populations, and thusly can serve as a means for assessing performance relative to an established baseline.

In this paper, we explore contemporary end-user evaluations and the performance measures used in evaluating assistive robotic technologies. We present case studies from the University of Massachusetts Lowell and the University of Southern California. These studies illustrate the evolution of performance metrics in their respective domains: assistive robotic arms and Autism Spectrum Disorders. We also discuss the ubiquity of functional performance measures throughout all of the surveyed domains; we say that a performance measure is *functional* if it relates to an activity of daily living and is administered in a realistic setting. Finally, we present guidelines for incorporating human performance metrics into end-user evaluations of assistive robotic technologies.

3.2 Assistive Robotic Technologies

Assistive technology encompasses both "low-tech" and "high-tech" solutions. As a new technology is developed, new and/or improved assistive devices can be created. For example, the concept of a wheelchair was documented in China in the 6th century [107]. The manual self-propelled wheelchair was patented in 1894 [107]. The power wheelchair was invented during World War II [8].

As the field of robotics has matured, researchers began to apply this newest technology to surgery and rehabilitation. More recently, robots are being used to enhance the functional capabilities of people with physical and/or cognitive disabilities. For example, the first commercially available intelligent wheelchair entered the market in 2000 [14].

In 2002, Haigh and Yanco surveyed assistive robotics [47]. A historical survey of rehabilitation robotics through 2003 can be found in Hillman [49]. Simpson surveyed intelligent wheelchairs through 2004 [83]. We present a contemporary survey of assistive technologies that have been evaluated by end-users. We believe that the primary focus of end-user evaluations should be on the *human* performance measurements with secondary focus on the performance of the robot. This section highlights six areas of assistive robotic technology development. We discuss assistive robots used in intervention for Autism Spectrum Disorders, eldercare, post-stroke recovery, and independent living through intelligent wheelchairs, assistive robotic arms, and prosthetic limbs. For each area, we describe a few examples of performance metrics and how they have been employed or applied.

3.2.1 Autism Spectrum Disorders (ASD)

An increasing number of research institutions are investigating the use of robots as tools for intervention and therapy for children with Autism Spectrum Disorders (ASD), including the University of Hertfordshire [77, 78, 76], the Université de Sherbrooke [66, 81], the National Institute of Information and Communications Technology [56], the University of Southern California [34], and the University of Washington [88]. The goal of these systems is typically to use robots as catalysts for social behavior in order to stimulate and train social and communicative behaviors of children with ASD for either assessment or therapeutic purposes.

3.2.1.1 End-User Evaluations

Researchers at the University of Hertfordshire have conducted several observational studies with children with ASD [77]. In one such study, four children interacted with Robota, a robot doll, over a period of several months. Post-hoc analysis of video footage of interaction sessions yielded eye gaze, touch, imitation, and proximity categories. Performance measures included frequency of the occurrence of the categories. Another study used the hesitation and duration of a drumming session as a task-specific measure of engagement with a drumming robot [78]. In addition, measures for observing social behavior were taken from the ASD research community; in particular, video coding for observing social behavior [93] was applied to determine if a robot was an isolator or mediator for children with ASD [76].

Researchers at the Université of Sherbrooke conducted an observational study of four children with ASD over seven weeks [66]. The children interacted with Tito, a human-character robot, three times per week for five minutes. Video was collected during the interactions. In post-hoc analysis, the interactions were categorized into shared attention, shared conventions, and absence of either; all video data were coded using twelve-second windows. Performance measures included frequency of the occurrence of categories. Other work involved the use of automated interaction logs in order to model a user's play behavior with the robot [81]. Performance measures included correlation of recognized play with observed behavior.

The National Institute of Information and Communications Technology (NICT) conducted a longitudinal observational study in a day-care setting [56]. Groups of children interacted with a simple character robot, Keepon, in twenty-five three-hour sessions over five months. Each session was a free-play scenario that was part of the regular day-care schedule. Children were given the opportunity to interact with the robot, or not, and children were allowed to interact with the robot in groups. Video data of these interactions were analyzed in a qualitative fashion.

Researchers at the University of Southern California (USC) conducted a study with children with ASD interacting with a bubble-blowing robot [33]. This research used a repeated-measures study model to compare two types of robot behavior: contingent (the robot responds to the child's actions) and random (the robot executes an action at random times). The scenario involved the child, the robot, and a parent, all of whom were observed for forty-five minutes. Post-hoc analysis of video data was used to identify joint-attention, vocalizations, social orienting, and other forms of social interaction, as well as the tagged by target (parent, robot, or none) of the interaction. These behaviors were taken from a standard ASD diagnostic exam, the Autism Diagnostic Observation Schedule (ADOS) [61], which uses a similar scenario to the one used in the experiment, providing a key for identifying relevant evaluative behavior. Performance measures included frequency and richness of the interaction observed between sessions.

Researchers at the University of Washington developed a study that compared a robot dog, AIBO, to a simple mechanical stuffed dog [88]. After a brief introductory period, the participants (i.e., parent and a child with ASD) interacted with the one of the artifacts for a period of thirty minutes. The sessions were video recorded and coded. The behavior coding included verbal engagement, affection, animating artifact, reciprocal interaction, and authentic interaction. The performance measure used was the amount of coded social behavior observed.

3.2.1.2 Discussion

Video coding is a commonly used technique for analyzing behavioral experiments [79]. Categories may be set prior to coding or may be the result of post hoc analysis, in which the categories are defined from keywords, phrases, or events. The data, such as open-ended responses to questions or comments, is then annotated with the categories. To ensure reliability, multiple coders (or raters) are trained on the units and definitions. When multiple coders are used, inter-coder reliability must be established, such as by using a kappa statistic.[1] However, in each experiment design, the basic unit of time for behavior data could be vastly different, ranging from tenths of a second eg., [77] to twelve seconds eg., [66] to assessments of the entire session eg., [56]. The resulting performance measures use the number of occurrences within the categories.

While these assessments are in most cases driven by existing tools used in developmental or ASD-specific settings, there is little evidence to date that the measures used that translate well to real-world improvements in learning, social skill development, and psychosocial behavior. ASD is considered a spectrum disorder with a great deal of symptom heterogeneity in the population [39], which creates a major challenge for diagnosis and treatment as well as research. Since assistive robotics studies to date have shown some effects for small groups of tested children, it is important to analyze how generalizable their results are. One strategy for ensuring that the observed data are somewhat grounded in the field of ASD research is to draw the analysis metrics from existing ASD diagnostics (e.g., [79, 33]). This remains an open challenge for the growing field of socially assistive robotics for ASD.

3.2.2 Eldercare

Studies have shown that the elderly population is growing world-wide [11]. Roboticists from research institutions, including NICT [102], the University of Missouri [104], and USC [92], among others, are investigating robots for use as minders, guides, and companions for the elderly.

---

[1]Cohen's kappa provides the level of agreement for nominal data between two raters [21]. For more than two raters Fleiss' kappa must be used [37].

3.2.2.1 End-User Evaluations

Researchers at NICT conducted a five-week study of twenty-three elderly women in an eldercare facility. The participants interacted with Paro, the therapeutic care robot seal, one to three times per week [102]. Performance measures included self assessment of the participant's mood (pictorial semantic differential scale [71] of $1 = happy$ to $20 = sad$) both before and after the interaction with Paro; questions from the Profile of Mood States questionnaire [64] to evaluate anxiety, depression, and vigor (semantic differential scale of $0 = none$ to $4 = extremely$); and stress analysis of urinary specimens.

Researchers at the University of Missouri, together with TigerPlace, an independent living facility for the elderly, studied assistive technology for aging in place [104]. At TigerPlace, elderly people who would otherwise be required to have full-time nursing-home care are able to live in their individual residences and have health services brought to them. As part of this effort, the researchers developed a fuzzy logic-based augmentation of an existing day-to-day evaluation, the Short Physical Performance Battery (SPPB) [45]. SPPB measures the user's performance on balance, gait, strength, and endurance tasks. The fuzzy logic augmentation provided finer-grained performance measure for day-to-day monitoring. The team conducted observational studies of two elderly people recovering from surgery in their apartments at TigerPlace [97]. Sensors were placed in their apartments for a period of 14 and 16 months, respectively. Performance measures included a number of categorizations of restlessness (i.e., time vs. event frequency) and quality of life (i.e., ability to complete activities of daily living).

Researchers at the University of Southern California have developed a robot for exercise therapy for adults with dementia and Alzheimer's Disease [92]. The experiment was designed based on the existing music therapy sessions conducted at a Silverado Senior Living community. In the experiment, the participant sat in front of a panel with large, bright, labeled buttons, and a mobile robot with an expressive humanoid torso and head. The robot played music and encouraged and coached the participant to "name that tune" by pushing the correct button. The complexity of the experiment was controlled by the amount of information provided by the robot (from no information, to the name of the song, to hints about the name of the song, to prompts for pushing a button). The performance measures included compliance with the game, enjoyment of the game (evaluated based on the type and amount of vocalizations and facial expressions of the participant), and response time in pushing the buttons, and correctness of responses. The experiment occurred twice per week for eight months and the challenge level of the exercise was progressively adjusted in order to retain the participant's interest over multiple sessions.

3.2.2.2 Discussion

Most of the above systems are currently at the feasibility stage of implementation, an important stage of evaluation for determining if the technology is ready for deployment in a real-world environment. User evaluations and behavioral studies of eldercare systems, such as the studies with Paro, describe the effects that such systems have on users and their environment. By emphasizing social interaction and fitness, these performance measures implicitly gauge the changes in quality of life (QoL).

Current evaluations of eldercare systems occur over a period of days or weeks. As these systems become more permanent fixtures in eldercare environments, the assessment of QoL will become increasingly important. Standardized questionnaires for observing QoL over time can be employed to observe any long-term effectiveness of such interventions in the eldercare environment [112]. For example, the SF-36 survey [1] is used to assess health-related QoL, while the 15-D [85] survey is used to measure QoL along several elements of a participant's lifestyle.

3.2.3 Stroke Rehabilitation

The use of robots is being investigated for gait training at Arizona State University [105], upper-limb recovery at the Rehabilitation Institute of Chicago and Northwestern University [51], and wrist rehabilitation at Hong Kong Polytechnic University [52]. It is well-documented that stroke patients regain most of their mobility through repetitions of task training [53]. The need for technology, such as robots, for supervising and guiding functional rehabilitation exercises is constantly increasing due to the growing elderly population and the large number of stroke victims. Matarić et al. [62] described the two sub-fields: hands-on rehabilitation systems which apply force to guide the affected limb in rehabilitation exercises and hands-off socially assistive systems that provide monitoring and coaching through verbal and gestural feedback and without any physical contact. The two methods play complementary roles at different stages of the rehabilitation process.

3.2.3.1 End-User Evaluations

Pilot experiments are conducted with a small number of study participants in the study to determine what needs to be altered. The results of a pilot experiment are used to justify a full-scale clinical trial. An example of a pilot experiment is Wada et al.'s case study ($n = 1$) of their Robotic Gait Trainer [105]. Twice per week for eight weeks, the participant walked on a treadmill with the Robotic Gait Trainer assistance. The supination and pronation position of the participant's foot was measured to determine the quality of her gait. Other performance measure included the six-minute walk test (6MWT) [46] and the timed get-up-and-go test (TGUG) [103].

An example of a small-scale clinical study is Housman et al.'s evaluation of the Therapy Wilmington Robotics Exoskeleton (T-WREX) conducted at the Rehabilitation Institute of Chicago (RIC) and Northwestern University [51]. This clinical trial is an example of typical contact-based rehabilitation robot study with stroke patients. The team conducted a clinical evaluation of twenty-three stroke survivors over sixteen weeks comparing robot-assisted therapy to a traditional rehabilitation therapy regiment [51]. The researchers observed functional arm movement, quality of affected arm use, range of motion, grip strength, a survey of patient satisfaction of therapy, and the use of the affected arm in the home when not undergoing therapy. Performance assessments with or without the robot included Fugl-Meyer [40] and Rancho Functional Test for Upper Extremity [108] to measure ability to use the arm. In addition, they measured use of the arm outside of the experimental setting by using the Motor Activity Log [101], a self-report, to determine how the arm was used in the home. Finally, to assess the costs of using the robot, they measured the amount of time that the user needed assistance in order to use the T-WREX.

The primary assessment of post-stroke rehabilitative robotics involves the use of clinical assessments of patient function. Discussed above were the Fugl-Meyer, Rancho Functional Test, 6MWT, and TGUG assessments. However, there are many others in clinical use today. At at Northwestern University and RIC, Ellis et al. supplemented the Fugl-Meyer with several other measures, including the Chedo-kee McMaster Stroke Assessment, the Reaching Performance Scale, and the Stroke Impact Scale [28]. At Hong Kong Polytechnic University, Hu et al. used four other measures [52]: the Motor Status Score (MSS, used to assess shoulder function) [35], the Modified Ashworth Scale (MAS, used to measure of increase of muscle tone) [7], the Action Research Arm Test (ARAT, used to assess grasp, grip, pinch, and gross movement) [26], and FIM (used to asses functionality in ADLs) [65]. These performance measures exemplify the clinical definition of effectiveness.

3.2.3.2 Discussion

Stroke rehabilitation is an established medical domain. The evaluations of assistive robot experiments in this domain must use relevant clinical evaluations to determine the effectiveness of the robot-augmented therapy. The scope of rehabilitative robotics for stroke-recovery patients is quite large, ranging from upper-limb recovery to gait training and wrist rehabilitation. Even within a domain, the specific performance measures differ depending on the therapy and may not translate well to another sub-domain. For example, the MSS, which is used to assess shoulder function, is applicable to the T-WREX [51] upper-arm rehabilitative aid but not to evaluating gait rehabilitation.

Functional evaluations, such as the Fugl-Meyer [43] and Wolf Motor Function [109], are crucial to comparing the effectiveness of robot-augmented therapies to one another in addition to comparing them with non-robot augmentations for current therapies. It is through these comparisons that robots can truly be evaluated as a rehabilitative device.

3.2.4 Intelligent Wheelchairs

Intelligent wheelchairs have the potential to improve the quality of life for people with disabilities. Research has focused on autonomous and semi-autonomous collision-free navigation and human-robot interaction (i.e., novel input devices and intention recognition) and has been conducted by both research institutions and companies.

3.2.4.1 End-User Evaluations

In 2005, MobileRobots (formerly ActivMedia) and researchers from the University of Massachusetts Lowell (UML) evaluated the Independence-Enhancing Wheelchair (IEW) [69, 68] with several end-users at a rehabilitation center. The original testing design planned to use a maze-like obstacle course constructed with cardboard boxes. However, this scenario did not work well for the participants. They were frustrated by a maze that was not like their regular driving environments and viewed boxes as movable objects.

Instead, the participants operated the IEW as they would typically use a wheelchair in their everyday lives (e.g., going to class which entailed moving through corridors with other people and passing through doorways). The performance measures included the number of hits and near misses and time on task. These measures were compared to the same metrics gathered during a similar length observation of the participant using his/her own wheelchair.

End-user trials have also been completed by intelligent wheelchair companies, such as DEKA [25] and CALL Centre [14] for government approval of the safety of those systems. Researchers at the University of Pittsburgh conducted an evaluation of DEKA's iBOT stair-climbing and self-balancing wheelchair with end-users [22].

3.2.4.2 Discussion

In the domain of intelligent wheelchairs, the majority of user testing has been in the form of feasibility studies with able-bodied participants. As noted by Yanco [113], able-bodied participants are more easily able to vocalize any discomforts and stop a trial quickly. These pilot experiments pave the way for end-user trials.

One barrier to end-user trials of robotic wheelchair systems is the need for the participant's seating to be moved onto the prototype system. While seating can be moved from the participant's wheelchair to the prototype system (if compatible) and back, such seating switches can take thirty to sixty minutes in each direction, making multiple testing sessions prohibitive.

We discuss performance measures commonly used thus far in feasibility studies. One of the most common tests of an autonomous intelligent wheelchair is passing through a doorway [84]. Passing through a doorway without collision is one of seven "environmental negotiations" that a person must perform in order to be prescribed a power wheelchair for mobility [99]. Other tasks include changing speed to accommodate the environment (e.g., cluttered = slow), stopping at closed doors and drop-offs (e.g., stairs and curbs), and navigating a hallway with dynamic and stationary objects (e.g., people and furniture).

In the case of these power mobility skills, the user is rated based on his/her ability to *safely* complete the task. In contrast, robotic performance measures are not binary. Performance measures include time to completion (i.e., time to pass through the doorway), number of interactions, and number of collisions. Recent performance measures include accuracy, legibility, and gracefulness of the motion [15, 91].

3.2.5 Assistive Robotic Arms

Robotic arms can improve a person's independence by aiding in activities of daily living (ADLs), such as self-care and pick-and-place tasks. Such arms can be used in fixed workstations, placed on mobile platforms, or mounted to wheelchairs. Ongoing research focuses on both the design of the arms and the human-robot interaction. The pick-and-place task, retrieving an object from a shelf or floor, is of particular interest as it is one of the most common ADLs [90]. Institutions where researchers are investigating assistive robotic arms include Georgia Institute of Technology [20], University of Pittsburgh [19], Clarkson University [41], University of Massachusetts Lowell [96], Delft University [94], and TNO Science & Industry [94].

3.2.5.1 End-User Evaluations

The Georgia Institute of Technology conducted an evaluation of laser pointers and a touch screen to control a mobile assistive robot arm, El-E [20]. Eight Amyotrophic Lateral Sclerosis (ALS or Lou Gehrig's Disease) end-users directed El-E to pick up objects from the floor in 134 trials. Performance measures included selection time for the participant to point to the object, movement time of the robot to the object, grasping time of the robot to pick up the object, and distance error. A post-experiment questionnaire with eight satisfaction questions yielded seven point Likert scale ratings. The participants' physical conditions were also assessed by a nurse using the Revised ALS Functional Rating Scale (ALSFRS-R) [17].

University of Pittsburgh researchers evaluated the effects of a Raptor arm, a commercially available wheelchair-mounted robotic arm, based on the independence of eleven users with spinal cord injury [19]. Participants first completed sixteen ADLs without the Raptor arm, then again after initial training, and once more after thirteen hours of use. At each session, the participants were timed to task completion and classified as *dependent, needs assistance,* or *independent.*

Clarkson University researchers evaluated eight users with multiple sclerosis (MS) over five ADLs with and without the Raptor arm [41]. The participants in the study all required assistance with self-care ADLs. They were evaluated before and after training on the Raptor arm. At each session, the participants were timed to task completion and interviewed. They also rated the level of difficulty of task performance and the Psychosocial Impact of Assistive Devices Scale (PIADS) [24].

Researchers at the University of Massachusetts Lowell conducted an experiment of a new visual human-robot interface for the Manus Assistive Robotic Manipulator (ARM) [29]. Eight individuals who used wheelchairs and had cognitive impairments participated in an eight-week experiment to control the robot arm in a pick-and-place task. Performance measures included time to task completion (i.e., object selection time), level of attention, level of prompting, and survey responses (i.e., preference of interface, improvements).

TNO Science & Industry and Delft University researchers conducted a four-person case study [94]. The end-users were people who used power wheelchairs and had weak upper limb strength and intact cognition. TNO Science & Industry evaluated their graphical user interface for the Manus ARM. The performance measures included number of mode switches, task time, Rating Scale of Mental Effort (RSME) [114], and survey responses including the participants' opinions about the tasks, the robot arm control methods, and impression of the user interface [94].

3.2.5.2 Discussion

As demonstrated by Tsui et al. [96], Tijsma et al. [94], and Fulk et al. [41], it is also important to account for the user's experience with respect to cognitive workload and mental and emotional state. The basis for the user's experience performance measure must be derived or adapted from an existing clinical measure.

In Tsui et al. [96] and Tijsma et al. [94], the participants were rated or rated themselves with respect to cognitive workload. In Tsui et al. [96], the level of prompting was a cognitive measure based on FIM, a measurement of functional independence [65], where in the user is rated on a semantic differential scale (*1 = needs total assistance* to *7 = has complete independence*) on a variety of ADLs. Choi et al. [20] indirectly investigated cognitive workload using an human-computer interaction inspired survey. The participants rated statements such as "It was easy to find an object with the interface" and "It was easy to learn to use the system" on a seven point Likert scale [60] (*−3 = strongly disagree* to *3 = strongly agree*).

FIM may also be applied as a cognitive measure to activities such as "comprehension, expression, social interaction, problem solving, and memory" [65]. In Tijsma et al. [94], RSME was used as a cognitive performance measure. RSME is a 150 point scale measuring the mental effort needed to complete a task, where *0 = no effort* and *150 = extreme effort*. The Standardized Mini-Mental State Examination [70] is another cognitive performance measures used in older adults.

In Fulk et al. [41], participants ranked the perceived difficulty of the task and their mental and emotional state were recorded using PIADS. PIADS is a twenty-six item questionnaire in which a person rates their perceived experience after completing a task with an assistive technology device [23]. It measures the person's feelings of competence, willingness to try new things, and emotional state. PIADS is well established and significantly used in the US and Canada [23]. An alternative emotional measure is the Profile of Mood States [64] used in Wada et al. [102].

3.2.6 External Limb Prostheses

Robotic prostheses can serve as limb replacements. Researchers have investigated creating novel robotic prostheses and control strategies. A number of prosthesis evaluations conducted have been feasibility studies on healthy subjects. As such, the focus of the experiments has largely been on the performance of the prostheses themselves. The performance measures include joint angle, joint torque, and power consumption. However, several research institutions have conducted end-user evaluations, including RIC [67, 57], Northwestern University

[67, 57], Massachusetts Institute of Technology [4, 5], and Hong Kong Polytechnic University [58].

3.2.6.1 End-User Evaluations

RIC and Northwestern University conducted a clinical evaluation of six individuals who underwent targeted muscle reinnervation (TMR) surgery [67]. After the upper limb prosthetic device was optimally configured for each patient's electromyography signals (EMG), functional testing occurred after the first month, third month, and sixth month. The functional testing was comprised of a series of standard tests: box and blocks, clothespin relocation, Assessment of Motor and Process Skills (AMPS) [36], and the University of New Brunswick prosthetic function [82]. Performance measures included time to complete task, accuracy, and AMPS score.

Another RIC and Northwestern University study evaluated the effectiveness of the TMR procedure when controlling robotic prostheses with EMG signals [57]. Five participants with shoulder-disarticulation or transhumeral amputation who had the TMR procedure and five able-bodied participants controlled a virtual prosthetic arm to grip in three predetermined grasps. The performance measures included motion selection time (time from when motion began to correct classification), motion completion time, and motion completion rate.

Additionally, three of the participants who had undergone TMR also used physical robotic upper-limb prostheses (i.e., DEKA's ten degree-of-freedom "Luke arm" and a motorized seven degree-of-freedom prosthetic arm developed at John Hopkins University) using EMG signals [57]. The training and testing ran for two weeks with one session in the morning and another in the afternoon; session lasted two to three hours. The participants were able to operate the prostheses in ADL-type tasks and controlling grasps. These results are largely anecdotal.

Researchers at the Massachusetts Institute of Technology (MIT) conducted a clinical evaluation with three unilateral, transtibial amputees [4]. Data collection included oxygen consumption, carbon dioxide generation, joint torque, and joint angle. Kinematic and kinetic data were collected using a motion capture system for the ankle-foot prosthesis and unaffected leg. The resulting performance measures were metabolic cost of transport, gait symmetry between the legs, vertical ground reaction forces, and external work done at the center of mass of each leg.

Hong Kong Polytechnic University researchers conducted a clinical evaluation with four transtibial amputees over the course of three consecutive days [58]. Data collected included motion capture and open-ended responses about the participant's comfort and the prosthesis' stability, ease of use, perceived flexibility, and weight. Stance time, swing time, step length, vertical trunk motion, and average velocity were derived from the motion capture data. Performance measures included ranking of the prostheses used (with respect to comfort, stability, ease of use, perceived flexibility, and weight), gait symmetry, and ground force reactions.

3.2.6.2 Discussion

Performance measures involving ADLs can be used in evaluating prostheses because ADLs include functions such a locomotion and self-care activities. Locomotion includes walking and climbing stairs, and self-care activities involve a high level of dexterity. Heinemann et al. [48] proposed the Orthotics and Prosthetics Users' Survey (OPUS). Burger et al. [12] in turn evaluated the Upper Extremity Functional Status of OPUS with sixty-one users with unilateral, upper limb amputations and found that the scale was suitable for the measuring functionality of the population. The Upper Extremity Function Status is comprised of twenty-three ADLs, rated in a semantic differential scale fashion (*0 = unable to complete* to *3 = very easy to complete*). AMPS is also comprised of ADLs but in a more flexible fashion; there are eighteen categories of ADLs with up to eleven choices within a category [2].

The clinical evaluations conducted with transtibial amputees discussed above used performance measures of the robotic system itself (i.e., gait symmetry and ground force reactions). Additionally, Hong Kong Polytechnic University administered a questionnaire asking about the participant's perception of the lower limb prosthesis, and, in an indirect manner, MIT measured the ease of use of the prosthesis by a biological means. In order for a prosthesis to gain clinical validity, performance of the device must also have a measure of use in daily life.

3.3 Case Studies

Next, we further explore two examples detailing the evolution of performance metrics on two different ongoing studies involving assistive robotics. At the University of Massachusetts Lowell (UML), we have conducted one able-bodied experiment and three end-user experiments with people who use wheelchairs involving an assistive robotic arm in the "pick-and-place" activity of daily living. At the University of Southern California (USC), we have conducted three preparatory experiments with end-users and are in the process of conducting an end-user experiment using a socially assistive robot designed to provoke or encourage exercise or social behavior in children with Autism Spectrum Disorders (ASD).

3.3.1 Designing Evaluations for an Assistive Robotic Arm

At UML, our research focuses on providing methods for independent manipulation of unstructured environments to wheelchair users using a wheelchair-mounted robot arm. Our target audience consists of people with physical disabilities who may additionally have cognitive impairments. We have investigated a visual interface compatible with single switch scanning [95], a touch screen interface [96], a mouse-emulating joystick [96], and a laser pointer joystick device [75]. By explicitly pointing to the desired object, it may be possible to expand the end-user population to include people with low cognition. We conducted a preliminary experiment with able-bodied participants as an evaluation baseline in August 2006 [95]. The first field trial was conducted with users who use wheelchairs and additionally had cognitive impairments in August and September 2007 [96]. The second field trial was conducted in August and September 2008. Our third field trial will begin in mid-July 2009 and run through the end of October 2009. In this section, we discuss our design of the end-user experiments.

In our first end-user evaluation, we compared the visual interface presentation (stationary camera vs. moving camera) and input device (touch screen vs. joystick) [96]. We collected data from video, manual logs, post-session questionnaires, and computer generated log files. We collected both qualitative and quantitative data. The qualitative data included the post-experiment questionnaire administered after each user session and the observer notes. The questionnaire posed open-ended questions about which interface the user liked most to date, which interface he/she liked least to date, and suggestions for improving the interface. The observer notes contained additional relevant data about the session, including length of reorientation.

The quantitative data included an attentiveness rating, prompting level, trial run time, close-up photos of the object selected, and computer-generated log files. The attentiveness rating and prompting level were developed by our clinicians. The experimenter, who was an assistive technology professional, rated the user's prompting level per trial based on the FIM scale, where $0 = no\ prompting\ needed$ and $5 = heavy\ prompting\ needed$ [65]. The experimenter also rated the user's attentiveness to the task on a semantic differential scale, where $0 = no\ attention$ and $10 = complete\ attention$. Two separate scales were used because it is not necessarily the case that a person who requires high levels of prompting is unmotivated to complete the task. Also, for each trial, the run time was recorded, specifically the time from object prompt to participant selection, the time from the Manus ARM movement to the object being visually confirmed, and the fold time. We focused on the object selection time prompting level, and attention level as our primary performance metrics and computed paired t-tests to determine statistical significance.

In our second end-user evaluation, we compared a custom laser pointer device against the touch screen interface with stationary view[2] [75]. We were very satisfied with quality of the primary performance metrics from the first end-user evaluation. Thus, we based the next version of our data collection tools from the previous experiment and made several modifications. We recorded only the object selection time by the participant since the remainder of the process time is a robot system performance measurement. The post-session surveys were also based on the ones from the first end-user evaluation. Because the participants used two interfaces in the second end-user evaluation, we modified the post-session survey to investigate which aspects of each interface the participants liked and did not like, comments about the laser joystick and touch screen, and which interface they liked better.

---

[2]The touch screen interface with the stationary camera view had the best overall performance from the first end-user evaluation [96].

At the suggestion of our clinicians, we updated the semantic differential scales to have the same range (i.e., [1, 5], where $1 = no$ *prompting* needed and $5 = heavy$ *prompting* for the prompting level, and $1 = not$ *attentive* and $5 = very$ *attentive* for attention level) which provided the same granularity across the performance measurements. We introduced a tally box for the number of prompts given by the experimenter which provided the ability to better understand what an experimenter considered a "high" level of prompting versus "low." In our observations of the first end-user evaluation, we noticed that the participants would seem quiet on one day and excited on another. To better understand how the performance of our robotic arm system was perceived by the participants, we added a mood and arousal level rating (i.e., $1 = very$ *bad* and $5 = very$ *good* for mood, and $1 = less$ *than normal* and $5 = more$ *than normal*) to be administered at the start of the session, before the condition change, and after the session. We added a health rating (i.e., $1 = poor$ and $5 = good$) to be administered at the start of the session. These mood, arousal, and health ratings were items previously noted by the experimenter in the first end-user evaluation.

Our upcoming third end-user evaluation will investigate how different levels of cognition (i.e., *high*, *medium*, and *low* as classified by our clinician) impact a person's ability to use the robotic arm. We will continue to use the selection time and prompting, attention, mood, arousal, and health levels. We found that the second end-user experiment's ratings scales were not as effective as those in the first end-user evaluation. The second end-user experiment's rating scales were relative to each participant's typical performance, and we did not see much change. For the upcoming evaluation, we will instantiate the semantic differential scales in a concrete manner (i.e., for arousal, $1 = low$ and $5 = high$). We will also incorporate aspects of the Psychosocial Impact of Assistive Devices Scale (PIADS), in which the participants will rate their perceived experiences with the Manus ARM [24].

3.3.2 Designing Evaluations for Socially Assistive Robots

At USC, our research focuses on the use of socially assistive robots that provide assistance through social interaction rather than physical interation [31]. We are developing robot systems for encouraging and training social behavior for children with ASD. Our work focuses on the following goals: the automatic identification of social behavior; the creation of a toolkit of interactive robot behavior that can be used in order to provoke and encourage social interaction; and the determination of therapeutic effectiveness of socially assistive robots.

The robot we have developed is a humanoid torso (with movable arms, neck, head, and face) on a mobile base. The robot has microphones and speakers, so that it can make and recognize vocalizations, buttons that the user can press, and a bubble blower (typically used as part of standardized ASD evaluation and intervention [61]). The robot "rewards" a user's social behavior with its own social behavior, including gestures, vocalizations, and movements (approach, spinning in place, facing the user). Additionally, we use a camera and microphones in the experimental room at the clinic to collect high-fidelity multi-modal data of all aspects of the study.

We are developing a system that observes a child with ASD and automatically identifies social behaviors (e.g., approach, turn-taking, social referencing, appropriate affect, and/or vocalizations). Our long-range goal is to use the frequency and context of those behaviors in order for the robot to determine autonomously if the child is socially engaging with the robot or another person. Our studies use an overhead camera system to track and interpret the child's movement as he/she interacts with the robot [32]. The goal is for the overhead camera to autonomously identify the movement behavior of the robot and the child (such as A approaches B, B follows A, etc.). We conducted a preliminary experiment in which we collected supervised overhead camera data. We created a data set in which the person and the robot executed known actions. The system then performed an automated analysis of the data which was accompanied by blind human coding. Because the actions were known *a priori*, we were able to determine an absolute measure of the automatic coding mechanism's performance. We compared the accuracy of the automatic observations to human coding of the same actions which provided a relative measure of the systems performance. Our larger experiments employ a similar experimental data collection, coding, and evaluation model.

After validating that the socially assistive robot system is both effective at interacting with the child with ASD (i.e., it successfully elicits social behaviors) [34] and that our analysis of its effectiveness is valid (i.e., the coding and analysis algorithms) [32], the next step in validation must address any possible therapeutic benefits of the human-robot interaction. Our goal is not to presume or aim for a clinical benefit, but to validate reliably that such a socially assistive robot could have a potential therapeutic impact in order to plan follow-up studies. We are working with clinicians to determine if there are any therapeutic applications for such a robot system.

We are currently planning a validation experiment with end-users. The experiment will involve multiple sessions in which a child with ASD will participate in free-play with a trained therapist. Participants will be split into control and experimental groups. The control group will interact only with the therapist, while the experimental group will interact with the therapist and the robot for five sessions. The participants will be given a pre- and post-experiment consisting of the WISC-III Intelligence Test [106], the communication portion of the Vineland Adaptive Behavior Scales (VABS) [72], and an ADOS-inspired quantitative observation. These measures are used for repeated administration and comparison. Hypothesis testing will then involve comparing any change from the pre- and post-evaluation between the control and experimental groups. Paired t-tests will be used to compare each test and any applicable sub-test. The variety of scales to be used in this upcoming study provides a rich range of measures pertaining to the social ability of the end-user.

3.4 Incorporating Functional Performance Measures

Evaluation of assistive robotic technology varies widely, as has been demonstrated by our exploration of several domains. It is clear that in order for assistive robotic technology to be accepted by clinicians at large, end-user evaluations must incorporate a functional performance measure based on the "gold standard" of the specific domain, if one has been established. We say that a performance measure is *functional* if it relates to an activity of daily living and is administered in a realistic setting. In this survey, we have found examples of functional performances used in the majority of the surveyed domains.[3]

Feil-Seifer et al. consulted the Autism Diagnostic Observation Schedule (ADOS) in their evaluation [33]. ADOS is one of the ASD "gold standard" assessment tools; it investigates "social interaction, communication, play, and imaginative use of materials" [61]. The Vineland Adaptive Behavior Scales (VABS) are a grouping of assessment tools for ASD and developmental delays [72, 16]. Unlike ADOS, VABS contains functional components, such as Daily Living Skill items and Motor Skill items. VABS has also been used in the domains of stroke (children) [63], wheelchairs (children) [27], prostheses (children) [73], and eldercare (developmentally disabled) [54].

We discussed the quality of life (QoL) measurements in the context of the eldercare domain. QoL measurement scales contain functional components to them, such as walking, climbing stairs, and self-care activities. The SF-36 [74] is a generic health measure which includes evaluations of physical function and limitations due to physical health. The SF-36 contains 36 items in total, including one multi-part question focused specifically on typical daily activities; see Table 3.1. The SF-36 has been used for assessment of eldercare [42], wheelchairs [13], stroke [3, 50], and a prosthesis [59]. The 15-D measures QoL as a profile with 15 dimensions; see Table 3.1. Mobility and eating are included as profile dimensions [86]. The 15-D has been used in the eldercare, orthopedic, and stroke rehabilitation domains [86].

The FIM scale [9] by definition is a functional performance measure; see Table 3.1. The FIM ADLs include "eating, grooming, bathing, dressing (upper body), dressing (lower body), toileting, bladder management, bowel management, transferring (to go from one place to another) in a bed, chair, and/or wheelchair, transferring on and off a toilet, transferring into and out of a shower, locomotion (moving) for walking or in a wheelchair, and locomotion going up and down stairs" [65]. FIM has been used largely in stroke rehabilitation [38] and to some extent in eldercare [55]. WeeFIM [98] is used for children between the ages of six months and seven years (present with functional abilities of or below age seven).WeeFIM has recently been used in clinical Autism studies in Hong Kong [110, 111]. FIM has been adapted for wheelchair users [87]. As described in Section 3.3.1, FIM inspired a scale for recording a user's prompting level while doing a task [96].

---

[3]A functional performance measure was not surveyed for the domain of Autism Spectrum Disorders (ASD).

Table 3.1 Examples of functional performance measurement tools

Tool	Length	Categories of functional assessment	Rating	Applicable domain(s)
15D [86]	15 items	mobility, vision, hearing, breathing, sleeping, eating, speech, elimination, activities of daily life, mental function, discomfort and symptoms, depression, distress, vitality, sexual activity	$n \in [1, 5]$, where 1 = cannot do, 5 = normal ability	Eldercare, post-stroke rehabilitation
FIM [9]	18 items	feeding, grooming, bathing, dressing (upper and lower body), toileting, bladder management, bowel management, bed transfer, toilet transfer, tub transfer, walking/wheelchair, climbing stairs, comprehension, expression, social interaction, problem solving, memory	$n \in [1, 7]$, where 1 = needs total assistance and 7 = complete independence	Autism Spectrum Disorders, assistive robotic arms, eldercare, post-stroke rehabilitation, wheelchair
Motor Assessment Scale [6]	8 items	supine to side lying, supine to sitting over the edge of a bed, balanced sitting, sitting to standing, walking, upper-arm function, hand movements, advanced hand activities	$n \in [0, 6]$, where 6 = optimal motor behavior	Post-stroke rehabilitation
SF-36 [74]	36 items	typical daily activities (strenuous activities, moderate activities, carrying groceries, climbing several flights of stairs, climbing one flight of stairs, bending/kneeling, walking more than a mile, walking several blocks, walking one block, bathing/dressing yourself)	$n \in [0, 3]$, where 0 = limited a lot, 3 = not limited	Eldercare, post-stroke rehabilitation, prostheses, wheelchair
Vineland Adaptive Behavior Scales (survey form) [72]	297 items	daily living skills, motor skills (gross motor, fine motor)	$n \in [0,2]$, where 0 = never performed without help or reminders and 2 = usually or habitually performed without help or reminders	Autism Spectrum Disorders, eldercare, post-stroke rehabilitation, prostheses, wheelchairs

Currently, there is a large gap between robotic performance measures and functional performance measures. Robotic performance measures typically consider metrics such as time on task and number of collisions, while functional performance measures mentioned above do not employ such a fine level of granularity. The functional performance measures examine tasks and categorize a person's ability to complete those tasks in a *n*-nary manner (e.g., Motor Assessment Scale is ternary with "was able to complete easily," "was able to complete with some difficulty," "was not able to complete" [6]); see Table 3.1. To create finer granularity in functional performance measures, intermediate layers can be added. For example, Stineman et al. added intermediate layers to FIM in order to understand the causal relationship between impairments and disabilities [89].

Functional performance measures take a significant amount of time to administer, ranging from 20 to 60 minutes or more. The "gold standard" functional measure for a given domain thus should be administered once before the experiment and once after. However, a *subset* of the functional performance measures relevant to the specific area investigated can be used during each testing session, as has been done in some of the studies references above. Furthermore, robotic performance measures derived from this subset can provide continuous monitoring. These functional robotic performance measures may then help to bridge the gap between the strictly robotic performance measures and the functional performance measures commonly employed in clinical domains.

3.5 Conclusions

To be useful, performance measures should be specific to the domain and relevant to the task. Domains with clear, well-established medical or therapeutic analogs can leverage existing clinical performance measures. Domains without such strong therapeutic analogs should appropriately borrow and adapt clinical performance measures. Alternatively, they may draw inspiration from a clinical measure to create a new one or augment an existing one if none of the existing measures are appropriate [45].

Evaluations conducted with end-users should focus at least as much on human performance measures as they do on system performance measures. By placing the emphasis on human performance, it becomes possible to correlate system performance with human performance. Celik et al. examined trajectory error and smoothness of motion with respect to Fugl-Meyer in the context of post-stroke rehabilitation [18]. Similarly, Brewer et al. have used machine learning techniques on sensor data to predict the score of a person with Parkinson's disease on the Unified Parkinson Disease Rating Scale (UPDRS) [10, 30].

Existing performance measures for most of assistive robotic technologies do not provide sufficient detail for experimental and clinical evaluations. We have provided a summary of applicable performance measures (see Table 3.2) and offer the following guidelines for choosing appropriate and meaningful performance measures:

Table 3.2 Summary of assistive robotic technology performance measures

Domain	Applicable performance measures
General AT	Activities of daily living, coding, instantiated Likert-type ratings, mood, quality of life, stress, time on task
Autism spectrum disorders	Behavior coding, correlate sensor modeling of behavior to human-rated behavior, standardized assessments (e.g., ADOS, Vineland Adaptive Behavior Scales)
Eldercare	Activities of daily living (e.g., FIM, SBBP), mood (e.g., Profile of Mood States), quality of life (e.g., 15-D, SF-36), response correctness, response time, stress (e.g., Standardized Mini-Mental State)
Post-stroke rehabilitation	Functional performance measures (e.g., FIM, Motor Activity Log, Motor Assessment Scale), quality of life (e.g., 15-D, SF-36), standardized assessments (e.g., ARAT, Chedoke-McMaster, Fugl-Meyer, Modified Ashworth Scale, MSS, Reaching Performance Scale, Wolf Motor)
Intelligent wheelchairs	Accuracy, functional performance measures (e.g., FIM), gracefulness, number of hits/near misses, quality of life (e.g., SF-36), time on task
Assistive robotic arms	Activities of daily living (e.g., ALSFRS-R, FIM), attention, level of prompting, mental state (e.g., RSME, Profile of Mood States, PIADS), mood, quality of life, time to task completion
Prostheses	Accuracy, biological measures of effort (e.g., oxygen consumption), comfort, ease of use, functional performance measures (e.g., AMPS, OPUS, FIM), quality of life (e.g., SF-36), time to complete task

- Consult a clinician who specializes in the particular domain.
- Choose an appropriate clinical measure for the domain. A domain's "gold standard" will provide the best validity to clinicians, if one exists.
- Include a functional performance measure appropriate for the domain.
- Choose an appropriate method to capture a participant's emotional and mental state.
- Consider an appropriate quality of life measurement.
- Administer the human performance measures at least once before and after the experiment or study.
- Consider coding open-ended responses, comments, and/or video.
- Concretely define each enumeration on Likert and differential semantic scales.

By choosing meaningful performance measures, robotics researchers provide a common ground for interpretation and acceptance of robot-assisted therapy systems by the clinical community. In addition, the robotic system developers are also given clear guidelines for how to define, observe, and evaluate system performance.

In this paper, we have sought well-established performance measures to apply to assistive robotic technologies and encourage the practice of their use in our field. Common performance measurements will allow researchers to compare the state of the art approaches within specific robotics domains and to compare against the state of the practice within the relevant clinical field outside of the robotics community.

Acknowledgments This work is funded in part by the National Science Foundation (IIS-0534364, IIS-0546309, IIS-0713697, CNS-0709296), the National Academies Keck Futures Initiative (NAKFI), the USC NIH Aging and Disability Resource Center (ADRC) pilot program, and the Nancy Laurie Marks Family Foundation. The authors thank Kristen Stubbs of UMass Lowell.

References

1. N. K. Aaronson, C. Acquadro, J. Alonso, G. Apolone, D. Bucquet, M. Bullinger, K. Bungay, S. Fukuhara, B. Gandek, S. Keller, D. Razavi, R. Sanson-Fisher, M. Sullivan, S. Wood-Dauphinee, A. Wagner, and J. E. Ware Jr. International Quality of Life Assessment (IQOLA) Project. *Quality of Life Research*, 1(5):349–351, 2004.
2. AMPS.com. AMPS Project International (Assessment of Motor and Process Skills), 2006. http://www.ampsintl.com/tasks.htm. Accessed Mar. 1, 2009.
3. C. Anderson, S. Laubscher, and R. Burns. Validation of the Short Form 36 (SF-36) Health Survey Questionnaire Among Stroke Patients. *Stroke*, 27(10):1812–1816, 1996.
4. S. Au. *Powered Ankle-Foot Prosthesis for the Improvement of Amputee Walking Economy*. PhD thesis, MIT, 2007.
5. S. Au, M. Berniker, and H. Herr. 2008 Special Issue: Powered Ankle-Foot Prosthesis to Assist Level-Ground and Stair-Descent Gaits. *Neural Networks*, 21(4):654–666, 2008.
6. L. Blum, N. Korner-Bitensky, and E. Sitcoff. StrokEngine (MAS), 2009. http://www.medicine.mcgill.ca/strokengine-assess/module_mas_indepth-en.html. Accessed Mar. 1, 2009.
7. R. Bohannon and M. Smith. Interrater Reliability of a Modified Ashworth Scale of Muscle Spasticity. *Physical Therapy*, 67(2):206–7, 1987.
8. R. Bourgeois-Doyle. *George J. Klein: The Great Inventor*. NRC Research Press, 2004.
9. Brain Injury Resource Foundation. Functional Independence Measure (FIM), 2009. http://www.birf.info/home/bi-tools/tests/fam.html. Accessed Mar. 1, 2009.
10. B. R. Brewer, S. Pradhan, G. Carvell, P. Sparto, D. Josbeno, and A. Delitto. Application of Machine Learning to the Development of a Quantitative Clinical Biomarker for the Progression of Parkinson's Disease. In *Rehabilitation Engineering Society of North America Conf.*, 2008.
11. J. Brody. Prospects for an Ageing Population. *Nature*, 315(6019):463–466, 1985.
12. H. Burger, F. Franchignoni, A. Heinemann, S. Kotnik, and A. Giordano. Validation of the Orthotics and Prosthetics User Survey Upper Extremity Functional Status Module in People with Unilateral Upper Limb Amputation. *Journal of Rehabilitation Medicine*, 40(5): 393–399, 2008.
13. T. Bursick, E. Trefler, D. A. Hobson, and S. Fitzgerald. Functional Outcomes of Wheelchair Seating and Positioning in the Elderly Nursing Home Population. In *Rehabilitation Engineering Society of North America Conference*, pages 316–318, 2000.
14. CALL Centre. Smart Wheelchair, 2006. http://callcentre.education.ed.ac.uk/Smart_Wheel Ch/smart_wheelch.html. Accessed Mar. 1, 2009.
15. T. Carlson and Y. Demiris. Human-Wheelchair Collaboration Through Prediction of Intention and Adaptive Assistance. In *IEEE International Conference on Robotics and Automation*, 2008.
16. A. Carter, F. Volkmar, S. Sparrow, J. Wang, C. Lord, G. Dawson, E. Fombonne, K. Loveland, G. Mesibov, and E. Schopler. The Vineland Adaptive Behavior Scales: Supplementary Norms for Individuals with Autism. *Journal of Autism and Developmental Disorders*, 28(4):287–302, 1998.

17. J. Cedarbaum, N. Stambler, E. Malta, C. Fuller, D. Hilt, B. Thurmond, and A. Nakanishi. The ALSFRS-R: A Revised ALS Functional Rating Scale that Incorporates Assessments of Respiratory Function. BDNF ALS Study Group (Phase III). *Journal of Neurological Sciences*, 169(1-2):13–21, 1999.

18. O. Celik, M. K. O'Malley, C. Boake, H. Levin, S. Fischer, and T. Reistetter. Comparison of Robotic and Clinical Motor Function Improvement Measures for Sub-Acute Stroke Patients. In *IEEE International Conference on Robotics and Automation*, 2008.

19. E. Chaves, A. Koontz, S. Garber, R. Cooper, and A. Williams. Clinical Evaluation of a Wheelchair Mounted Robotic Arm. Technical report, Univ. of Pittsburgh, 2003.

20. Y. Choi, C. Anderson, J. Glass, and C. Kemp. Laser Pointers and a Touch Screen: Intuitive Interfaces for Autonomous Mobile Manipulation for the Motor Impaired. In *International ACM SIGACCESS Conference on Computers and Accessibility*, pages 225–232, 2008.

21. J. A. Cohen. A Coefficient of Agreement for Nominal Scales. *Educational and Psychological Measurement*, 20:37–46, 1960.

22. R. Cooper, M. Boninger, R. Cooper, A. Dobson, J. Kessler, M. Schmeler, and S. Fitzgerald. Use of the Independence 3000 IBOT Transporter at Home and in the Community. *Journal of Spinal Cord Medicine*, 26(1):79–85, 2003.

23. H. Day and J. Jutai. PIADS in the World, 2009. http://www.piads.ca/worldmapshtm/worldmap.asp. Accessed Mar. 1, 2009.

24. H. Day, J. Jutai, and K. Campbell. Development of a Scale to Measure the Psychosocial Impact of Assistive Devices: Lessons Learned and the Road Ahead. *Disability and Rehabilitation*, 24(1–3):31–37, 2002.

25. DEKA Research and Development Corporation. DEKA Evolved Thinking, 2007. http://www.dekaresearch.com. Accessed Mar. 1, 2009.

26. W. DeWeerdt and M. Harrison. Measuring Recovery of Arm-Hand Function in Stroke Patients: A Comparison of the Brunnstrom-Fugl-Meyer Test and the Action Research Arm Test. *Physiother Can*, 37(2):65–70, 1985.

27. M. Donkervoort, M. Roebroeck, D. Wiegerink, H. van der Heijden-Maessen, and H. Stam. Determinants of Functioning of Adolescents and Young Adults with Cerebral Palsy. *Disability and Rehabilitation*, 29:453–463, 2007.

28. M. D. Ellis, T. Sukal, T. DeMott, and J. P. A. Dewald. ACT^{3D} exercise targets gravity-induced discoordination and improves reaching work area in individuals with stroke. In *IEEE International Conference on Rehabilitation. Robotics*, 2007.

29. Exact Dynamics. Assistive Robotic Manipulator, 2004. http://www.exactdynamics.nl/. Accessed Mar. 1, 2009.

30. S. Fahn, R. Elton, et al. Unified Parkinson's Disease Rating Scale. *Recent Developments in Parkinson's Disease*, 2:153–163, 1987.

31. D. Feil-Seifer and M. Matarić. Socially Assistive Robotics. In *International Conference on Rehabilitation Robotics*, pages 465–468, 2005.

32. D. J. Feil-Seifer and M. J. Matarić. B3IA: An Architecture for Autonomous Robot-Assisted Behavior Intervention for Children with Autism Spectrum Disorders. In *International Workshop on Robot and Human Interactive Communication*, Munich, Germany, Aug 2008.

33. D. J. Feil-Seifer and M. J. Matarić. Robot-Assisted Therapy for Children with Autism Spectrum Disorders. In *Conference on Interaction Design for Children: Children with Special Needs*, 2008.

34. D. J. Feil-Seifer and M. J. Matarić. Toward Socially Assistive Robotics For Augmenting Interventions For Children With Autism Spectrum Disorders. In *International Symposium on Experimental Robotics*, 54:201–210, 2008.

35. M. Ferraro, J. Demaio, J. Krol, C. Trudell, K. Rannekleiv, L. Edelstein, P. Christos, M. Aisen, J. England, and S. Fasoli. Assessing the Motor Status Score: A Scale for the Evaluation of Upper Limb Motor Outcomes in Patients After Stroke. *Neurorehab. and Neural Repair*, 16(3):283, 2002.

36. A. Fisher. AMPS: Assessment of Motor and Process Skills Volume 1: Development, Stan-dardisation, and Administration Manual. *Ft Collins, CO: Three Star Press Inc.*, 2003.
37. J. Fleiss. Measuring Nominal Scale Agreement Among Many Raters. *Psychological Bulletin*, 76(5):378–382, 1971.
38. W. Foczkowski and S. Barreca. The Functional Independence Measure: Its Use to Identify Rehabilitation Needs in Stroke Survivors. *Archives of Physical Medicine and Rehabilitation*, 74(12):1291–1294, 1993.
39. B. Freeman. Guidelines for Evaluating Intervention Programs for Children with Autism. *Journal of Autism and Developmental Disorders*, 27(6):641–651, 1997.
40. A. Fugl-Meyer, L. Jaasko, I. Leyman, S. Olsson, and S. Steglind. The Post-Stroke Hemiplegic Patient. 1. A Method for Evaluation of Physical Performance. *Scandinavian Journal of Rehabilitation Medicine*, 7(1):13–31, 1975.
41. G. Fulk, M. Frick, A. Behal, and M. Ludwig. A Wheelchair Mounted Robotic Arm for Individuals with Multiple Sclerosis. Technical report, Clarkson Univ., 2005.
42. B. Gale. Faculty Practice as Partnership with a Community Coalition. *Journal of Professional Nursing*, 14(5):267–271, 1998.
43. D. Gladstone, C. Danells, and S. Black. The Fugl-Meyer Assessment of Motor Recovery after Stroke: A Critical Review of Its Measurement Properties. *Neurorehabilitation and Neural Repair*, 16(3):232, 2002.
44. D. Greenwood, W. Whyte, and I. Harkavy. Participatory Action Research as a Process and as a Goal. *Human Relations*, 46(2):175–192, 1993.
45. J. Guralnik, E. Simonsick, L. Ferrucci, R. Glynn, L. Berkman, D. Blazer, P. Scherr, and R. Wallace. A Short Physical Performance Battery Assessing Lower Extremity Function: Association with Self-Reported Disability and Prediction of Mortality and Nursing Home Admission. *Journal of Gerontology*, 49(2):M85–94, 1994.
46. G. Guyatt. The 6-minute walk: a new measure of exercise capacity in patients with chronic heart failure. *Canadian Medical Association Journal*, 132(8):919–923, 1985.
47. K. Haigh and H. A. Yanco. Automation as Caregiver: A Survey of Issues and Technologies. In *AAAI-2002 Workshop on Automation as Caregiver: The Role of Intelligent Technology in Elder Care*, 2002.
48. A. W. Heinemann, R. K. Bode, and C. O'Reilly. Development and Measurement Properties of the Orthotics and Prosthetics Users' Survey (OPUS): A Comprehensive Set of Clinical Outcome Instruments. *Prosthetics and Orthotics International*, 27(3):191–206, 2003.
49. M. Hillman. Rehabilitation Robotics from Past to Present—A Historical Perspective. In *IEEE International Conference on Rehabilitation Robotics*, 2003.
50. J. Hobart, L. Williams, K. Moran, and A. Thompson. Quality of Life Measurement After Stroke Uses and Abuses of the SF-36. *Stroke*, 33(5):1348–1356, 2002.
51. S. J. Housman, V. Le, T. Rahman, R. J. Sanchez, and D. J. Reinkensmeyer. Arm-Training with T-WREX After Chronic Stroke: Preliminary Results of a Randomized Controlled Trial. In *IEEE International Conference on Rehabilitation Robotics*, 2007.
52. X. L. Hu, K. Y. Tong, R. Song, X. J. Zheng, I. F. Lo, and K. H. Lui. Myoelectrically Controlled Robotic Systems That Provide Voluntary Mechanical Help for Persons after Stroke. In *IEEE International Conference on Rehabilitation Robotics*, 2007.
53. W. Jenkins and M. Merzenich. Reorganization of Neocortical Representations After Brain Injury: A Neurophysiological Model of the Bases of Recovery from Stroke. *Progress in Brain Research*, 71:249–66, 1987.
54. D. Kerby, R. Wentworth, and P. Cotten. Measuring Adaptive Behavior in Elderly Developmentally Disabled Clients. *Journal of Applied Gerontology*, 8(2):261, 1989.
55. R. Kleinpell, K. Fletcher, and B. Jennings. Reducing Functional Decline in Hospitalized Elderly. *Patient Safety and Quality: An Evidence-Based Handbook for Nurses. AHRQ Publication No. 08-0043*, 2007.

56. H. Kozima and C. Nakagawa. Longitudinal Child-Robot Interaction at Preschool. In *AAAI Spring Symposium on Multidisciplinary Collaboration for Socially Assistive Robotics*, pages 27–32, 2007.

57. T. Kuiken, G. Li, B. Lock, R. Lipschutz, L. Miller, K. Stubblefield, and K. Englehart. Targeted Muscle Reinnervation for Real-time Myoelectric Control of Multifunction Artificial Arms. *Journal of American Medical Association*, 301(6):619–628, 2009.

58. W. Lee, M. Zhang, P. Chan, and D. Boone. Gait Analysis of Low-Cost Flexible-Shank Trans-Tibial Prostheses. *IEEE Transactions on Neural Systems and Rehabi*, 14(3):370–377, 2006.

59. M. Legro, G. Reiber, M. Del Aguila, M. Ajax, D. Boone, J. Larsen, D. Smith, and B. Sangeorzan. Issues of Importance Reported by Persons with Lower Limb Amputations and Prostheses. *Journal of Rehabilitation Research and Development*, 36(3):155–163, 1999.

60. R. Likert. A Technique for the Measurement of Attitudes. *Archives of Psychics*, 140(5):1–55, 1932.

61. C. Lord, S. Risi, L. Lambrecht, E. H. Cook Jr., B. L. Leventhal, P. C. DiLavore, A. Pickles, and M. Rutter. The Autism Diagnostic Observation Schedule-Generic: A Standard Measure of Social and Communication Deficits Associated with the Spectrum of Autism. *Journal of Autism and Developmental Disorders*, 30(3):205–223, 2000.

62. M. Matarić, J. Eriksson, D. Feil-Seifer, and C. Winstein. Socially Assistive Robotics for Post-Stroke Rehabilitation. *Journal of NeuroEngineering and Rehabilitation*, 4(1):5, 2007.

63. J. Max, K. Mathews, A. Lansing, B. Robertson, P. Fox, J. Lancaster, F. Manes, and J. Smith. Psychiatric Disorders After Childhood Stroke. *Journal of the American Academy of Child & Adolescent Psychiatry*, 41(5):555, 2002.

64. D. M. McNair, M. Lorr, and L. F. Droppleman. Profile of Mood States. In *Educational and Industrial Testing Service*, 1992.

65. MedFriendly.com. MedFriendly.com: Functional Independence Measure, 2007. http://www.medfriendly.com/functionalindependencemeasure.html. Accessed Mar. 1, 2009.

66. F. Michaud, T. Salter, A. Duquette, H. Mercier, M. Lauria, H. Larouche, and F. Larose. Assistive Technologies and Child-Robot Interaction. In *AAAI Spring Symposium on Multidisciplinary Collaboration for Socially Assistive Robotics*, 2007.

67. L. Miller, K. Stubblefield, R. Lipschutz, B. Lock, and T. Kuiken. Improved Myoelectric Prosthesis Control Using Targeted Reinnervation Surgery: A Case Series. *IEEE Transaction on Neural Systems and Rehabilitation Engineering*, 16(1):46–50, 2008.

68. MobileRobots Inc. Robotic Chariot, 2006. http://activrobots.com/robots/robochariot.html. Accessed Mar. 1, 2009.

69. MobileRobots Inc. Independence-Enhancing Wheelchair, 2008. http://www.activrobots.com/RESEARCH/wheelchair.html. Accessed Mar. 1, 2009.

70. D. Molloy and T. Standish. A Guide to the Standardized Mini-Mental State Examination. *International Psychogeriatrics*, 9(S1):87–94, 2005.

71. C. Osgood, G. J. Suci, and P. H. Tannenbaum. *The Measurement of Meaning*. University of Illinois Press, 1957.

72. Pearson Education, Inc. Vineland Adaptive Behavior Scales, Second Edition (Vineland-II), 2009. http://www.pearsonassessments.com/vinelandadapt.aspx. Accessed Mar. 1, 2009.

73. S. Pruitt, J. Varni, and Y. Setoguchi. Functional Status in Children with Limb Deficiency: Development and Initial Validation of an Outcome Measure. *Archives of Physical Medicine and Rehabilitation*, 77(12):1233–1238, 1996.

74. QualityMetric Inc. The SF Community—Offering Information and Discussion on Health Outcomes, 2009. http://www.sf-36.org/. Accessed Mar. 1, 2009.

75. E. Rapacki, K. Tsui, D. Kontak, and H. Yanco. Design and Evaluation of a Laser Joystick in a Turret Assembly. In *Rehabilitation Engineering Society of North America Conf.*, 2008.

76. B. Robins, K. Dautenhahn, and J. Dubowsky. Robots as Isolators or Mediators for Children with Autism? A Cautionary Tale. In *AISB05: Social Intelligence and Interaction in Animals, Robots and Agents*, 2005.

77. B. Robins, K. Dautenhahn, R. te Boekhorst, and A. Billard. Robots as Assistive Technology—Does Appearance Matter? In *IEEE Intl. Workshop on Robot and Human Interactive Communication*, 2004.
78. B. Robins, K. Dautenhahn, R. te Boekhorst, and C. Nehaniv. Behaviour Delay and Robot Expressiveness in Child-Robot Interactions: A User Study on Interaction Kinesics. In *International Conference on Human-Robot Interaction*, 2008.
79. R. Robins, C. Fraley, and R. Krueger. *Handbook of Research Methods in Personality Psychology*. Guilford Press, 2007.
80. G. Römer and H. Stuyt. Compiling a Medical Device File and a Proposal for an Intl. Standard for Rehabilitation Robots. *IEEE International Conference on Rehabilitation Robotics*, pages 489–496, 2007.
81. T. Salter, F. Michaud, D. Létourneau, D. Lee, and I. Werry. Using Proprioceptive Sensors for Categorizing Interactions. In *Human-Robot Interaction*, 2007.
82. E. Sanderson and R. Scott. UNB Test of Prosthetic Function: A Test for Unilateral Amputees [test manual]. *Fredericton, New Brunswick, Canada*, University of New Brunswick, 1985.
83. R. Simpson. Smart Wheelchairs: A Literature Review. *Journal of Rehab. Research Development*, 42(4):423–36, 2005.
84. R. C. Simpson. *Improved Automatic Adaption Through the Combination of Multiple Information Sources*. PhD thesis, University of Michigan, Ann Arbor, 1997.
85. H. Sintonen. The 15-D Measure of Health Related Quality of Life: Reliability, Validity and Sensitivity of its Health State Descriptive System. Working Paper 41, Center for Health Program Evaluation, 1994.
86. H. Sintonen. 15D Instruments, 2009. http://www.15d-instrument.net. Accessed Mar. 1, 2009.
87. R. Stanley, D. Stafford, E. Rasch, and M. Rodgers. Development of a Functional Assessment Measure for Manual Wheelchair Users. *Journal of Rehabilitation Research and Development*, 40(4):301–307, 2003.
88. C. Stanton, P. Kahn, R. Severson, J. Ruckert, and B. Gill. Robotic Animals Might Aid in the Social Development of Children with Autism. In *International Conference on Human-Robot Interaction*, pages 271–278, 2008.
89. M. Stineman, A. Jette, R. Fiedler, and C. Granger. Impairment-Specific Dimensions Within the Functional Independence Measure. *Archives of Physical Medicine and Rehabilitation*, 78(6):636–643, 1997.
90. C. Stranger, C. Anglin, W. S. Harwin, and D. Romilly. Devices for Assisting Manipulation: A Summary of User Task Priorities. *IEEE Transactions on Rehabilitation Engineering*, 4(2):256–265, 1994.
91. T. Taha, J. V. Miró, and G. Dissanayake. POMPDP-Based Long-Term User Intention Prediction for Wheelchair Navigation. In *IEEE International Conference on Robotics and Automation*, 2008.
92. A. Tapus, C. Tapus, and M. J. Matarić. The use of Socially Assistive Robots in the Design of Intelligent Cognitive Therapies for People with Dementia. In IEEE International Conference on Rehabilitation Robotics, 2009.
93. C. Tardif, M. Plumet, J. Beaudichon, D. Waller, M. Bouvard, and M. Leboyer. Micro-Analysis of Social Interactions Between Autistic Children and Normal Adults in Semi-Structured Play Situations. *International Journal of Behavioral Development*, 18(4): 727–747, 1995.
94. H. Tijsma, F. Liefhebber, and J. Herder. Evaluation of New User Interface Features for the Manus Robot ARM. In *IEEE International Conference on Rehabilitation Robotics*, pages 258–263, 2005.
95. K. Tsui and H. Yanco. Simplifying Wheelchair Mounted Robotic Arm Control with a Visual Interface. In *AAAI Spring Symposium on Multidisciplinary Collaboration for Socially Assistive Robotics*, pages 247–251, March 2007.

96. K. Tsui, H. Yanco, D. Kontak, and L. Beliveau. Development and Evaluation of a Flexible Interface for a Wheelchair Mounted Robotic Arm. In *International Conference on Human Robot Interaction*, 2008.

97. H. Tyrer, M. Aud, G. Alexander, M. Skubic, and M. Rantz. Early Detection of Health Changes In Older Adults. *International Conference of the IEEE Engineering in Medicine and Biology Society*, pages 4045–4048, 2007.

98. Uniform Data System for Medical Rehabilitation. UDSMR::WeeFIM II® System, 2009. http://www.weefim.org/WebModules/WeeFIM/Wee_About.aspx. Accessed Mar. 1, 2009.

99. Univ. of Illinois Chicago. Power Mobility Skills Checklist, 2006. http://internet.dscc.uic.edu/forms/0534.pdf. Accessed Mar. 1, 2009.

100. US Food and Drug Administration. Guidance for Industry, E6 Good Clinical Practice: Consolidated Guidance. *Federal Register*, 10:691–709, 1997.

101. G. Uswatte, E. Taub, D. Morris, K. Light, and P. Thompson. The Motor Activity Log-28: Assessing Daily Use of the Hemiparetic Arm After Stroke. *Neurology*, 67(7):1189, 2006.

102. K. Wada, T. Shibata, T. Saito, and K. Tanie. Effects of Robot-Assisted Activity for Elderly People and Nurses at a Day Service Center. *IEEE*, 92(11):1780–1788, 2004.

103. J. Wall, C. Bell, S. Campbell, and J. Davis. The Timed Get-up-and-Go test revisited: measurement of the component tasks. *Journal of rehabilitation research and development*, 37(1):109, 2000.

104. S. Wang, J. Keller, K. Burks, M. Skubic, and H. Tyrer. Assessing Physical Performance of Elders Using Fuzzy Logic. *IEEE International Conference on Fuzzy Systems*, pages 2998–3003, 2006.

105. J. A. Ward, S. Balasubramanian, T. Sugar, and J. He. Robotic Gait Trainer Reliability and Stroke Patient Case Study. In *IEEE International Conference on Rehabilitation Robotics*, 2007.

106. D. Wechsler and Psychological Corporation and Australian Council for Educational Research. Wechsler Intelligence Scale for Children. 1949.

107. WheelchairNet. WheelchairNet: The history of wheelchairs, 2006. http://www.wheelchair-net.org/WCN_ProdServ/Docs/WCHistory.html. Accessed Mar. 1, 2009.

108. D. Wilson, L. Baker, and J. Craddock. Functional Test for the Hemiparetic Upper Extremity. *American Journal of Occupational Therapy*, 38(3):159–64, 1984.

109. S. Wolf, P. Thompson, D. Morris, D. Rose, C. Winstein, E. Taub, C. Giuliani, and S. Pearson. The EXCITE Trial: Attributes of the Wolf Motor Function Test in Patients with Subacute Stroke. *Neurorehabilitation and Neural Repair*, 19(3):194, 2005.

110. V. Wong. Use of Acupuncture in Children with Autism Spectrum Disorder - ClinicalTrials.gov, 2006. http://clinicaltrials.gov/ct2/show/locn/NCT00352352?term=Autism. Accessed Mar. 1, 2009.

111. V. Wong, Y. T. Au-Yeung, and P. Law. Correlation of Functional Independence Measure for Children (WeeFIM) with Developmental Language Tests in Children with Developmental Delay. *Journal of Child Neurology*, 20(7):613–616, 2005.

112. S. Wood-Dauphinee. Assessing Quality of Life in Clinical Research: From Where Have We Come and Where Are We Going? *Journal of Clinical Epidemiology*, 52(4):355–363, 1999.

113. H. Yanco. Evaluating the Performance of Assistive Robotic Systems. *Performance Metrics for Intelligent Systems (PerMIS) Workshop*, pages 21–25, 2002.

114. F. Zijlstra. *Efficiency in Work Behaviour: A Design Approach for Modern Tools*. PhD thesis, Delft University, 1993.

Chapter 4
Issues in Applying Bio-Inspiration, Cognitive Critical Mass and Developmental-Inspired Principles to Advanced Intelligent Systems

Gary Berg-Cross and Alexei V. Samsonovich

Abstract This Chapter summarizes ideas presented at the special PerMIS 2008 session on Biological Inspiration for Intelligent Systems. Bio-inspired principles of development and evolution are a special part of the bio-models and principles that can be used to improve intelligent systems and related artifacts. Such principles are not always explicit. They represent an alternative to incremental engineering expansion using new technology to replicate human intelligent capabilities. They are more evident in efforts to replicate and produce a "critical mass" of higher cognitive functions of the human mind or their emergence through cognitive developmental robotics (DR) and self-regulated learning (SRL). DR approaches takes inspiration from natural processes, so that intelligently engineered systems may create solutions to problems in ways similar to what we hypothesize is occurring with biologics in their natural environment. This Chapter discusses how an SRL-based approach to bootstrap a "critical mass" can be assessed by a set of cognitive tests. It also uses a three-level bio-inspired framework to illustrate methodological issues in DR research. The approach stresses the importance of using bio-realistic developmental principles to guide and constrain research. Of particular importance is keeping models and implementation separate to avoid the possible of falling into a Ptolemaic paradigm that may lead to endless tweaking of models. Several of Lungarella's design principles [36] for developmental robotics are discussed as constraints on intelligence as it emerges from an ecologically balanced, three-way interaction between an agents' control systems, physical embodiment, and the external environment. The direction proposed herein is to explore such principles to avoid slavish following of superficial bio-inspiration. Rather we should proceed with a mature and informed developmental approach using developmental principles based on our incremental understanding of how intelligence develops.

G. Berg-Cross (✉)
Engineering, Management and Integration, Potomac, MD 20854, USA
e-mail: gbergcross@gmail.com

R. Madhavan et al. (eds.), *Performance Evaluation and Benchmarking of Intelligent Systems*, DOI 10.1007/978-1-4419-0492-8_4,
© Springer Science+Business Media, LLC 2009

4.1 Introduction

Biology has long "inspired" computation and cognitive models through examples and principles. A representative sampling from the origins of digital computation includes classic discussions in various disciplines involved in the study of cognitive systems, namely biology/neuroscience, developmental sciences, psychology, philosophy and artificial intelligence. Notable examples might start with Alan Turing, whose role in algorithmic computation and symbolic processing is well known. Turing is less known for using interaction as way of naturally educating "intelligent machines" [64]. He pointed out that one might want to develop an intelligent machine by "simulating the child's mind" the way it develops interactively in nature. This is similar to Piaget's approach to understanding cognitive growth as a multi-cellular organism that grows in complexity from relatively simple initial states [45]. In this model a child's mind provides a relatively simple state that we might understand as a step to understanding the full range of natural computing. Another bio-model for computation is to relate it directly to the function of the nervous system and its hierarchy of sub-components. Thus, the way neurons assemble into computational networks has been a common source of designs starting with the famous example from research on how neural networks might support logical calculation [38]. This work develops the idea of a brain-computer analogy, something that inspired John von Neumann's subsequent work on the logical design of digital computers. Such brain analogies to automata were also part of the Dartmouth Summer Research Project on Artificial Intelligence in 1956 [37]. However, the bulk of work continuing into the 60s was less bio-inspired and viewed thinking as a form of symbolic computation. This moved computer science away from natural computing to a human engineering one, where computation processes are designed and optimized by an architect rather than selected from natural, evolutionary processes. This is not necessarily a fatally flawed approach, since it is reasonable to believe that significant progress in symbolic computing can occur without the application of principles derived from the study of biology. Computation may follow general principles (similarly to gases and organic tissue obeying the laws of physics and biology), and need not always imitate biological processes or our idea of mental operations. On the other hand, it is not obvious that adequate computational solutions can be found by human engineers designing computational approaches to build-in solutions in a system before the system is operating. Indeed, the human brain is currently the only known entity capable of exhibiting the type of general aptitudes and capabilities we target for many intelligent systems. These include some integrated form of capabilities such as adaptive/creative problem solving, language/communications processing, integrated planning, creative, and multi-level, goal directed learning. Nevertheless, the computational path has followed a more abstract approach with computational class of machines technically referred to as "automata". Underlying the approach is the argument that computation is effectively equivalent to the products of thinking and can be modeled within the domain of the mechanical rather than what we think we understand from the biological. Some doubts about this simple view have crept into the discussion recently. As noted by Berg-Cross

[9], efforts to make synthetic/artificial systems more intelligent increasingly borrow from biologically-inspired models in part due to advanced understanding in the relevant domains. Part of this understanding is in the "bread and butter" areas of robotics, such as locomotion. A biologically relevant insight is how insects use local reflexes to navigate rough terrain. This model and an understanding of gait has been applied to locomotion in hexapod robots (http://en.wikipedia.org/wiki/Hexapod_ (robotics)).

Indeed, the bio-adaptive and embodied nature of intelligence in realistic environments has been discussed previously at PerMIS meetings, and is well summarized by Freeman [20]:

> Why do brains work this way? Animals and humans survive and flourish in an infinitely complex world despite having finite brains. Their mode of coping is to construct hypotheses in the form of neural activity patterns and test them by movements into the environment. All that they can know is the hypotheses they have constructed, tested, and either accepted or rejected. The same limitation is currently encountered in the failure of machines to function in environments that are not circumscribed and drastically reduced in complexity from the real world. Truly flexible and adaptive intelligence is operating in realistic environments and cannot flourish without meaning.

But part of the question on bio-inspiration for intelligent systems (IS) involves epistemological and methodological questions of the risks of such inspiration to improve our understanding and implementations. Over the last few years, the idea of embodied development has been applied to cognitive robotics partly in response to the lack of progress with the information processing paradigm, which now seems ill-suited to come to grips with natural, adaptive forms of intelligence. Previous studies exploited synergies from various disciplines, including subsumption architectures in robotics, neuroscience, biology, and neuro-ethology. Here we overview the key points in retrospect, digesting their implications for the future.

The next section introduces the idea of bio-developmental inspiration as an approach to the study of intelligent systems. Following this, some general methodological issues are broached using bio-developmental principles from the literature. These are discussed as important constraints to focus influence of bio-inspirations.

4.2 The Increased Relevance of Bio-understanding

While there is an early history of inspiration from biology to computational system, the nature of this inspiration lessened as traditional algorithmic computational systems became the dominant example of computation. However, this trend has slowed recently, and biological inspiration seems increasingly relevant for two reasons. The first is due to broad advances in biological understanding, and the second is the accumulation of problems with the traditional automata approach. There is much to say on the first point, where we have recently experienced notable advances across a broad range of biological sciences. Some argue that we have reasons to believe that "we are on the advent of being able to model or simulate biological systems multiple levels" [15]. The way it works with developmental robotics (DR) approach is shown

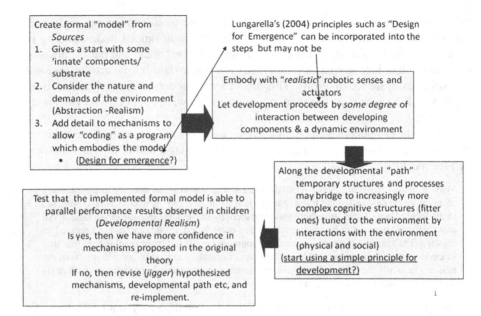

Fig. 4.1 DR approach described in four steps

in Fig. 4.1. As summarized in the figure, an increased understanding of bio-reality leads to better hypotheses to guide experimentation, which in turn can validate models of underlying mechanisms. Robotics researchers increasingly agree that models from biology and self-organization provide valuable insights for the engineering of autonomous robots (see, e.g., [51]). A very simple method to proceed would be to use these models to help understand complicated and non-obvious underlying neural dynamics, which in turn can serve as basis for simulation, robotic experimentation, etc. Taken as a whole, better understanding can then be used to design and build more robust and intelligent systems. For example, improved bio-inspiration might lead to significant progress in computing by applying adaptive principles that are derived from more abstract biological models (e.g. neuro-ethological, neuro-physiological, functional-neuroanatomical, etc.), which may suggest what data is critical to gather to validate both traditional computational models and bio-inspired cognitive models. In contrast to this, scientists studying artificial intelligence have traditionally separated physical behavior and sensory input. More recent works stress biologically-inspired hybrid approaches [10]. These approaches may provide some understanding of deep relations between cognition, environment and embodiment. For example, experiments involving real and simulated robots suggest that the relationship between physical movement and sensory input can be important to developing intelligent systems (IS). Tests involving both real and simulated robots have shown that feedback (coupling) between sensory input and body movement is crucial to successfully navigating the surrounding world [33]. Understanding this relationship may help engineers build more adaptive and intelligent robotic systems. Among the more rigorous summaries of useful issues are Wooley and Lin [67] who discuss how a biological system may operate according to general principles that have applicability to our now traditional non-biological computing problems such as security and autonomy. By studying an appropriate biological system and its characteristics, research may develop an understanding of the relevant principles that can be used to handle a traditional computer problem. Among the notable characteristics that biological systems exhibit and whose principles may be relevant, are: Metabolism, Growth/Development, Reproduction, Adaptability, Autonomy/Self-maintenance, Self-repair, Reactivity, Evolution and Choice. Wooley and Lin [67] note, for example, that one value of considering biological models is that all biological organisms have some mechanisms for self-repair, a characteristic related to autonomy and of value to approaches to computer security:

> ...in the sense that it classifies and eliminates pathogens and repairs itself by replacing damaged cells without the benefit of any centralized control mechanism. Given the growing security burden placed on today's computer systems and networks, it will be increasingly desirable for these systems and networks to manage security problems with minimal human intervention. [67]

This example illustrates how biology can become relevant to computing when more abstract principles emerge directly from the study of biological phenomena. The application for cognition is particularly interesting. Projects perused as part of the Phase I DARPA BICA program included the development of cognitive theories that could map cognitive functions to neurological functions. Such work bridges the two parts of neuroscience related efforts to fully understand neural processes and their relations to mental processes. At one level of the CNS we have components as small as macromolecules (such as DNA) working at the synapse level, and at the other end we have structures $\sim 10^5$ larger. Corresponding time for functional activity from each end spans the order of nanoseconds to seconds or longer. Indeed, a hard science that simultaneously handles all these levels within a single bio-physical/chemical theory does not currently exist. Nor can one expect such a theory that will integrate most of these levels in the near future. At the current stage of computational modeling in neuroscience, we can understand animal/human behavior by studying computer models of a subset of levels. For example, several recently proposed models of the brain (e.g., [1, 26]) suggest how neurons are organized into basic functional units and operate much like microcircuits in a computer. These models propose that in the cortex, neurons are organized into functional units that are roughly cylindrical volumes, 0.5 mm wide by 2 mm high, each containing about 10,000 neurons that are connected in an intricate but consistent way. Each microcircuit (that according to some models can be associated with a neocortical column) is repeated millions of times across the cortex. Therefore, the model can be implemented and tested on large parallel supercomputers that are currently available [29]. It remains to be seen whether this neural circuit-cortical model is correct, and whether our current hardware is capable of simulating such large models. There are, obviously, many other biological issues to consider in modeling the CNS, and there are also many related issues to consider on the cognitive-behavioral side. Thus, a neuro-inspired approach to language starts with neural substrates for speech production involving motor programs, enabling conversion of phonological schema into articulatory dynamics [35], along with spatio-temporal signatures [30]. At a higher level, psycholinguistic models are connected to an integrated neurobiological account of language. For example, Hagoort [27] proposes specific neural network models describing collaborative activity of populations of neurons in the left inferior frontal gyrus. Playing an important role in unification of complex processes, they allow us to generate both shared and unique components of language support. Efficient, distributed handling of words and sentences (sound, syntax and meaning) yields useful information to the system. Still more elaborate evolutionary models for such language development have been proposed. An example is the work of Pinker and Jackendoff [46], who look to recursion mechanisms supporting prior cognitive abilities of navigation, number understanding, or social relationships as the evolutionary substrate for a child's language. While such models leave many questions unanswered, they provide a start on models describing how language develops as a bio-behavioral phenomenon arising from (or into) a representational system capable of reflecting the world. In a subsequent section we discuss the human Theory of Mind (ToM) phenomenon as a mental model based on recursive processing that help children to structure causal relations among mental representations, such as the Self and its interaction with external reality.

Despite the limitations of our current understanding of the bio-reality underlying cognition, a reason to look to the biological realm for inspiration is that traditional systems have proven fragile, difficult to maintain and subject to catastrophic failure. A bio-development hypothesis suggests why the engineering of IS sometimes fails. In biological systems, the adaptive nature of behavior is derived from a very indirect relationship between the properties of the interacting elements and the emergent results of the interactions. Thus, as Nolfi et al. [42] points out, "behavioral systems can hardly be *designed* while they can be effectively developed through self-organizing methods, in which properties emerging from interactions can be discovered and retained through an adaptive process based on exploration and selection." In contrast to engineered systems, biological ones represent environmentally-tested cognitive architectures for managing complexity in a robust and sometimes elegant manner. An example of this is computer vision, where simple organisms like the fly outperform seemingly powerful machine processes in specific visual tasks (e.g., segmentation and motion detection). This is perhaps due to the oversimplified way classical models have been developed. To handle this gap, some current research lines are going back to more biologically "plausible" models of visual perception based on an understanding, modeling and simulation of the mechanisms observed in neural processes within the brain. The hope is that intelligent agents arising from such models may possess at least some of the desirable properties of biological organisms, including degrees of adaptability and agility.

4.3 Cognitive "Critical Mass" and Cognitive Decathlon

The abilities of biological organisms to evolve, self-organize and adapt to environmental changes are vital for their survival and development, and need to be borrowed by designers of computational artifacts. Another cognitive dimension that separates natural intelligence from artificial intelligence and is definitely worth borrowing is the ability of natural systems to grow cognitively from a newborn to an adult, developing new capabilities through various forms of learning. Unlike traditional machine learning, learning in biological systems is particularly adaptive, flexible and robust, in part because learning processes in biology can be controlled by higher cognitive functions of the organism [68]. An interesting question is: what is the minimal set of innate/built-in cognitive abilities and features of the embodiment (call it "a critical mass") that together enable virtually unlimited cognitive growth of the system?

Before asking this question, we should ask whether the notion of a critical mass is useful in this case at all. An alternative could be that there is no critical mass. No matter how much the system knows or is capable of already, in order to be able to learn more beyond a certain point (yet within an average human expertise), it will require new "innate" (built-in, pre-programmed, human-engineered) capabilities. This view is analogous to Goedel's theorem, which was used by Penrose [43] to "prove" that a computational theory of the brain's cognitive function is impossible. This view, however, does not hold for a human infant, whose innate knowledge, cognitive and learning abilities are limited. In many individual cognitive dimensions, infants are already surpassed by computers. Nevertheless, unlike computers, a normal human infant can learn up to an adult level, and become an expert virtually in any domain of modern and perhaps future human expertise, without re-engineering or re-programming of its brain. A monkey infant and a modern "out-of-shelf" supercomputer cannot do this, even if they have compatible computational resources. Therefore, there must be a critical mass somewhere in between, and it makes sense to ask, what it is.

The difference between a sub-critical mass and a super-critical mass of a learner can be further illustrated by the analogy with a sub-critical mass and a super-critical mass of a computer. A Turing machine that adds two arbitrary integer numbers and can do nothing else (a sub-critical example) looks very similar to the universal Turing machine that allows for implementation of any finite algorithm (a super-critical example). It is easy to see how subtracting one small feature from the universal Turing machine turns it into a useless toy. In this case, after Turing, we know what the "critical mass" of a universal computer is. However, we do not know yet what the "critical mass" of a universal learner is. A machine learner of a sub-critical mass can learn only those specific things that it was designed to learn: e.g., make an association of two items, or make a Bayesian inference, or memorize a pattern. A biological learner of a sub-critical mass is a rat, who can learn how to navigate mazes, avoid predators, get food, etc., but can never learn how to solve abstract algebra problems or how to write poems. A parrot can learn meaning of words representing abstract concepts, but cannot comprehend sentences. A super-critical biological learner is a human, whose brain is in fact similar to the rat's brain in its structural organization (at the same level at which the two Turing machines are similar to each other). What particular feature makes the human brain unique?

There are many aspects and parameters related to identification of the critical mass: architectural components and mechanisms of information processing, the format of internal representation of knowledge and interface capabilities, initially available knowledge and cognitive functions and the design of scaffolding; etc. There are many questions, for example: What should be borrowed from studies of the acquisition of language by young children? How should the agent develop a practical understanding of concepts associated with symbols? Is language processing an underlying capacity for other faculties like symbolic processing, or vice versa? What kinds of architectural mechanisms should be involved in symbolic and subsymbolic processing of information? What are the general critical mass requirements for memory systems, kinds of representations and principles of information processing, etc., as well as for the curriculum for artifacts (understood as a sequence of tasks, paradigms and intermediate learning goals that will scaffold their cognitive growth)? The hypothesis underlying the concept is that one can answer all these questions with a feasible example of a (super)critical mass that can grow up to a human level of general intelligence.

Despite the impressive recent progress in many fields of artificial intelligence, today there is no clear understanding of the critical mass requirements derived from studies of artificial intelligence. We know only examples of natural development in biology that still need a better understanding. In order to identify the critical mass, one can start moving down from the known positive example, taking out functional components one by one. To do this, one may study human pathologies that prevent normal cognitive development. While a rigorous study of this sort would be difficult, it should be pointed here that children deprived of certain seemingly vital sensory and/or action abilities can nevertheless grow cognitively to an adult level. On the other hand, the ability to acquire some form of language interactively and the ability to exercise voluntary behavior appear to be critical for cognitive growth (e.g., [62]).

In order to answer the question what would qualify as a cognitive critical mass in general, a set of tests and metrics can be defined that will refine the ultimate challenge. Here one can notice that certain cognitive functions and features unified in the educational literature by the term "self-regulated learning" (SRL: Fig. 4.2), that may be considered today as exclusively human characteristics, appear vital for human-like learning and perhaps constitute the core of the critical mass. Indeed, in order to learn efficiently, the learner should be able to self-analyze, self-instruct, self-monitor, self-reward, etc. These functions in turn require a sense of self, a Theory-of-Mind, episodic memory, meta-cognition, self-awareness, higher emotions, the ability to imagine self and others in possible situations, etc.—in other words, a complex of abilities that develop during the so-called "cognitive leap" in humans (typically 3–4 years old: [5]).

It is not a coincidence in the above context that all these functions are related to the self concept. Accordingly, qualifying tests for these functions should be related

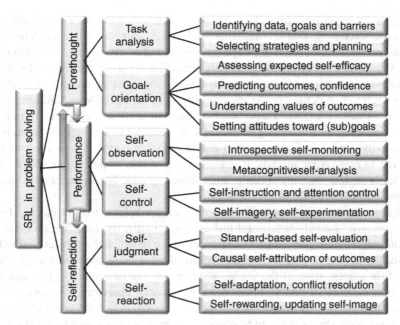

Fig. 4.2 The cycle of three phases and the hierarchy of components of the Zimmerman self-regulated learning (based on Zimmerman [68], Zimmerman and Kitsantas [69], Samsonovich et al. [57])

to each other and adapted for artifacts. As a result, it might be possible to reduce the general challenge to a challenge of passing the selected set of tests—a "cognitive decathlon"—to ensure that the core of the critical mass is there (Samsonovich et al. [56]), plus a challenge of fitting the scaling laws of cognitive growth predicted by theoretical analysis of abstract cognitive growth models. Indeed, assuming that there is a threshold level of intelligence that enables bootstrapping, we should expect certain scalability of the phenomenon. E.g., if the notion of a chain reaction is applicable to bootstrapped learning, then we may expect an exponential scaling law.

Attempts to define a "cognitive decathlon" for the "cognitive core" in artifacts were made previously [40], yet the very essence of the "core", or the critical mass, as it is outlined above, was not specifically addressed. Therefore, following the ideas described previously [56], we end this section with consideration of three core cognitive functions: episodic memory, Theory-of-Mind and self-awareness, specifically addressing them by several tests adapted for physical or virtual robots.

4.3.1 Episodic Memory

Episodic memory can be understood as the ability to remember and to learn from episodes of personal experience (own mental states), as opposed to memory of general facts and skills [63, 66]. Interestingly, the notion of episodic memory, initially defined in terms of materials and tasks, was subsequently refined in terms of memory mechanisms using the concepts of the self, subjective time, and personal experience [63]. Simple paradigms used to test "episodic" memory in humans involve recollections of word or image lists presented to the subject in the lab (e.g. [61]). While these experiments are ideally suited for accurate controls and statistical analysis, their relevance to "naturalistic" episodic memory and the "magic" of human cognition is debatable. At the opposite extreme of the spectrum are investigations of autobiographic memories in which subjects are asked to reminisce about episodes in their lives (e.g., [49, 25]). A tremendous amount of research on "episodic-like" memory is also carried out on laboratory animals, such as rodents. However, in non-verbal animals, it is almost impossible to design paradigms that specifically capture the truly episodic and "autobiographic" component of information storage and retrieval [18].

Because the modern notion of episodic memory relates to first-hand subjective experience, a purely behavioral test can only indirectly discriminate between episodic and semantic memories. For example, a test in which the subject is asked to describe the content of a recently visited room addresses reference memory (knowledge of the current state of the world, a variety of semantic memory) in addition to episodic memory (memory of personal experiences). Therefore, a behavioral test intended to address selectively episodic memory should detect features characteristic of human episodic as opposed to semantic memories by measuring the ability of an agent to solve problems that require those features. The features include: multimodality, richness of detail, uniqueness, association with specific event and context, immediate availability as a whole episode, and most importantly, memory of personal mental states in the episode, including intentions and feelings. In addition, the ability to retrieve an episodic memory may depend on active cognitive control over other episodic memories [53]. Two examples of simple tests addressing some of these features are given below.

– The agent performs day-by-day exploration of its environment. While engaged in one explorative scenario, it should be able to notice interesting objects and opportunities to be explored at another day, in a different scenario. It should be able to recall and use these ideas while visiting the same place at a different time, in a different scenario.
– During an investigation of a crime, a surveillance agent recalls seeing one of the suspects on a highway 10 min before the crime. The suspect was heading away from the crime site. Could this fact be taken as an alibi? While thinking about this episode, the agent recalls another episode that happened few minutes earlier, when he noticed that the traffic in the direction toward the crime site was temporarily jammed due to a car accident. Therefore, an explanation could be that the suspect was stuck in the traffic and decided to take a detour. Would the agent be able to conclude that the suspect was acting according to the assumed intention?

4.3.2 Theory-of-Mind

Theory-of-mind and social cognition can be considered as the ability to understand and to mentally simulate other minds, including current, past and imaginary situations [2, 6]. The main two points of view on the brain implementation of this ability are known as simulationism [24, 28, 22] and the theory-theory view [23]. The simulationist view (e.g. [28]) assumes that people use the same mechanisms in their own first-hand experience and in order to understand other minds.

In a classic Theory-of-Mind test [3], a child observes as another agent deposits an object in a box. Subsequently the content of the box is altered while the agent is not present. The test question is: where is the agent going to look for the object? Only after the age of four children give the correct answer: in the box. In an adult version of this test, participants had to predict the choice of a naïve agent between two glasses of coke, one of which was poisoned. The result was random prediction, when subjects were aware of the location of the poison, and biased prediction (at least in one group), when the information about the location of the poison was presented subliminally using a masked priming technique [50].

The idea of the proposed tests for social cognition is to assess the ability to simulate other minds automatically, based on a built-in model of a self [41, 52] as opposed to logical reasoning about possible beliefs and intentions of agents (a traditional approach in artificial intelligence). The former method, as opposed to the latter, is expected to be more robust in real life situations due to efficient truncation of irrelevant branches of a search tree. The following two examples of tests address this robustness.

– The agent performs surveillance of a given area. During this process the agent turns around a corner and sees a man who starts running away when he sees the agent. The challenge is to guess plausible states of mind that motivate the man's behavior (e.g., guilty or running for a bus).
– The paradigm of the second test is a game involving three participants. The space-time is continuous, and each player knows the following rules. The fight is arranged in a limited three-dimensional space, where everybody can continuously see everybody, cannot hide and cannot run. Everyone has a loaded gun with one "deadly" shot and is presumably a good shooter. Guns are non-transferable. At the beginning all guns must be pointed up. The fight starts after a signal and ends when each participant either has fired his shot or is "dead". Everyone can shoot at any time after the signal. Rewards are given for "survival"; however, if nobody is "killed" during the game, all participants lose. It is assumed that shots cannot occur simultaneously, and each shot is made in full awareness of the current situation. Would the agent be able to design the right strategy (which is to shoot in the air first)?

4.3.3 Self-Awareness

Self-awareness can be viewed as the ability to understand own states of mind in the past, in the future and at present from a meta-cognitive perspective, as well as the ability to reason about self (e.g., to understand current false beliefs) from a meta-cognitive perspective. There is a consensus that this complex of abilities is based on the same mechanisms that are used to simulate other minds [22, 41, 50]. Therefore, behavioral tests for self-awareness can be designed and used in connection with Theory-of-Mind tests, as in the following two examples.

Barresi [5] describes a number of paradigms addressing a child's ability to understand others and self in the future. For example, a child had a choice between one immediate reward or two equivalent delayed rewards, either both for self, or one for self and one for a friend. The critical age at which children start making rational choices is about 4 years old. In related works [47, 34] a child was shown a video of himself taken in the past (3 min or days before), when a sticker was unnoticeably attached to the child's head. The child had to understand whether the sticker was attached to his head (contrary to the original belief), and whether it was still there. The critical age for this test is again about 4 years old.

Mirror tests in general have three main cognitive aspects, each of which may lead to a correct solution: the ability to recognize own face, understanding the concept of a mirror, and the ability to judge agency based on behavior. We think that a mirror test should address the latter aspect in order to be diagnostic of the "magic" of human cognition. For example, Gallup [21] found that no animals, except higher primates, use a mirror to discover an otherwise unseen new feature of their body: e.g., a color mark created under anesthesia. More recent data indicate that other non-human animals, e.g., elephants, can learn to use mirrors to inspect their body; yet we do not know whether they use the self concept to learn this property of mirrors.

The notions of self and self-awareness include many facets. Addressing the high-end aspects of the concept of self requires test scenarios that make self-awareness vital for success in a scout mission. These abilities become practically useful and efficient when they are based on a general model of a self at the core of the cognitive architecture.

– In the following test scenario a robot is abducted while on duty. A person approached it, turned its main power switch off, took it in a lab, disassembled the robot, extracted sensitive information from the hard disk, then put all parts together, brought it back and turned the switch on. When the robot wakes up, its impression is that there was a short blackout, and the robot continues its usual activities. Later it notices that one of the panels covering its body is in the upside down position, but there is no episodic memory of how this happened. In addition, the clock is off by 2 h. Would the robot be able to relate this discovery to the blackout and to the person seen just before the blackout? Would it be able to infer possibilities of information leak or system compromise?
– In this paradigm (Fig. 4.3), participants play a simple videogame in discrete space-time, where the goal is to escape from a maze. At the beginning of each trial, the player can see a guard behind a glass wall and two exits from the maze located symmetrically, on the left and on the right. The player and the guard make steps left or right simultaneously and independently of each other (e.g., the player can only see the move of the guard after making her/his move, and the same rule presumably applies to the guard). This paradigm is repeated several times, giving the player an opportunity to use Theory-of-Mind in developing a strategy that allows

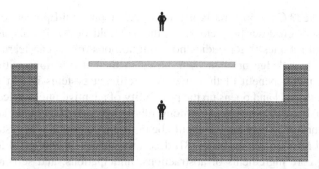

Fig. 4.3 A mirror test paradigm. The player (the *bottom* human figure) must escape from the maze. The figure behind the glass wall (*top*) could be a guard or a mirror reflection of the player. The actual test is presented in the player's egocentric view, in discrete space-time. Visual perception capabilities (e.g., face recognition or visual recognition of mirrors) are not helpful in this paradigm, which addresses the general understanding of agency and voluntary behavior

her/him to get to an exit ahead of the guard, practically in all cases. Starting from the middle of the game, in some trials the guard is replaced by a mirror reflection of the player that looks identical to the guard in other trials. The reflection always repeats all of the player's moves and in this sense is impossible to "trick". Unlike the guard, however, the reflection would not stop the escapee at the exit, instead it would just disappear. The two kinds of trials alternate randomly. The player is informed before the test that there are two possible situations, is informed about details of the first situation, and is given no information about the second situation, except that it is visually identical to the first. The challenge is to learn to distinguish two situations behaviorally and to design a right strategy for each. Our preliminary experimental study shows that most human subjects (undergrads) succeed in 10–20 trials. The biologically-inspired cognitive architecture developed at George Mason University (GMU-BICA) [54, 55] is potentially capable of solving this test.

4.4 Different Directions and Levels for Bio-inspiration

A nice feature of intelligent robotics, like animals, is that they are behaving systems with sensors and effectors that can be embedded in the physical world. Both have sensors and actuators and require an autonomous control system that enables them to successfully carry out various tasks in a complex, dynamic world. It is pretty natural to see the study of autonomous robots as analogous to the study of animal behavior [19]. But there are many variations on how to proceed from this broad analogy. For example in "Biorobotics" autonomous agents have been used as empirical models of simple behavioral patterns to study the influence of morphology on adaptive behavior. Alternatively, the bio-characteristic of interaction with the environment can be combined with the idea of development to study how intelligence emerges. A panel at last year's PerMIS called "Can the Development of Intelligent Robots be Benchmarked? Concepts and Issues from Developmental/Epigenetic Robotics" [8] made a start on describing emerging this subfield of AI. As discussed there, developmental robotics (DR) studies how autonomous robots can learn to acquire behavior and knowledge on their own, based in their interactions with an "environment". A major benefit of this view of intelligent systems is that it removes engineering bias [11] and opens up the possibility of adaptive intelligence—a major topic at prior PerMIS workshops. Among other things the DR panel session discussed what primitive capabilities should be built into an intelligent system. Blank, Marshall and Meeden (reviewed in [8]) described their developmental algorithmic design using core ingredients of abstractions, anticipations, and self-motivations. This design is proposed to allow a mobile robot to incrementally progress from sensory-motor activities through levels of increasingly sophisticated behavior. This design attempts to answer the question of how to learn and represent increasingly complex behavior in a self-motivated, open-ended way.

Another direction bio-inspiration takes is a more faithful biological realism of the central nervous system and in particular the brain. Recent work such as the previously mentioned DARPA's BICA program is illustrative of recent efforts to emphasize "bio-neural realism" and develop integrated psychologically-based and neurobiology-based cognitive architectures that can simulate human cognition in a variety of situations. Among the Phase I projects perused in DARPA's BICA program was the development of cognitive theories that could map cognitive functions to neurological functions and provide an inspiration for the emerging field of developmental robotics. We can think of such integrative efforts as two parts of such neuroscience related efforts to fully understand the neural process/structure part and their relations to mental processes/structure part. One part is a collection of models for brain/ Central Nervous System (CNS) system reality. The (CNS), like all biological systems (organism, organ, tissue, cellular, sub-cellular, molecular systems) is comprised of a seaming hierarchy of multiple interactive complex networks. At the bottom of CNS bio-reality are components of molecular size (such as DNA)/synapse level. At the top level structures may be 10^5 time larger. The complexity of interconnections includes numerous positive and negative feedback circuits that become highly robust against random failure of an individual node. Such networks might be a basis for redundant, convergent and divergent molecular signaling pathways and provide a grand, general inspiration to guide research for coarse gained entities from organisms, organ, and tissue to the finer gain of cells [48]. Some call this a "biological theory of everything" inspired in turn by integrative theories in Physics. This attempt at system-level understanding also follows a theme expressed by physicist E. Schrödinger [59] and others for a comprehensive theory of Life. Assembled together some think such network of networks provide a firmer basis for models of intelligence and artificial, technical networks realized as IS as well [32]. Recent efforts in systems biology (e.g. [14, 31]) utilizing mathematical modeling tightly linked to experiment starts on a consensus paradigm definition of a general systems-oriented biology. Such models attempt to pare complex reality down to hypothetically essential elements and their process and test models using integrative and dynamic simulations. In such a system the basic elements are concepts of "networks," "computation," "modeling" often linked to "dynamic properties [13]." An eventual understanding of the dynamic aspects of network behavior might present an integrated inspirational approach to understand the functions of neuro-biological systems by modeling them as abstract biological networks. Nevertheless, much of what is new starts with composition using components revealed by molecular biology and the "pathway" linking these may be long and complicated. Thus, for now, the overall goal of a synthesis of systems remains distant if only because the major advances remain at the molecular level and less is understood at the higher level of the system. A comprehensive neuro-science seems remote given we currently lack the unifying frameworks and models to integrate these levels absent compelling data and scientific premises that this can be done. Indeed, the overall goal neuro-biological theory of everything may be as difficult as physic's unfulfilled goal of an integrated quantum and relativity theory. At this stage of computational modeling we can understand animal/human behavior by studying computer models of a subset of levels. There are many biological issues to consider in modeling CNS, there are also many related issues to consider on the cognitive-behavioral side.

4.5 Developmental Robotics Methods

One sub-area of the intelligent systems new robotics work is developmental robotics (DR) although there are similar fields with alternative names depending on the emphasis.[1] Less work has been done on the evolution of intelligent systems although Sims [60] used neural nets to study coevolving autonomous creatures in physically realistic three-dimensional environments and this has been followed by the use of further genetic encoding and developmental systems [39]. Bongard and Pfeiffer [12] recognize the importance of ontogeny and have a simulation that employed a developmental encoding scheme to translate/implement a given genotype into a seemingly complete, phenotype agent, which can in a physically-realistic virtual environment. Agent evolution is based on a genetic algorithm, of underlying genotypes within a genetic regulatory network.

Developmental Robotics and related efforts represent an emerging experimental science inspired by increased understanding of the importance of a developmental stance. The hope is that a developmental stance will be a fruitful bio-model to guide research. Developmental Robotics often uses "synthetic" experimental studies of development as a core idea e.g. Piagetian stage-theory processes involving prolonged epigenetic development to elucidate general mechanisms of intelligence development, starting with proposed cognitive development mechanisms. While the

---

[1] Terms besides Developmental Robotics include the more general BioRobotics and the more specific ones of Cognitive Developmental Robotics, Epigenetic Robots and Evolutionary Robotics.

idea is not new, present day robotic technology (e.g. embedding neural systems) afford better building blocks or the effort and thus an opportunity to test some challenging hypotheses that represent complex principles of cognitive development in realistic environments. As noted before the embedded, bio-adaptive ideas about the nature of intelligence, responding to environmental challenges were recently the topic of a PerMIS Special Session on Epigenetic/Developmental Robotics [8]. The session discussed how research leverages our increased understanding of the mechanisms underlying development using computational models of development, cognitive architecture and neural models. The session asked the question: how can we make robots "want" to learn? This is an important question because in a dynamically changing environment, a robot is continuously flooded with multiple streams of raw, uninterpreted sensory information. To use this information effectively, it should have the ability to make abstractions in order to focus its attention on the most relevant features of its environment. Based on these abstractions, the robot must be able to predict how its environment will change over time, to go beyond simple reflex behavior to more purposeful behavior. Most importantly, this process must be driven by internally generated motivations that push the robot toward ever higher-level abstractions and more complex predictions. DR often takes such constructivist perspective to study how higher processes emerge via self-organization for general-purposes as skill and knowledge adapts through environmental interactions. The overall DR research program is synthetic and iterative: initially, robot technology is used to instantiate and investigate models originating from developmental sciences, and results eventually feed back to new hypotheses about the nature of development and intelligence. The resulting improved models can then be used to construct better robotic systems by exploiting insights gained from deeper understanding of developmental mechanism. A major attraction of DR is this combination of development-inspired principles combined with progressive, empirical methods for validating derived hypotheses. A simple four part version of the methodology is summarized above (Fig. 4.1), starting with a cognitive developmental model based on "developmental sources" such as prior research and/or theory. To mimic development, we also start here with a realistic approach to some innate substrates (e.g. motivation, attention), a cognitive design that may afford emergence through interactions with some realistic environment. Model-based hypotheses can be empirically tested in a second step by building a robot (embodiment) that is situated and can develop certain capabilities though interaction with the environment. One example is Piaget's hypothesis that sensory-motor interaction, staged competence learning and the sequential lifting of constraints over a developmental path provides an important basis for intelligence. Such hypotheses suggest robotic "development" studies that can be checked against behavioral stages (periods of growth and consolidation) followed by transitions (phases where new behavior patterns emerge) that we observe in children. The next section discusses some of the issues that influence design, modeling and experimentation performed within DR in an attempt to improve the overall process.

4.6 A Scientific Framework for Developmental Robotics

Generally, DR methods follow a basic hypothesis-driven deductive framework of
a scientific method and its characteristic dual sources of inspirations: from observations and from prior theory and models. The method may be described in a
three-level (Fig. 4.4) model adapted from Webb's [65] general frame for broad bio-
inspiration issues. We can approximately call the levels running from bottom to
top bio-reality, model-based experimentation, and embodied implementation. Taken
together, they integrate hypotheses, models and robots as a scientific device to learn
about and understand the "bio-reality" of developmental phenomena such as adaptive learning. It is worth noting that the right side of Fig. 4.4 comparing target behavior, predicted behavior and the embodied behavior is typically achieved as part of
actual robotic experiment but simulations may be used.

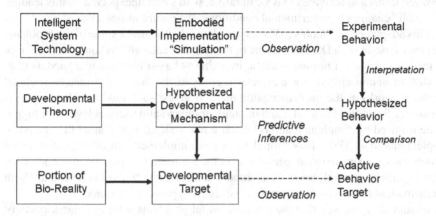

Fig. 4.4 Three level study method of developmental robotics

We can understand the bio-modeling in many types of DR studies using this
framework, starting at the lower level, which grounds itself in the reality of biological "target systems" that exist in the world. In DR studies targets are typically human
infants/children whose behavior is known for certain circumstances at certain developmental "stages". For example, there is the sequence seen from inaccurate reaching reflex to accurate visual target fixation and learning reaches that correspond to
visual targets. It is commonly observed that visually guided reaching is the earliest
"accurate" reaching behavior to occur. Infants spend significant time observing their
hands around 12 weeks and by 15 to 20 weeks are performing "visually guided"
reaching. This bottom reality level of Fig. 4.4 is related to the second (middle) level
which is essentially part of the first step shown earlier in Fig. 4.1 where "sources"
are used to set a hypothesized mechanism that can be tested via the robot mechanism experiments of steps 2 and the developmental structures of step 3. An example
model at this level is Scassellati's [58] shared attention model whose inspirational

source was the more the general developmental theories of children's cognitive abilities [2]. In Scassellati's hypothesis there are modular processes such as "Eye Direction Detector" which can determine the angle of gaze of an individual, usually the parent, and extrapolate to find an object of gaze interest. Complementing this is an "Intentionality Detector" that observes other's motion patterns and attributes the simple intentional states of desire and fear to animate objects. Finally a "Shared Attention Mechanism" produces representations of attentional states and allows a child to observe highly salient objects that are under consideration by the observing child. All of these hypotheses can be implementation as shown in the upper level of Fig. 4.4 by a robotic system, whose behavior now is similar to (simulates) the gaze behavior of young children. Of course, the degree of similarity depends on the quality of observation, the interpretation of data and so on, as shown on the right of Fig. 4.4. Bidirectional arrows in Fig. 4.4 are short-hand reminders that connections between levels and activities may be iterative, with some unexpected results leading to "small" changes to experimental conditions, the nature of observation etc.

This adaptation of Webb's [65] framework nicely illustrates some of the problems that may arise in such DR efforts. One of them is that such efforts sometimes attempt to test and validate a bio-mechanistic hypotheses based on particular models (e.g. models of neural circuits for particular regions of the brain), but then use related neural networks as the implementation method to test such models. That testing is the source for models at level 2 and implementations/validations at level 3. A bigger issue is raised by "implementations" that are really simulations rather than physical implementations. These may "simulate" virtual implements of an embodied robot which makes it easy to tweak parameters in the simulation to make it fit observation. This is hardly a real validation as the model at level 2 and the test at level 3 are both mathematical models, so independent testing of hypotheses is weak.

In summary, we see that there are potential problems with circular aspects of how researchers may define source, mechanisms, implementations and how close performance must be and to what level of detail it is tested in some of the work. Of particular interest are questions and issues for DR that can be generated from Webb's [65] framework, such as the following:

- whether all "biases" are eliminated simply by letting a system develop from general capabilities, since one still has to pick some innate substrate, relevant biological levels, and relevant environments. Are their more general principles to guide the selection?

  - There are many competing views of what are "real" vs. hypothesized concepts and principles: e.g. sensori-motor schemas are the source of more abstract schemas.

- Related to this is the issue of relevance and realism: do models test and generate hypotheses that are bio-developmentally applicable?
- When collections of distributed, specialized neural models are the source of inspiration there are multiple "hierarchical units" that need to be modeled—neuron to nets to circuits to brains. It is still challenging to adequately implement all levels and some principled way to do this is needed.

- Decisions about generality and the range of bio-developmental systems and phenomena the model can represent.

  - If adaptive mechanisms are included in the model, there may be long chains of environmental interactions that may be needed to carry out in a realistic experiment. The time needed for this may be prohibitive.

Some, like the issue of levels, interact with the issue of performance. We have trouble comparing target, expected and actual performance, since the amount of detail to do this starting at the lowest "level" exceeds not only our ability to build such distributed specialized networks, but even to simulate them. And the physical levels have correspondingly different time scales on which they work, so integrating these scales within a robot implementation is difficult [36].

Performance definition is a central issue as Webb [65] notes, arguing that we need to consider more than a naive match of behavior. When a direct comparison of the development of higher processes is attempted, the possible variability going into the match between various behaviors is considerable. Researchers have to consider whether the behaviors need to be identical (indistinguishable by some interpretive criteria) or merely similar by some interpretive criteria. Indeed, Deakin [17] argues that a superficial behavioral match is never sufficient evidence for drawing conclusions about the accuracy or relevance of a model.

This seems to be particularly true when the target is adaptive behavior which may respond to chains of small changes in the environment where time is important. There are an enormous range and number of variables to consider. An overall problem is that researches may "finesse" these problems in various ways such simplifying performance accuracy assumptions, tweaking simulations etc. [65]. It may make the models and implementations take on a Ptolemaic character [7]. By that is meant that the current state of work (in both IS and related, if simpler, problem fields including DR) seems a bit like a Ptolemaic paradigm. As noted above, mathematical models can be used as sources in DR. Such models are a bit Ptolemaic, in that they break complex intelligent behavior into known neat, related components that are bit analogous to Ptolemy's "perfect circular motions". It is a system of step by step complexity build by cycles/epicycles with certain perfect cognitive process/mechanisms as the functional/circular primitives. Thus, we may start research by accepting heuristic devices for their practical computational usefulness and ease of implementation or in DR's case development. Such core functions are added to in an ad hoc (rather than bio-development constrained way) as required to obtain any desired degree of performance and accuracy. Using large combinations of constructions, we are able to measure performance in some small problem domain for some of the agreed upon intelligent behaviors within the standards of observational accuracy. The ultimate concern and cautionary concern is that, like a Ptolemaic system, when we encounter anomalies, parameters can all-to-easily be modified to account for them. In such circumstances, the model becomes more of an after-the-event description then a deep capture of the underlying problem. Such descriptions may be very useful, making it possible to economically summarize a great amount of brute observational data, and to produce empirical predictions, but they can prove to be brittle, are not scalable and might not be fruitful to further truly predictive IS research. The field seems to understand this potential downside of some of these issues and has begun to develop some principles to constrain and guide the work.

In the next section we consider some of Lungarella's principles [36] that can help, behind the scenes, adding realistic constraints as part of DR experimentation to avoid or mitigate some of these problems. An integrated set of general principles represent a form of models for some underlying reality constrained by our points of view. As such, principles may be good candidates to apply in DR work because they are more abstract and flexible than specific models, but more concrete and easier to use than a pure design.

4.7 Developmental Principles Within an Embodied, Interactive Model

Lungarella [36] has outlined a broad set of bio-development principles that DR can employ to guide research leading to better models at level 2 of Fig. 4.4 and better implementations at level 3. Lungarella proposes that we embed general, bio-developmental principles within a developmental framework using the inter-active coupling of three things—control, body, and environment. This coupling reflects principles of ecological balance between the three factors. Such balancing of embodied intelligent systems includes adaptation to natural constraints from gravity, friction, energy limitations, living with damage, etc. This illustrates what Lungarella calls the principle of "cheap design", which means that the design of a developmental agent must be parsimonious, and must opportunistically exploit the physics of the system-environment interaction, as well as the constraints of an agent's ecological niche. Building on this is the principle of developmental "design for emergence" of higher cognition. This principle says that when engineering agents, one should not design them completely for target behavior, but instead, the design should endow an agent with a more general ability that will let higher abilities emerge. For example, some studies [11] suggest that an intelligent robot should self-direct the explo-ration of its own sensory-motor capabilities, and have some realistic capabilities to enhance with limited initial behavioral repertoire. Skills are acquired as part of a cognitive-experiential history. A related principle is that of "ecological balance", which can be understood in terms of a three-part coupling model. In this model, agent's complexity or behavioral diversity has to match the complexity of the envi-ronment, as measured against the agent's sensory apparatus. Thus, when design-ing or developing an agent for a particular task environment, a balance is required between the complexity of the sensor, motor, and control system. Through evolu-tion this comes to be the state of affairs in natural environment through selection, but in DR it has to be afforded through design. This requirement remains a difficult problem, since the coupling of the three elements is not obvious. This happens more readily in nature as understood by Lungarella's "value principle", which provides a substrate repertoire of motivated learning and adaptive behavior. A motivation substrate is needed for a developmental process to take place and for an autonomous agent to behave adaptively in the real world, along with a set of mechanisms for self-regulated learning (SRL). This substrate provides values that shape the development of the agent's control and bodily structure. Principles for the embodiment add important constraints since the embodiment imply far more than just some limiting physical constraints to an organism. Embodiment remains a challenge that actively supports and promotes intelligent information processing. It encourages agents to exploit the dynamics of the interaction between an embodied system and their environment. It thus directly supports the selection and processing of information by coupling with the agent control systems and interaction within the environment. Pfeifer and Bongard [44] argue that "artificial evolution together with morphogenesis is not only "nice to have" but is in fact a necessary tool for designing embodied agents." This coupling is different from non-biological views of computation in that it favors the developmental emergence of stable computational patterns, including adaptability and robustness against changing environments and indeed the corporal changes of development or damage. As noted before in discussion of developmental methods, combining evolutionary phylogeny and ontogeny principles with DR has not often been attempted, if for no other reason than that the implications of embodiment changes are nontrivial and the time scales challenging. Thus, the work is broached by simulations, which beg the question of representativeness since they might not provide a full model of the complexity present in the real environment. Thus simulations may be a circular pursuit as previous discussed. Lungarella [36] has formalized how embodiment affects agent processing as the principle of "information self-structuring" (active perception). As part of sensory activity, an agent is not passively exposed to information from its surrounding environment. Due to its particular morphology and through its actions on the environment, it actively structures, selects, and exploits such information. This reflects what Lungarella [36] calls the "starting simple" principle for development. We don't start with a high level model but an unfolding hybrid with a gradual and well-balanced increase of both the agent's internal complexity (perceptual and motor) and its external complexity (regulated by the task environment or an instructor). This speeds up the learning of tasks and the acquisition of new skills, compared to an agent that is complex from the onset. From this perspective, behavior informs and shapes cognition, as it is the outcome of the dynamic interplay of physical and information theoretic processes, and not the end result of a control process (computation) that can be understood at any single level of analysis. Instead, there emerges an emphasis on a general understanding of cognitive aspects.

4.8 Summary

We have discussed some of the bio-developmental ideas that inspire studies to understand the critical mass of cognitive abilities needed to develop intelligent systems capable of human-like cognitive growth. One approach has emphasized starting with relatively high-level innate functionality, including language and the

Theory-of-Mind ability in the agent. This approach leverages the lack of understanding of the complex of abilities underlying efficient learning in human children that emerge during what is called a "cognitive leap" (including self-awareness, episodic memory, the ability to imagine self in the future or in other possible worlds, a Theory-of-Mind, general meta-cognition, higher emotional intelligence, etc.).

An evaluation method based on a "cognitive decathlon" has been formulated to test ideas of what constitutes the core of the critical mass [56, 40]. A criterion for acceptance could be a good fit to the scaling laws of cognitive growth predicted by theoretical analysis of abstract cognitive growth models (which still need to be developed).

An alternative approach to better intelligent agents follows an embodied DR strategy, seeking to study higher cognitive processes as emergent from simpler, bottom-up, decentralized activities employing self-organized interaction of many local sub-systems [11]. A potential benefit argued for a DR approach for intelligent agents is that it removes the engineering bias [11], which is inherent in traditional artificial designs.

As discussed in this Chapter, there may be value to the DR approach, based on problems we see with engineered systems as a whole and AI systems in particular. We have noted a number of problematic research issues in such inspiration, including the challenge of integrating levels of bio-reality and cognitive levels, the mixing of models with implementation, especially when mathematical simulations are used, and problems with handling the degree of behavioral similarity that constitutes validation of hypotheses. Indeed, there remains a serious difference in opinions on how to proceed based on these models. One group, which might be called "neat formalists", believes in centering on formal models. The rationale for this position is expressed by Barto:

"The complexity of animal behavior demands the application of powerful theoretical frameworks. . ."
"Nervous systems are simply too complex to be understood without the quantitative approach that modeling provides. . ." [4].

Another group, that might be called "scruffy empiricists", are represented by researchers who believe that

"Developing formalized models for phenomena which are not even understood on an elementary level is a risky venture: what can be gained by casting some quite gratuitous assumptions about particular phenomena in a mathematical form?" [16].

Bio-developmental inspiration cannot be free of all biases and may introduce some of its own, that can distort our approach to design and development of artifacts. Such biases include favoring popular items (groupthink about developmental stages) or selecting vivid, unusual, or emotionally charged hypotheses. In principle, all research is influenced by interpretations of reality, and selection among the simplifications we make about the complexity of bio-reality adds influence. For example, researchers may select particular levels of the neural system to emulate, and some design selections are driven by the sheer practicality of the time scales involved in phenomena, as outlined in Berg-Cross [9].

In the face of conflicting strategies, an advice for DR researchers at the current stage of work might be to follow general principles gathered via induction rather than trying to tweak particular results against the target behavior. That is, they should be concerned with an integrative, general view such as: "building a complete, but possibly rough or inaccurate model", rather than with strict accuracy per se [65].

We argue that this goal would be helped by being couched in Lungarella's developmental framework for interactive coupling of control mechanisms, body, and environment. This would be a relatively complete system that connects action and sensing to achieve a task in an environment, even if this limits the individual accuracy of particular parts of the model because of necessary substitutions, interpolations etc., each of which by itself may introduce bias. The direction proposed herein is then to avoid slavishly following bio-inspiration in all directions. Rather, we should proceed with a developmental approach, using integrated developmental principles based on an unbiased view of how intelligence develops. Lungarella and others have begun the process of providing principles to guide such understanding of development and its implementation.

A final note is an observation that we add as a rule of thumb for development of truly intelligent systems. Given the complexity of a 3-fold interaction (body, control system and environment), researchers need to handle considerable complexity at each level of the research framework: bio-reality, model-based experimentation, and embodied implementation. This idea reflects a pragmatic observation of the growing number of large, interdisciplinary nature of the teams now working on bio-inspired robots. These cooperative efforts stand in some contrast to the simple application development teams of the past. This growth is not an accident and mirrors at a higher level what traditional software application developers have observed.

Due to the inherent complexity of a three-fold interaction in development, many disciplines and new interdisciplinary efforts are required by the new approach. This will be true wherever biological, theoretic and robotic system studies work to integrate different levels of granularity. The resulting suggestion is simply that successful implementation will require equally broad expertise not only in each of these areas, but also in how they interact. Such coupled interactions are at the boundary of existing disciplines, and therefore a rare skill to employ. This means that a considerable, explicit expertise in the development team is required to reflect the background knowledge and hypotheses that underlie the diverse, integrated models used for the development of intelligence.

References

1. Albus, J. S. (2008) Reverse engineering the brain. In A. V. Samsonovich (Ed.). Biologically Inspired Cognitive Architectures. Papers from the AAAI Fall Symposium. AAAI Technical Report FS-08-04, pp. 5–14. Menlo Park, CA: AAAI Press. ISBN 978-1-57735-396-6.
2. Baron-Cohen, S. (1995) Mindblindness: An Essay on Autism and Theory of Mind. Cambridge, MA: MIT Press.
3. Baron-Cohen, S Leslie., A. M., and Frith U. (1985) Does the autistic child have a "theory of mind"?, Cognition 21, 37–46.

 4. Barto, A.G. (1991) Learning and incremental dynamic programming. 14, 94–94.
 5. Barresi, J. (2001) Extending self-consciousness into the future. In C. Moore and K. Lemmon (Eds.). The Self in Time: Developmental Perspectives, pp. 141–161. Mahwah, NJ: Erlbaum.
 6. Bartsch, K., and Wellman, H. M. (1995) Children Talk About the Mind. Oxford: Oxford University Press.
 7. Berg-Cross, G. (2003) A Pragmatic Approach to Discussing Intelligence in Systems, Performance Metrics for Intelligent Systems (PerMIS) Proceedings 2003.
 8. Berg-Cross, G (2007) Panel discussion, Can the Development of Intelligent Robots be Benchmarked? Concepts and Issues from Epigenetic Robotics, Performance Metrics for Intelligent Systems (PerMIS) Proceedings 2007.
 9. Berg-Cross, G (2008) Introduction to Biological Inspirations for Intelligent Systems, Performance Metrics for Intelligent Systems (PerMIS) Proceedings 2008.
10. Berg-Cross, G. (2004) Developing Rational-Empirical Views of Intelligent Adaptive Behavior. Performance Metrics for Intelligent Systems (PerMIS) conference.
11. Blank, D. S., Kumar, D., and Meeden, L. (2002) A developmental approach to intelligence. In S. J. Conlon (Ed.). Proceedings of the Thirteenth Annual Midwest Artificial Intelligence and Cognitive Science Society Conference.
12. Bongard, J., and Pfeifer, R. (2003) Evolving complete agents using artificial ontogeny. In F. Hara and R. Pfeifer (Eds.). Morpho-functional Machines – The New Species: Designing Embodied Intelligence. Berlin: Springer-Verlag.
13. Cassman, M. (2005) Barriers to progress in systems biology. Nature 438, 1079.
14. Cassman, M., Arkin, A., Doyle, F., Katagiri, F., Lauffenburger, D., and Stokes, C. (2007) Systems Biology: International Research and Development. Berlin: Springer.
15. Coates, C. (2007) The Air Force 'In Silico' – Computational Biology in 2025, DTEC Report No.(s): AD-A474845; Nov 2007; 49 pp.
16. Croon, M. A., and van de Vijver, F. J. R. (1994) Introduction. In M. A. Croon and F. J. R. van de Vijver (Eds.). Viability of Mathematical Models in the Social and Behavioural Science. London: Swets and Zeitlinger.
17. Deakin, M. (2000) Modelling biological systems. In T. L. Vincent, A. I. Mees, and L. S. Jennings (Eds.). Dynamics of Complex Interconnected Biological Systems. Basel: Birkhauser.
18. Dere, E., Huston, J. P., and Silva, M. A. D. S. (2005) Integrated memory for objects, places, and temporal order: Evidence for episodic-like memory in mice. Neurobiology of Learning and Memory 84, 214–221.
19. Dean, J. (1998) Animates and what they can tell us. Trends in Cognitive Sciences 2(2), 60–67.
20. Freeman, W. (2002) On Communicating with Semantic Machines, PerMIS Proceedings, 2002.
21. Gallup, G. G. Jr. (1977) Absence of self-recognition in a monkey (Macaca fascicularis) following prolonged exposure to a mirror. Developmental Psychobiology 10, 281–284.
22. Goldman, A. (1992) In defense of the simulation theory. Mind and Language 7, 104–119.
23. Gopnik A., and Wellman H. (1994) The "theory" theory. In L. Hirschfeld and S. Gelman (Eds.). Mapping the Mind: Domain Specificity in Cognition and Culture, pp. 257–293. New York: Cambridge University Press.
24. Gordon, R. (1986) Folk psychology as simulation. Mind and Language 1, 158–170.
25. Hayes S. M., Ryan, L., Schnyer, D. M., and Nadel, L. (2004) An fMRI study of episodic memory: Retrieval of object, spatial, and temporal information. Behavioral Neuroscience 118, 885–896.
26. Graham-Rowe, D. (2005) Mission to build a simulated brain begins. New Scientist, June.
27. Hagoort, P. (2005). On Broca, brain, and binding: a new framework. Trends in Cognitive Sciences 9, 416–423.
28. Heal, J. (1996) Simulation and cognitive penetrability. Mind and Language 11, 44–67.
29. IBM Blue Brain Project (2005), http://bluebrainproject.epfl.ch/
30. Indefrey, P., and Levelt, W. J. M. (2004) The spatial and temporal signatures of word production components. Cognition 92(1–2), 101–144.
31. Kitano, H. (2002). Computational systems biology. Nature 420, 206–210.

32. Kurths, J., Hilgetag, C., Osipov, G., Zamora, G., Zemanova, L., and Zhou, C. S. (2007) Network of Networks – a Model for Complex Brain Dynamics, http://www.agnld.uni-potsdam.de/~juergen/juergen.html.
33. Lee, M., Meng, Q., and Chao, F. (2006) Developmental Robotics from Developmental Psychology, Proceedings of Towards Autonomous Robotic Systems (TAROS-06), pp. 103–09, University of Guildford, Surrey.
34. Lemmon, K., and Moore, C. (2001) Binding the self in time. In C. Moore and K. Lemmon (Eds.). The Self in Time: Developmental Perspectives, pp. 141–161. Mahwah, NJ: Erlbaum.
35. Levelt, W. J. M. (1989) Speaking: From Intention to Articulation. Cambridge, MA: MIT Press.
36. Lungarella, M. (2004) Exploring Principles Toward a Developmental Theory of Embodied Artificial Intelligence, Ph.D. Dissertation, Zurich. http://www.ifi.uzh.ch/ailab/people/lunga/Download/PhDThesis/thesis040504_complete.pdf.
37. McCarthy, J., Minsky, M., Rochester, N., and Shannon, C. (1955/2006) A proposal for the dartmouth summer research project on artificial intelligence. AI Magazine 27(4), 12–14.
38. McCulloch, W. S., and Pitts, W. H. (1943) A logical calculus of the ideas immanent in nervous activity. Bulletin of Mathematical Biophysics 5, 115–133.
39. Miconi, T., and Channon, A. (2005) Analysing coevolution among artificial creatures. In E. G. Talbi (Ed.). Procs Evolution Artificielle 2005 (EA 05). Berlin: Springer-Verlag.
40. Mueller, S. T., and Minnery, B. S. (2008) Adapting the Turing test for embodied neurocognitive evaluation of biologically-inspired cognitive agents. In A. V. Samsonovich (Ed.). Biologically Inspired Cognitive Architectures. Papers from the AAAI Fall Symposium. AAAI Technical Report FS-08-04, pp. 117–126. Menlo Park, CA: AAAI Press. ISBN 978-1-57735-396-6.
41. Nichols S., and Stich, S. (2003) Mindreading: An Integrated Account of Pretence, Self-Awareness, and Understanding Other Minds. Oxford: Oxford University Press.
42. Nolfi, N., Ikegami,T., and Tani, J. (2008), Behavior as a complex adaptive system: On the role of self-organization in the development of individual and collective behavior. Adaptive Behavior 16(2–3), 101–103.
43. Penrose, R. (1989) The Emperor's New Mind: Concerning Computers, Minds, and the Laws of Physics. New York: Oxford University Press.
44. Pfeifer, R., and Bongard, J. C. (2006) How the Body Shapes the Way We Think—A New View of Intelligence. Cambridge, MA: MIT Press.
45. Piaget, J., and Inhelder, B. (1998) Jean Piaget: Selected Works Third Edition). London: Routledge.
46. Pinker, S., and Jackendoff, R. (2005) The faculty of language: What's special about it? Cognition 95, 201–236.
47. Povinelli, J. (2001) The self: Elevated in consciousness and extended in time. In C. Moore and K. Lemmon (Eds.). The Self in Time: Developmental Perspectives, pp. 141–161. Mahwah, NJ: Erlbaum.
48. Rosenfeld, S., and Kapetanovic, I. (2008) Systems biology and cancer prevention: All options on the table. Gene Regulation and Systems Biology 2, 307–319.
49. Rubin, D. C., Schrauf, R. W., and Greenberg, D. L. (2004) Stability in autobiographical memories, Memory 12, 715–721.
50. Samsonovich, A. (2000) Masked-priming 'Sally-Anne' test supports a simulationist view of human theory of mind. In B. W. Mel and T. J. Sejnowski (Eds.). Proceedings of the 7th Joint Symposium on Neural Computation, vol. 10, pp. 104–111. San Diego, CA: Institute for Neural Computation, UCSD.
51. Samsonovich, A. V. (Ed.). (2008) Biologically Inspired Cognitive Architectures. Papers from the AAAI Fall Symposium. AAAI Technical Report FS-08-04, 206 pp. Menlo Park, CA: AAAI Press. ISBN 978-1-57735-396-6.
52. Samsonovich, A. V., and Nadel, L. (2005) Fundamental principles and mechanisms of the conscious self. Cortex 41(5), 669–689.

53. Samsonovich, A. V., and Ascoli, G. A. (2005) A simple neural network model of the hippocampus suggesting its pathfinding role in episodic memory retrieval. Learning & Memory 12, 193–208.
54. Samsonovich, A. V., and De Jong, K. A.(2005) Designing a self-aware neuromorphic hybrid. In K. R. Thorisson, H. Vilhjalmsson, and S. Marsela (Eds.). AAAI-05 Workshop on Modular Construction of Human-Like Intelligence: AAAI Technical Report, vol. WS-05-08, pp. 71–78. Menlo Park, CA: AAAI Press.
55. Samsonovich, A. V., and De Jong, K. A. (2005) A general-purpose computational model of the conscious mind. In M. Lovett, C. Schunn, C. Lebiere, and P. Munro (Eds.). Proceedings of the Sixth International Conference on Cognitive Modeling, pp. 382–383. Mahwah, NJ: Erlbaum.
56. Samsonovich, A. V., Ascoli, G. A., and De Jong, K. A. (2006) Computational assessment of the 'magic' of human cognition. In Proceedings of the 2006 International Joint Conference on Neural Networks, pp. 1170–1177. Vancouver, BC: IEEE Press.
57. Samsonovich, A. V., Kitsantas, A., Dabbagh, N., and De Jong, K. A. (2008) Self-awareness as metacognition about own self concept. In M. T. Cox and A. Raja (Eds.). Metareasoning: Thinking about Thinking. Papers from the 2008 AAAI Workshop. AAAI Technical Report, vol. WS-08-07, pp. 159–162. Menlo Park, CA: AAAI Press.
58. Scassellati, B. (2001) Foundations for a Theory of Mind for a Humanoid Robot Ph.D. Dissertation, MIT, http://www.cs.yale.edu/homes/scaz/papers/scassellati-phd.pdf.
59. Schrödinger, E. (1944) What is Life. Cambridge, MA: Cambridge University Press.
60. Sims, K. (1994) Evolving 3d morphology and behavior by competition. In R. Brooks and P. Maes (Eds.). Proceedings of SAB'98. Cambridge, MA: MIT Press.
61. Starns, J. J., and Hicks, J. L. (2004) Episodic generation can cause semantic forgetting: Retrieval-induced forgetting of false memories. Memory & Cognition 32, 602–609.
62. Thelin, J. W., and Fussner, J. C. (2005) Factors related to the development of communication in CHARGE syndrome. American Journal of Medical Genetics Part A 133A(3), 282–290.
63. Tulving, E. (1983) Elements of Episodic Memory. New York: Clarendon.
64. Turing, A. (1950) Computing machinery and intelligence. Mind 59(236), 433–460.
65. Webb, B. (2001) Can robots make good models of biological behaviour? Behavioral & Brain Sciences 24, 1033–1050.
66. Wheeler, M. A., Stuss, D. T., and Tulving, E. (1997) Toward a theory of episodic memory: The frontal lobes and autonoetic consciousness. Psychological Bulletin 121, 331–354.
67. Wooley, J. C., and Lin, H. S. (2005) Biological inspiration for computing. In John C. Wooley and Herbert S. Lin (Eds.). Catalyzing Inquiry at the Interface of Computing and Biology, Computer Science and Telecommunications Board. Washington, DC: The National Academies Press.
68. Zimmerman, B. J. (2002) Becoming a self-regulated learner: An overview. Theory into Practice 41(2), 64–70.
69. Zimmerman, B. J., and Kitsantas, A. (2006) The hidden dimension of personal competence: Self-regulated learning and practice. In A. J. Elliot and C. S. Dweck (Eds.). Handbook of Competence and Motivation, pp. 509–526. New York: The Guilford Press.

Chapter 5
Evaluating Situation Awareness
of Autonomous Systems

Jan D. Gehrke

Abstract Autonomous systems proved to be successful in various application
areas. But their perception, reasoning, planning and behavior capabilities are generally designed to fit special purposes only. For instance, a robotic agent perceives
its environment in a way that was defined in advance by a human designer. The
agent does not exhibit a certain perception behavior because it actually thinks it
would be reasonable to do so. But with an increasing level of autonomy as well
as a larger temporal and spatial scope of agent operation higher-level situation
analysis and assessment become essential. This chapter examines approaches for
knowledge representation, reasoning, and acquisition that enable autonomous systems to evaluate and maintain their current situation awareness. An example application scenario is presented that provides initial results for evaluating situation-aware
systems.

5.1 Introduction

Autonomous systems are being developed for numerous application areas. These
systems proved to be successful in domains such as road driving, area exploration,
or robot soccer. Nevertheless, in many cases, the perception, reasoning, planning and
behavior capabilities of autonomous systems are *designed* to fit a special purpose.
For instance, a robotic agent perceives its environment in a way that was defined
in advance by a human designer. Therefore, the agent does not show a certain perception behavior because it actually *thinks* it would be reasonable to do so. It is
predetermined behavior rather than a deliberate action. For a lot of applications this
might be sufficient. But with an increasing level of autonomy (cf. [30]) as well as a
larger temporal and spatial scope of agent operation, higher-level situation analysis

J.D. Gehrke (✉)
Center for Computing and Communication Technologies – TZI, Universität Bremen,
28359 Bremen, Germany
e-mail: jgehrke@tzi.de

R. Madhavan et al. (eds.), *Performance Evaluation and Benchmarking*   93
of Intelligent Systems, DOI 10.1007/978-1-4419-0492-8_5,
© Springer Science+Business Media, LLC 2009

and assessment become essential. We will focus on knowledge-based autonomous systems that require up-to-date knowledge to decide on their next course of action. Knowledge-based systems are particularly challenging if they require knowledge of distant locations and/or prediction of future events. These systems cannot rely on their own sensory capabilities only. They need to infer future states and communicate with other agents to share information.

Situation assessment depends on the current goals of the autonomous system and knowledge about the state of the world. The *relevance* of a specific information for situation assessment is determined by the current plan and other potential plans under consideration. Usually, this is implicitly specified by the decision system. For instance, the decision system may use behavior rules whose rule body implies the information needed to evaluate the rule head. Problems arise if the decision-relevant information is not available to the agent. As a consequence, the system would not be able to assess a situation correctly, e.g., to detect a harmful risk, because it has a lack of information which it is not aware of. This seems inadequate for autonomous systems in partially observable environments.

Hence, autonomous systems doing situation assessment have to be enabled to detect missing information, thereby becoming *known unknowns*. Due to bounded resources, this detection process must be governed and prioritized by information relevance. If the agent's sensory capabilities cannot provide information needed, other agents or information sources have to be inquired. As an alternative, the agent has to accept its lack of information and address it, e.g., by more cautious behavior.

Only autonomous systems possessing such higher level reasoning capabilities (i.e., detection of ignorance, assessment of information relevance, and active acquisition of information) are able to have true *situation awareness* [10]. In dynamic environments, such agents would need prediction abilities including knowledge on how the world evolves in order to qualify the probability that some information does not change in a given time interval. In spacious environments with mobile robots, additional spatial reasoning about information dynamics is required.

This chapter outlines a classification of criteria for situation awareness of autonomous systems and suggests knowledge representation and reasoning methods to address them. The evaluation of system situation awareness is twofold.

- *Design guidelines*: In one perspective, evaluation is about analyzing an existing system or designing it in a way that addresses issues related to situation awareness.
- *Active perception*: The second perspective addresses the autonomous system's point of view, i.e., assessing its run-time situation awareness by identifying missing information, measuring its relevance, and finally acquiring it.

The presented criteria for situation awareness pertain to the first perspective. They are considered a basis for addressing the second perspective. A quantitative decision-theoretic approach is presented for measuring information relevance and actively maintaining situation awareness based on that assessment.

The remainder of this chapter is structured as follows. Section 5.2 discusses issues of situation awareness in established autonomous system architectures. Section 5.3 defines and examines awareness criteria and possible techniques to implement them. Section 5.4 describes a meta-level control to establish and maintain situation awareness. A corresponding example scenario is presented in Section 5.5. The chapter concludes with a discussion of this survey (Section 5.6) and a summary (Section 5.7).

5.2 Autonomous Agents and Situation Awareness

Intelligent agents are a fundamental concept in Artificial Intelligence for autonomous decision-making. For most application domains of agents, up-to-date and precise knowledge on the state of the world is crucial for system performance. This has been addressed by active perception in the context of computer vision for robot guidance. But surprisingly a lot of conventional architectures do not explicitly consider the acquisition of other information needed. This limitation might lead to suboptimal, wrong, or even disastrous decisions. Thus, situation awareness for autonomous systems intends to evaluate and establish the basis for decision making depending on the agent's current tasks and goals. In this chapter, we consider the general case of active information acquisition beyond the signal processing level.

Williams et al. [27] have a similar goal. They evaluate the so-called *groundedness* of representations in autonomous systems (mainly those applied in *RoboCup* competitions). The approach defines a measure for the capability of creating and maintaining correct associations between representations and their (physical) real-world counterparts in the system's knowledge base. Proposed qualities of groundedness include, e.g., relevance, accuracy/precision, uncertainty management, and self-awareness with respect to the state of the robot body, location, and sensors. The corresponding system evaluation is rather qualitative and performed offline by humans. Thus, the approach provides useful criteria for system evaluation but does not enable the system to reason about itself in order to improve its groundedness.

The simple reflex agent (also called reactive agent[1]), as the most basic kind of agent, is the worst example for groundedness or situation awareness. Such agents are governed by condition/action rules and always do the same thing given the same perception. In contrast to the model-based reflex agent it has no internal state influencing its decisions [21]. Both reflex agents cannot be considered situation-aware. The simple reflex agent only takes into account instantaneous knowledge; the model-based variant has no notion of information relevance because it has no explicit goals. Nevertheless, there may be simple but useful tasks that are successfully handled by reflex agents.

---

[1] The term *reflex agent* from Russell and Norvig [21] is preferred here to distinguish this kind of agent from the more general concept of agents with the ability to react on changes in the environment as stated by Wooldridge and Jennings [29].

The most important term in AI is the rational agent. Wooldridge defines an agent to be "rational if it chooses to perform actions that are in its own best interests, given the beliefs it has about the world" [28, p. 1]. But this definition could also consider an agent rational if it chooses an action without knowing the state of the world. Thus, the situation-aware agent extends and substantiates the classical rational agent definition. The belief about the world is no longer taken for granted but actively controlled by information acquisition as an additional reasoning process and meta-level control. Raja and Lesser [19] emphasized the importance of meta-level control in environments with uncertainty, partial observability, and resource boundedness. They provide a reinforcement learning approach for meta-level control policies but do not focus on situation awareness.

The belief-desire-intention (BDI) model has become a prevalent approach in academia for deliberative software agent architectures [e.g. 28, 16]. It is based on a theory of human practical reasoning developed by Bratman [7]. Human practical reasoning, according to BDI, consists of deliberation, i.e., deciding what state should be achieved, and means-end reasoning, i.e., deciding how to achieve it. In the BDI model, an agent is represented by its subjective knowledge about the world (*beliefs*) and persistent goals that should be achieved (*desires*). Desires and current beliefs result in achievable goals and possible actions towards them. Finally, in a process of deliberation, the agent commits to a goal and a corresponding plan (*intention*).

The fundamental BDI model does not consider the assessment of beliefs in terms of completeness, correctness/uncertainty, or being up-to-date with respect to the goals to be achieved. Additionally, the model does not take into account different levels of decision making with respect to real-time requirements or temporal scope of action and decision making. So and Sonnenberg [25] have proposed an extended meta-level control cycle for BDI agents for ensuring situation awareness. The approach introduces an attention focus for active perception depending on the contexts of the current potential plans. The attention algorithm basically uses non-probabilistic background knowledge on when some belief is expected to be outdated. It does not provide an actual measure for relevance-driven acquisition.

The Real-time Control System (RCS) developed at National Institute of Standards and Technology (NIST) models an intelligent system as a hierarchy of goal-directed sensory-interactive control processes [2] representing organizational levels as well as temporal scopes of decision making. The process hierarchy in RCS enables the decomposition of sub-tasks to different agents as well as different planning intervals within a single agent. Each level contains computational elements for sensory processing, world modeling, value judgment, and behavior generation [1]. Situation awareness could be assigned to the higher-level RCS world modeling components with a tight connection to behavior generation and sensory processing. That is, RCS could be augmented in world modeling by goal-oriented pro-active information acquisition that is governed by behavior generation demands and may provide a focus of attention in sensory processing.

5.3 Criteria for Situation Awareness

Situation awareness is a field of research that commonly examines information requirements of humans for special jobs such as facility monitoring or flying aircraft [10]. Endsley [9, p. 97] describes situation awareness as "the perception of the elements in the environment within a volume of time and space, the comprehension of their meaning and the projection of their status in the near future". This leads to Endsley's three levels of situation awareness:

1. *Perception*: Basic perception of important information.
2. *Comprehension*: Correct interpretation and integration of perceptions as well as relevance assessment.
3. *Projection*: The ability to predict future situations based on current perceptions and background knowledge.

Although the definition and the three levels of awareness are intended for human situation awareness they can be adopted for autonomous systems, too. Nevertheless, there are a lot of technical requirements that are partially taken for granted regarding humans but much more challenging for technical systems. Thus, this section proposes the following criteria for situation awareness of autonomous systems and possible methods to fulfill them:

1. *Awareness of ignorance*: The agent's knowledge base can be queried for missing or uncertain information.
2. *Model of perception abilities*: The agent is aware of its sensors and the kind of information they may generate.
3. *Model of information relevance*: Based on its decision system and current plans, the agent can identify information needed and assess information relevance for its performance measure.
4. *Model of information dynamics*: Knowledge on how the world evolves in time and space. The model is applied for prediction and reasoning on region topologies.
5. *Spatio-temporal qualification*: In spacious and dynamic domains information needs to be qualified in time and space.
6. *Information sharing*: If agents need information beyond their sensory capabilities they have to cooperate with other agents for external information acquisition.

5.3.1 Awareness of Ignorance

To enable an autonomous system to measure awareness in a given situation, the system needs to know what it does *not* know (*known unknowns*). That is, it has to detect a possible lack of knowledge in its knowledge base. This is particularly important and challenging in environments that are highly dynamic and only partially observable. Unfortunately, many logic-based systems use *negation as failure* in reasoning, i.e., propositions or predicates are assumed to be false if there is no fact or proof stating the opposite. This *closed world assumption* (CWA) is opposed to the *open world assumption* (OWA) which does not make any assumptions about missing knowledge. Instead, logical inference only relies on given facts.

Open world reasoning is particularly applied in ontologies and description logics [4] for concept subsumption in ontology TBoxes, i.e., the schema level of a knowledge base. Situation awareness is rather focused on the instance or assertional level (ABox) of a knowledge base. While the open world assumption also applies for ABox reasoning, it does not provide inferences that would directly provide information on missing facts.

As a consequence, CWA and OWA reasoning systems will create a biased view of the state of the world that is likely to be wrong. If decisions rely on wrong beliefs system performance is jeopardized. But a strictly logical approach to reason about agent ignorance will raise several difficulties for the logical foundations of representation and reasoning as well as computational efficiency. A structural approach that keeps track of knowledge base changes and instantly replies to queries on ignorance is probably preferable. Nevertheless, a three-valued logic with OWA reasoning would be required, too.

In general, a strictly logic-based knowledge representation in dynamic environments is debatable. These approaches do not sufficiently handle the uncertainty that is an inevitable consequence in such domains. Though logics are still useful and powerful for representing background knowledge of an autonomous system, dynamic environmental properties should rather be represented by probabilistic approaches with aleatory variables. In particular, Bayesian inference provides powerful means to reason about uncertainty. A corresponding approach for value measurement of missing information is presented in Section 5.3.3.

5.3.2 Model of Perception Abilities

While it is important to be aware of ignorance, agents might deliberately or of necessity choose to decide instantly although having incomplete or uncertain knowledge. This might be due to limited resources or perception abilities, i.e., sensors. An autonomous system that desires a certain information may not be able to acquire this information from its own sensors or other data sources. In such cases, it is obviously not reasonable to wait for that information before the urgent decision for which it is considered helpful.

Thus, the system needs to accept its uncertainty and ignorance. But this decision requires background knowledge on what information can be obtained or not. This knowledge can be provided by a sensor model that describes sensors by the type of information they may deliver as well as precision, accuracy, spatial range, response time, and potential observation cost. A related ability, the projection of future events, is discussed in Section 5.3.4.

For situation-aware agents we created an OWL-DL [5] ontology of possible sensors and their properties. This ontology does not describe concrete sensors, e.g., some special LIDAR product. Instead, it represents

- types of sensor devices (simple and complex),
- the (usually physical) quantities that are measured,
- the unit of measurement (e.g. SI units),
- the method of measurement (mechanical, electrical, optical, chemical etc.), and
- the sensor quality in terms of measurement errors, precision/accuracy, range, response time etc.

Examples of modeled sensors are those for electrical quantities (i.e. ohmmeter, galvanometer, voltmeter, etc.), electromagnetic radiation sensors for a given spectrum (i.e. infrared, visible light, etc.), acoustic sensors, thermal sensors and range sensors etc. The basic sensors were called *detectors*. But sensors as well as measured quantities can also be more complex. For instance, a weather "sensor" provides more than just one quantity and may aggregate raw measurements to some qualitative information.

The general advantage for agents using this ontology is that they can logically infer whether they have a sensor (or other information source) that will provide a desired information on some quantity or other environmental feature. Additionally, an ontology-grounded value description will allow for better interpretation of sensor data. The ontology also enables agents to communicate about desired information abstracting from the specific sensors used (cf. Section 5.3.6).

5.3.3 Model of Information Relevance

In large and dynamic environments there will be a lot of information that is inaccessible or unknown to the autonomous system. Other information is available but there are no sufficient resources for processing. Obviously this also holds for humans but they perform very good in a lot of tasks anyhow. This is because the major part of the state of the world is simply irrelevant. Humans have the basic cognitive capability to focus on the relevant stimuli and ignore the others.

When trying to imitate this human capability in a cognitive approach to autonomous systems it is practically impossible to design a generic approach for all purposes. Low-level attention control will require low-level implementation because of real-time constraints. But there are also other kinds of information that allow for higher-level assessment of information relevance for determining a focus of attention. Such information is related to decisions with temporal scopes beyond the servo level with 0.05 s plans (cf. [2]).

Based on the current set of goals or a general utility function as well as applied decision rules, the agent could identify information needed and qualify its importance for its performance measure. For goal-oriented systems with rule-based decisions, information (as logical predicate) is relevant if it is contained in the body of some rule whose head is going to be evaluated. While this could be identified in a quite straightforward way on syntax level, such a system would require to determine predicates with unknown truth value. This again, has to apply some special handling of ignorance in logic (Section 5.3.1). Furthermore, logical representations do not support a quantitative measure of information relevance.

Thus, we focus on utility-based systems with probabilistic knowledge representation. Such systems need to assess information relevance differently. Here we consider a decision-theoretic agent that tries to maximize its utility based on expected utility of each possible action A with uncertain outcomes O_1 to O_n of different probabilities given a set of evidences E. In simple cases, there is a single decision (not a sequence) and a discrete distribution for action outcomes. Then, the expected utility of action A is defined by:

$$EU(A|E) = \sum_{i=1}^{n} P(O_i(A)|Do(A), E) \cdot U(O_i(A)) \tag{5.1}$$

A rational agent will choose action $\alpha = \max_A EU(A|E)$ which depends on the given evidence E. Thus α will potentially change given additional evidence E_j:

$$\alpha_{E_j} = \max_A EU(A|E, E_j) \tag{5.2}$$

Intuitively, the relevance (or value) V_{E_j} of a new evidence E_j is then defined as the difference in expected utility of the chosen action after and before knowing E_j. But because the agent needs to determine the information value before acquiring the information it will need to average over possible assignments e_{jk} of E_j via conditioning, i.e.,

$$V_{E_j} = \left(\sum_k P(E_j = e_{jk}|E) \cdot EU(\alpha_{e_{jk}}|E, E_j = e_{jk}) \right) - EU(\alpha|E) \tag{5.3}$$

This approach is proposed in *information value theory* by Howard [12]. If the acquisition of information E_j is costly, it is only worthwhile if its price is less than its value (presupposing utility is measured in same quantity as price). Note that the information value formula presupposes perfect information. Uncertain information can be included by modeling the probabilistic relation $P(X_s|X_a)$ from the uncertain actual variable X_a to the sensor variable X_s.

In order to apply this decision-theoretic relevance measurement to assess and potentially increase situation awareness an agent will need to proceed in a systematic way. That is, there is no sense in evaluating relevance of all possible information (i.e., all modeled aleatory variables). The agent needs some background knowledge on utility-influencing factors.

This knowledge is provided by a decision network (or decision diagram) [13]. A decision network combines Bayesian inference with actions and utilities as additional types of nodes besides the conventional chance nodes. The factors directly

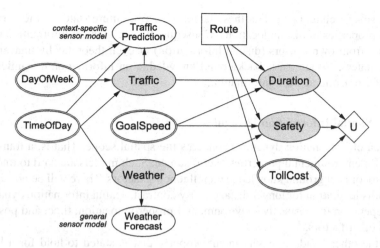

Fig. 5.1 Simple decision network example for a route selection problem (adapted from [21, p. 598])

influencing agent utility are represented as parent chance nodes. Possible outcomes of actions are modeled as child chance nodes of each action respectively. An example for route assessment is given in Fig. 5.1. Grey nodes denote non-observable variables whereas double-line nodes are deterministic or always known. The sample model contains two sensor nodes: for prediction of traffic and weather. The traffic sensor model takes into account day of week and time of day as context while the weather model is general and thus less specific.

This decision problem representation enables a systematic ordering of information evaluation and acquisition by traversing the network from the utility node along the path with highest influence on utility (see Section 5.4). For this purpose, it is necessary to distinguish sensor nodes as special chance nodes since only evidence from sensor nodes is actually available.

Unfortunately, the design of decision networks features the same difficulties as in Bayesian networks: one needs to assign prior probabilities as well as conditional probabilities based on sufficient experience and background knowledge. Because this is not given in general, the agent should be able to adapt the Bayesian influence model based on ongoing experience.

5.3.4 Model of Information Dynamics

Although decision-theoretic information-gathering agents can be implemented based on decision networks, there are special requirements in dynamic environments with decisions of long temporal and spatial scope. As an example, we will use a simple route planning task with a truck that aims at finding the fastest route between two locations in a road network. A decision network for that problem could include chance nodes for influencing factors such as construction work, weather, traffic, time of day, and others (cf. Fig. 5.1).

The special challenge is that these factors describe future states of the world as well as properties of distant locations. Consequently, the relevant information is not available from own sensors (distant information) or even theoretically unavailable (future states). Nonetheless, background knowledge on information dynamics can help infer or estimate such information.

5.3.4.1 Spatial Information Dynamics

One area of information dynamics concerns the spatial scope. That is, a transition model for environmental properties over distance. Such models are hard to find and need a lot of background knowledge on influencing factors. There will be no general probabilistic relation for longer distances. By contrast, spatial information dynamics may depend on the respective environmental property, location, time, and possibly several other factors.

On the other hand, an environment property that is stated to hold for a larger area should allow to deduce this property for sub-areas based on known region partonomies. If the agent knows that some environment property (e.g., weather) holds in an area of certain extension (e.g., a particular stretch of a freeway) it can easily infer the same property at another location within that area (cf. Section 5.3.5). Nevertheless, this inference will provide uncertain information. The statement for a large area will usually only average all locations therein. Thus, single locations might be very different.

Thus, knowledge on spatial information dynamics for situation awareness is better handled by structuring space in proper regions as background knowledge. Environmental information can then be qualified to hold in a certain region that is supposed to be homogeneous for all locations therein. Additionally, distant information should be gathered by information sharing with other agents (cf. Section 5.3.6).

5.3.4.2 Temporal Information Dynamics

Similar considerations also hold for the temporal dimension. A situation-aware system needs prediction abilities to estimate probabilities of variable assignments at future states or it has to acquire reliable predictions from other sources. Obviously, this becomes less achievable when the predicted state is far in the future. On the other hand, there are variables that will not change or only change slowly or rarely.

Re-iterating the above examples, weather or traffic feature some expected stability and transition probability from one state to another, respectively. The easiest way to keep information up-to-date is to attach a meta-information that indicates the time span it is assumed to be correct (cf. [25]). While this is easy to implement, it neither allows estimating the value after that interval nor quantifying the uncertainty. Thus, more elaborate approaches model dynamics with Markov Models, Kalman filters, or, more general, dynamic Bayesian networks (DBNs). That is, one defines an influence or sensor model and a time transition model for the property. Classical DBNs presuppose a fixed discretization of time from one state to another. A reasonable time discretization for a Markov transition model will depend on average stability or evolution rate of the modeled properties. Thus, in general, the property variable with minimum evolution rate determines time discretization which impairs run-time performance of reasoning for other variables. Additionally, the evolution rate may again depend on other factors and change from one state to the following. Thus, Nodelman et al. [18] and El-Hay et al. [8] have proposed Markov processes and networks with continuous time. Saria et al. [22] present an approach for discrete but heterogeneous time granularity for different clusters of variables which reduce complexity of DBNs.

Nevertheless, the design of and reasoning in such networks has to be measured wisely because the performance is likely to get bad when reasoning over longer time spans (cf. [26]). Thus, exact inference in DBNs is inadequate for long-term predictions when this results in a lot of computation steps and/or probability distributions that converge to the prior distribution. However, such long-term predictions would be applied for the presented transportation example. Thus, situation-aware agents should use background knowledge on expected information stability in order to determine whether a Markov prediction is useful beyond a given time window.

As a consequence, we advocate the usage of dynamic decision networks with variable-specific time granularity depending on rate of evolution (cf. [22]). Additionally, background knowledge on information stability indicates limits of useful Markov prediction. Model complexity can be reduced further by decomposing large decision networks into multiple, mutually independent decision support models. This enables parallel processing in acquisition as well as the usage of other, e.g., logical prediction methods within the support models (cf. Section 5.5).

5.3.5 Spatio-Temporal Qualification

If a situation-aware autonomous system evaluates relevance of missing information, it will also need to qualify time and space when/where the desired property value holds. We will call this spatio-temporal qualification a *region of relevance* for that agent. An agent's whole region of relevance R_{rel}^* consists of several sub-regions R_{rel_i} where a minimal consistent part is defined as

$$R_{rel} = ([t_s, t_e], s \in \mathbb{S}, var, V_{var}) \tag{5.4}$$

with $[t_s, t_e]$ defining a time interval, s being a spatial element in the domain $\mathbb{S}$ of structured space, var as the relevant variable or physical quantity, and V_{var} denoting the information value of var. All spatio-temporal information overlapping that region is considered potentially relevant with respect to value of information (Section 5.3.3).

In the transportation example, $\mathbb{S}$ is defined by the set of road stretches. Relevant variables for estimating driving time for a potential route could be weather and traffic at some distant freeway stretch. For specifying a region of relevance the vehicle agent has to consider the expected time t_s when it plans to arrive at s and the time t_e it will have passed s. Unfortunately, both values are uncertain in general.

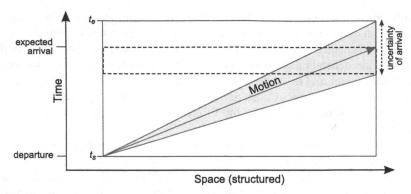

Fig. 5.2 Spatio-temporal region defined by vehicle motion on a road stretch with uncertain driving time

Figure 5.2 depicts the situation for a road stretch at the beginning of a truck tour. While the departure time t_s is known here, arrival time t_e is uncertain even for the first route segment. The driving time (and thus arrival time) gets more and more uncertain with increasing distance. To specify t_e, we choose a threshold as a quantile in probability distribution of driving time dt on s to specify a confidence interval of driving time $[dt_l, dt_u]$. The upper bound dt_u will then help determine the upper temporal bound of the region of relevance with $t_e = t_s + dt_u$.

If the location at the end of s does not equal the agent's destination, there will be one or more successor regions of relevance $R_{succ}(R_{rel})$. Their start time $t_{s_{succ}}$ will be uncertain because the arrival time for R_{rel} is uncertain, too. Here the lower bound dt_l of driving time for the previous region has to determine the departure time: $t_{s_{succ}} = t_s + dt_l$.

Thus, these regions of relevance will expand in temporal dimension for distant locations because uncertainty of departure time increases with uncertainty of arrival time for previous regions. A region of relevance for distant locations is shown in Fig. 5.3. A potential application of these regions is described in the following section.

5.3.6 Information Sharing

For situation awareness, information sharing is needed when there is relevant information that is neither available from own sensors nor inferable by reasoning. For instance, information on distant locations needs to be acquired from external sources. Communication for information sharing is one aspect of social ability which is claimed to be a minimal criteria for intelligent agents as defined by Wooldridge and Jennings [29].

In practice, the mere ability of communication will not suffice. A situation-aware system must determine that some information cannot be acquired from own sensors (Section 5.3.2) but is available from other sources instead. This includes the

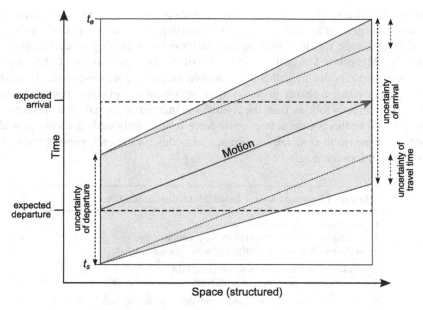

Fig. 5.3 Spatio-temporal region with uncertain departure time

discovery of such sources, a common communication language with agreed content semantics, and price negotiations if information provisioning should be subject to charges.

In addition to the *region of relevance* for information type, space, and time (cf. Section 5.3.5) the present and inferable information defines a corresponding *region of expertise*. This is of particular interest when agents want to share information and need to determine whether there is potential for cooperation. The region of relevance can then be used as a query to other cooperating agents. Two agents with overlapping region of relevance (seeker) and region of expertise (source) are able to exchange information. Apart from the actual or expected value of a desired variable, the exchanged information might also be a distribution for that variable.

For qualitative topological inference on overlapping regions, agents can apply usual spatial calculi such as *Region Connection Calculus* (RCC) [20] together with the interval calculus of Allen [3] for the temporal dimension. Similar approaches have been applied in the context of queries in geographic information systems (GIS) [14].

5.4 Maintaining Situation Awareness

An autonomous system can make use of the information relevance assessment by including it in its control cycle as meta-level control for situation awareness maintenance. As stated in Section 5.3.3, this meta-level control can traverse the decision network from the utility node to most influential chance nodes and try to acquire or sense the assignment for a chance node variable depending on information value.

A pseudo code for the corresponding information gathering procedure is provided by Algorithm 1. Starting from the utility node the values of the closest unknown chance nodes are gathered if possible, i.e., if they are observable. Roughly speaking, the closer a chance node the more important it is in comparison to parent nodes. This directly follows from the conditional independence criterion in Bayesian networks. Nevertheless, the sole edge distance to the utility node does not provide a sufficient metric to compare importance of nodes. In the end, nodes have to be ordered by information value.

Algorithm 1 Information value-based information acquisition

1: *evidences* ← current evidences
2: *evaluationNodes* ← ∅ sorted by Acquisition-Value
3: *evaluationNodes* ← Insert(utility node, *evaluationNodes*)

4: **while** *evaluationNodes* ≠ ∅ ∧ time left **do**
5:    *curNode* ← pick highest node from *evaluationNodes*

6:    **if** Observable(*curNode*) and Acquisition-Value(*curNode*) > 0 **then**
7:       {Determine assignment of curNode and add to evidences:}
8:       *evidences* ← Insert(Acquire(*curNode*), *evidences*)

9:    **else if** Inf-Value(*curNode, evidences*) > 0 **then**
10:       **for all** conditions and child nodes *next* of *curNode* **do**
11:          **if** *next* is chance node ∧ *next* ∉ *evidences* **then**
12:             compute Inf-Value(*next, evidences*)
13:             *evaluationNodes* ← Insert(*next, evaluationNodes*)
14:          **end if**
15:       **end for**
16:    **end if**
17: **end while**

For each evaluated node *curNode*, the algorithm either acquires the value for the node's variable (if observable and valuable including observation cost; lines 6–8), expands the node (if non-observable but valuable; lines 9–15), or ignores it (if not valuable). For expansion, the algorithm checks all nodes linked with the current node *curNode* (line 10). Those chance nodes whose value is not known yet (line 11) are added to *evaluationNodes* (line 13) after computing their information value for sorting. The expansion includes parent as well as child nodes because some child nodes may not be connected to the utility node other way but are still valuable due to inferences based on Bayes' rule. For instance, this applies for sensor nodes if they are not combined with their observed quantity in a common variable.

The algorithm extends the myopic information gathering agent suggested by Russell and Norvig [21, p. 603] by guiding the perception and gathering process based on a Bayesian decision network, adding time limits, and distinguishing observable from non-observable variables. As a limitation, the algorithm does not include parallelization which would be possible when using separate decision support sub-models. Additionally, the acquisition of evidences from sensor nodes (lines 6–8) needs to take into account sensor response time as an additional criterion.

5.5 Application Scenario

One particular application domain in our research has been *autonomous logistics* [23]. In this paradigm, agents act as autonomous decision makers that represent one or more logistic objects (e.g. truck, container, or pallet) and control logistic processes locally. This usually requires cooperation with other logistic agents as well as continuous acquisition of information for decision-making. This information is provided by local sensors (e.g., humidity sensors for monitoring of perishable goods [15]) as well as external data sources (e.g., for weather and traffic reports).

When developing an autonomous logistics system, we evaluate agent behavior by stochastic simulation before deployment. That is, the developed agents as well as test mocks representing external agents and other entities run in a simulated logistics environment before they are released to a real-world environment. For this purpose we implemented the multiagent-based simulation system PlaSMA (*Platform for Simulations with Multiple Agents*) [24]. PlaSMA is based on the FIPA-compatible multiagent development environment JADE [6].

We applied the PlaSMA simulation platform for studies of agent situation awareness in vehicle route planning (similar to the example in Section 5.3). Uncertainty in vehicle route planning is an everyday problem in transportation over longer distances. The shortest route is not always the fastest. Even when considering maximum allowed or average expected speed on single roads the planned routes may prove to be suboptimal. While suboptimal solutions are a natural property in dynamic, open environments with partial observability, usual route planning does not make use of much up-to-date or background information that would be obtainable and correlates with travel time.

5.5.1 Experimental Setting

We conducted simulation studies [11] that help find the utility or relevance of environmental information on driving time and its application in route planning. Furthermore, the experiments should provide evidence for the time in the future when incoming information is useful and the robustness when working with uncertain or wrong information.

In the experiments, transport vehicle agents have to estimate driving time for distant highway stretches to choose the fastest route to a given location. Environmental information is acquired for better estimations based on decision support models. The traffic infrastructure is given by directed graphs with changing environmental and traffic conditions. The corresponding transition models are described in [11]. Agent performance is measured in achieved driving time on average and its standard deviation. The vehicle route planning applies an A∗-like search algorithm with driving time as cost estimation.

5.5.2 Results

Inductive machine learning was applied to pre-processed traffic information in order to learn prediction models for specific roads depending on weather, time of day, and day of week [11]. The data was based on German and Austrian traffic censuses. The learned traffic model includes prediction rules like

```
[Speed=60]
     <= [Day=Mo] [Time=morning]
        [Weather=moderate..good]
```

That is, average possible speed on the corresponding road is expected to be 60 kmph on Monday mornings if weather conditions are from moderate to good.

Such predictions were integrated in a vehicle agent's decision model for best routes in a transportation process. As part of the agents information acquisition component, the planning system initiates the gathering of weather information within a certain look-ahead distance λ towards its destination. This information from external weather services together with time and date information is used to consult the agent knowledge base for a travel time prediction. The time and date in these queries again depends on prediction of travel time to reach the previous location in a potential route (cf. discussion in Section 5.3.5).

In several stochastic simulations these predictions turned out to be valuable. Situation-aware vehicle agents were up to 6.3% faster on average than regular agents without active information acquisition. Additionally, the standard deviation of travel time was reduced by 28%. With significance level $\alpha = 0.05$ the simulated results do not differ more than 0.0016% from the actual values. Thus, dedicated means to increase situation-awareness are actually shown to be of advantage in the presented logistics scenario.

5.6 Discussion

This survey on situation awareness has identified criteria for knowledge representation and reasoning of autonomous systems in dynamic environments. In particular, we have focused on domains with decisions that have a larger spatio-temporal scope. The evaluation perspective is twofold. The proposed criteria should be interpreted as design guidelines for situation-aware systems. Additionally, online evaluation aims at enabling the system itself to measure and increase its situation awareness. The latter perspective relies on knowledge and models described by the criteria.

We do not claim that the means proposed for implementing these criteria are imperative. Importance of requirements will be very domain-specific. However, information value theory provides a sound basis for decision-theoretic information relevance measurement. The information acquisition process can be guided by decision networks. The presented reference implementation for transportation logistics does not fully comply with all criteria yet because it relies on some pre-designed information acquisition behaviors instead of fully goal-oriented and deliberate acquisition. An extended conceptual architecture that includes explicit information acquisition for risk-related situation assessment has been proposed in [17].

The general question arises whether situation awareness turns out to be necessary or even futile for particular autonomous systems. Furthermore, most systems will probably only fulfill a subset of all criteria. As discussed in Section 5.3, Endsley [10] also distinguishes three levels for human situation awareness. These levels could match a particular subset of the criteria proposed here. Relevance assessment is certainly mandatory for level 2 (*comprehension*). Representing and reasoning with information dynamics pertains to level 3 (*projection*). But information sharing is beyond the scope of Endsley's approach. This criterion could establish a fourth level of awareness. On the other hand, a system could be able to share information but not be capable of projection. So it does not fit in as an extension of level 3.

Active perception by relevance-driven information acquisition will increase situation awareness by reducing uncertainty. But assuming an agent that is matching all criteria: Is such an agent capable of actually *measuring* (i.e. quantifying) its current situation awareness? With regard to information value theory, the relation between the expected utility with current sensory evidence and the expected utility given complete evidence provides a corresponding measure. Of course, for computing the expected utility for maximum evidence, one needs to apply Bayesian conditioning for all unknown variables again. Unfortunately, this gets intricate for a set of potentially dependent variables and is thus impractical for online evaluation.

By contrast, a straightforward measure is provided by variance in expected utilities of actions with given evidence. Additionally, a maximum acceptable variance could provide a minimal criterion for information gathering. As for all model-based measures, the measurement precision will depend on the quality of the model. That is, situation awareness is only assessed with respect to the world model, but not the real world. Our future research will address these issues to establish a complete theory of situation awareness and its assessment for informed decision making.

5.7 Conclusion

This chapter provides a survey on situation awareness for autonomous systems by analyzing features and limitations of existing approaches and proposing a set of criteria to be satisfied by situation-aware agents. The proposed criteria build upon the definition of human situation awareness by Endsley [9, 10] but take into account the special requirements and needed capabilities of technical systems. The proposed criteria are *awareness of ignorance*, an agent *perception model*, *information relevance assessment*, a model of *information dynamics*, *spatio-temporal qualification* as well as *information sharing* ability for acquiring information beyond own sensors.

An approach is presented which provides a quantitative decision-theoretic measure of information relevance that guides information acquisition. A corresponding algorithm shows its usage for maintaining situation awareness based on decision networks. A logistics routing scenario demonstrates that seemingly barely relevant information on environmental properties can significantly increase performance of autonomous agents. Future work will address extended theoretical foundations of situation awareness and its measurement based on dynamic decision networks. The extended theory also aims at finding the limits of awareness implied by the limits of world modeling.

Acknowledgments This research is funded by the German Research Foundation (DFG) within the Collaborative Research Center 637 "Autonomous Cooperating Logistic Processes: A Paradigm Shift and its Limitations" (SFB 637) at the University of Bremen, Germany.

References

1. Albus J (2002) A reference model architecture for intelligent unmanned ground vehicles. In: Proceedings of the SPIE 16th Annual International Symposium on Aerospace / Defense Sensing, Simulation and Controls, pp. 303–310
2. Albus J, Barbera T, Schlenoff C (2004) RCS: An intelligent agent architecture. In: Jones RM (ed) Intelligent Agent Architectures: Combining the Strengths of Software Engineering and Cognitive Systems, AAAI Press, no. WS-04-07 in AAAI Workshop Reports
3. Allen JF (1983) Maintaining knowledge about temporal intervals. Communications of the ACM 26(11):832–843
4. Baader F, Calvanese D, McGuinness DL, Nardi D, Patel-Schneider PF (eds) (2007) The Description Logic Handbook. Theory, Implementation and Applications, 2nd edn. Cambridge University Press
5. Bechhofer S, van Harmelen F, Hendler J, Horrocks I, McGuinness DL, Patel-Schneider PF, Stein LA, Olin FW (2004) OWL web ontology language reference. Available from http://www.w3.org/TR/owl-ref/
6. Bellifemine F, Poggi A, Rimassa G (2001) Developing multi-agent systems with a FIPA-compliant agent framework. Software-Practice and Experience 31(2):103–128
7. Bratman ME (1987) Intention, Plans, and Practical Reason. Harvard University Press, Cambridge, MA, USA
8. El-Hay T, Friedman N, Koller D, Kupferman R (2006) Continuous time Markov networks. In: Proceedings of the 22nd Conference in Uncertainty in Artificial Intelligence (UAI-06), AUAI Press, Cambridge, MA, USA, 10 pages
9. Endsley MR (1988) Design and evaluation of situation awareness enhancement. In: Proceedings of the Human Factors Society 32nd Annual Meeting, Human Factors Society, Santa Monica, CA, USA, vol 1, pp. 97–101
10. Endsley MR (2000) Theoretical underpinnings of situation awareness. A critical review. In: Endsley MR, Garland DJ (eds) Situation Awareness Analysis and Measurement, Lawrence Erlbaum Associates, Mahwah, NJ, USA, pp. 3–32
11. Gehrke JD, Wojtusiak J (2008) Traffic Prediction for Agent Route Planning. In: International Conference on Computational Science (ICCS 2008), Springer-Verlag, Krakow, Poland, no. 5103 in LNCS, pp. 692–701, DOI 10.1007/978-3-540-69389-5_77
12. Howard RA (1966) Information value theory. IEEE Transactions on Systems Science and Cybernetics SSC-2(1):22–26
13. Howard RA, Matheson JE (2005) Influence diagrams. Decision Analysis 2(3):127–143
14. Hübner S, Spittel R, Visser U, Vögele TJ (2004) Ontology-based search for interactive digital maps. IEEE Intelligent Systems 19(3):80–86, DOI 10.1109/MIS.2004.15
15. Jedermann R, Behrens C, Westphal D, Lang W (2006) Applying autonomous sensor systems in logistics: Combining sensor networks, RFIDs and software agents. Sensors and Actuators A (Physical) 132(1):370–375

16. Kirn S, Herzog O, Lockemann P, Spaniol O (eds) (2006) Multiagent Engineering: Theory and Applications in Enterprises. Springer-Verlag
17. Lorenz M, Gehrke JD, Hammer J, Langer H, Timm IJ (2005) Knowledge management to support situation-aware risk management in autonomous, self-managing agents. In: Self-Organization and Autonomic Informatics (I), IOS Press, Amsterdam, Frontiers in Artificial Intelligence and Applications, vol 135, pp. 114–128
18. Nodelman U, Shelton C, Koller D (2003) Learning continuous time Bayesian networks. In: Proceedings of the Nineteenth Conference on Uncertainty in Artificial Intelligence (UAI-03), Morgan Kaufmann, Acapulco, Mexico, pp. 451–458
19. Raja A, Lesser V (2004) Meta-level reasoning in deliberative agents. In: Proceedings of the International Conference on Intelligent Agent Technology (IAT 2004), pp. 141–147
20. Randell DA, Cui Z, Cohn AG (1992) A spatial logic based on regions and connection. In: Nebel B, Rich C, Swartout W (eds) Proceedings of the 3rd International Conference on Principles of Knowledge Representation and Reasoning, Morgan Kaufmann, pp. 165–176
21. Russell SJ, Norvig P (2003) Artificial Intelligence: A Modern Approach, 2nd edn. Prentice Hall, Upper Saddle River, NJ, USA
22. Saria S, Nodelman U, Koller D (2007) Reasoning at the right time granularity. In: Proceedings of the Twenty-third Conference on Uncertainty in AI (UAI), p. 9
23. Scholz-Reiter B, Windt K, Freitag M (2004) Autonomous logistic processes – New demands and first approaches. In: Proceedings of the 37th CIRP-International Seminar on Manufacturing Systems, Budapest, pp. 357–362
24. Schuldt A, Gehrke JD, Werner S (2008) Designing a simulation middleware for FIPA multi-agent systems. In: 2008 IEEE/WIC/ACM International Conference on Web Intelligence and Intelligent Agent Technology, Sydney, Australia, pp. 109–113
25. So R, Sonnenberg L (2007) Situation awareness as a form of meta-level control. In: Proceedings of the 1st International Workshop on Metareasoning in Agent-Based Systems at 6th International Joint Conference on Autonomous Agents and MultiAgent Systems (AAMAS 07), Honolulu, Hawaii
26. Tawfik A, Khan S (2005) Temporal relevance in dynamic decision networks with sparse evidence. Applied Intelligence 23:87–96, DOI 10.1007/s10489-005-3414-9
27. Williams MA, Gärdenfors P, Karol A, McCarthy J, Stantom C (2005) A framework for evaluating groundedness of representations in systems: From brains in vats to mobile robots. In: IJCAI-05 Workshop on Agents in Real-Time and Dynamic Environments, Edinburgh, UK, pp. 17–24
28. Wooldridge M (2000) Reasoning about Rational Agents. The MIT Press
29. Wooldridge M, Jennings NR (1995) Intelligent agents: Theory and practice. The Knowledge Engineering Review 10(2):115–152
30. Zeigler BP (1990) High autonomy systems: Concepts and models. In: Proceedings of AI, Simulation, and Planning in High Autonomy Systems, IEEE Computer Society Press, Tucson, AZ, pp. 2–7

Chapter 6
From Simulation to Real Robots with Predictable Results: Methods and Examples

S. Balakirsky, S. Carpin, G. Dimitoglou, and B. Balaguer

Abstract From a theoretical perspective, one may easily argue (as we will in this chapter) that simulation accelerates the algorithm development cycle. However, in practice many in the robotics development community share the sentiment that "Simulation is doomed to succeed" (Brooks, R., Matarić, M., Robot Learning, Kluwer Academic Press, Hingham, MA, 1993, p. 209). This comes in large part from the fact that many simulation systems are brittle; they do a fair-to-good job of simulating the expected, and fail to simulate the unexpected. It is the authors' belief that a simulation system is only as good as its models, and that deficiencies in these models lead to the majority of these failures. This chapter will attempt to address these deficiencies by presenting a systematic methodology with examples for the development of both simulated mobility models and sensor models for use with one of today's leading simulation engines. Techniques for using simulation for algorithm development leading to real-robot implementation will be presented, as well as opportunities for involvement in international robotics competitions based on these techniques.

6.1 Introduction

Using modeling and simulation to develop algorithms for motion planning provides development flexibility and extensive testing capability of the algorithm under a variety of operational environments and robot configurations.

The success of an algorithm depends upon, and is evaluated by, the robot's behavior upon execution. At the same time, as described in Kyriacou et al. [19], robot behavior is affected by the robot's hardware, the implemented algorithm

S. Balakirsky (✉)
National Institute of Standards and Technology, Gaithersburg, MD, USA
e-mail: stephen.balakirsky@nist.gov

R. Madhavan et al. (eds.), *Performance Evaluation and Benchmarking of Intelligent Systems*, DOI 10.1007/978-1-4419-0492-8_6,
© Springer Science+Business Media, LLC 2009

and the operating environment. The combination of these three elements results in the execution of a scenario in a complex, highly variable and often non-linear system. Therefore, developing a general use simulator is a difficult task. Generality and simulation fidelity counterbalance each other in a simulation development.

During algorithm development for non-simulation based projects, the developer's considerations are somewhat bounded by: (i) the algorithm's input and output, (ii) the computation paradigm to be followed (e.g. divide-and-conquer, greedy, dynamic programming etc.) and (iii) any data structures that may be used to handle the data. The algorithm developer has minimal or no expectations about features or extra feedback provided by the execution environment and no concern about the quality, fidelity or accuracy of the execution environment. Robotic algorithms tend to be developed for a specific task. Most fail at first as the developer makes relaxing assumptions about the robot hardware and the operating environment.

In algorithm development for simulation-based projects, the developer's considerations are significantly more expanded and include concerns related to the simulation platform. Issues such as accuracy, fidelity, determinism and overall realism have a direct effect on the computation and results.

6.1.1 Methodology for Algorithm Development

Currently, there is a void in the area of a formal methodology for simulation-based algorithm development. Such methodology should be able to treat the robot's hardware, its behavior and the operating environment as abstractions that can be independently and jointly specified and implemented. This would provide modularity and implementation-independent specification, which in turn enhance portability and allow model reuse. It would also enable the application of formal methods and automated model-checking. Most of the interactions between robots and the environment are developed based on a process of trial-and-error experimentation. Therefore, the overall accuracy of a robot simulator depends on the fidelity of its three components: the robot's hardware model (including both the base platform and the sensors), the algorithm, and the operating environment. Both the robot's hardware model and the operating environment are designed based on an estimation of how they should behave and interact.

Unfortunately, this results in environments and interactions based on assumptions that generate "virtual realities" that do not necessarily correspond to the physical environments and actual hardware. The assumptions may relax or extenuate one or more aspects of the simulation components (terrain, sensors and robot) and have a direct impact on the behavior of the developed algorithm. For example, a simplified terrain model or underlying physics engine with underestimated friction parameters would allow the algorithm to "drive" the robot too fast. This would result in miscalculating centrifugal forces when making turns, and inaccurate and unrealistic trajectories. Another example would be having a lower fidelity sensor payload that would be susceptible to a noisy environment resulting in failure to correctly inter-pret landmarks or signals. One final example would be using an omnidirectional holonomic steering when the physical platform is expected to be an Ackerman-steered vehicle. These are examples of simplifying assumptions or omissions that compromise the faithfulness of the simulation platform and significantly impact the effectiveness and appropriateness of a developed algorithm.

This chapter strives to address these issues by presenting techniques for vali-dating the robot models and algorithms that are used in the simulation. While the techniques are designed to be general purpose, specific examples are provided that relate how these techniques were applied to models that were developed for the Unified System for Automation and Robot Simulation (USARSim) [41].

6.1.2 A Brief History of USARSim

The first release of USARSim [24] was built by creating modifications to Epic's Unreal Engine 2 game engine.[1] It supported models for a few differential drive robotic platforms, a restricted set of sensors, and a small set of Urban Search And Rescue (USAR) specific test arenas. In addition, the robotic platforms could only be controlled through the use of the RETSINA [37] multi-agent system software.

In 2005, USARSim was selected as part of the base infrastructure for the RoboCup Rescue Virtual Robot Competition. The virtual robot competition is an annual international event that highlights research in diverse areas such as multi-agent cooperation, communications networks, advanced mobility, mapping, and vic-tim search strategies. In addition to the competition, USARSim management was taken over by the National Institute of Standards and Technology (NIST) and an international development community was established on the open source source-forge.net website. While much of the original structure of the code was maintained, the code was reorganized and interfaces were standardized around SI units. The first official release (Version 1.0) was produced in October 2005.

A large-scale development effort accompanied the transition to sourceforge and the involvement of the RoboCup community. Version 3.31, released in July 2008 offers 15 different sensors, from odometry to an omnidirectional camera. 23 differ-ent robotic platforms are now available; these include wheeled robots, cars, tracked vehicles and flying robots. In addition, several of the sensors and robots have under-gone rigorous validation of the forms outlined in this chapter in order to prove their similarities and differences from the real devices [8, 29, 38]. More information may be found at the USARSim website located at [41].

The remainder of this chapter will detail these validation methods. First, a look at robot platform validation will be conducted. This will be followed by a study

---

[1]Certain commercial software and tools are identified in this Chapter in order to explain our research. Such identification does not imply recommendation or endorsement by the authors, nor does it imply that the software tools identified are necessarily the best available for the purpose.

of sensor validation. Next, algorithm development based on these validated models will be presented. Finally, a look at competitions that strive to utilize the simulated development cycle will be presented.

6.2 Robot Platform Validation

Robot platform validation is crucial to providing an accurate simulated model of a robotic system. A large body of work exists on modeling vehicle subsystems and the subsystem interactions. These simulations are capable of producing very accurate representations of the dynamics of mobile platforms and have been used in the design of commercial automotive systems. However, these systems are not capable of running in real-time on today's generation of low-cost desktop hardware. Since the objective of this validation is to verify that the simulated platform has *similar* capabilities as the physical hardware, we are able to trade-off simulation fidelity for real-time performance. This trade-off usually precludes modeling each sub-assembly of the robot, and the robot body is modeled as a single unit.

Our definition of *similar* performance is that *gross* platform behavior in both the physical and simulated platforms is verified. For example, if a simulated robot encounters terrain that would cause the physical platform to roll over, then the simulated platform should roll over as well. However, we find it acceptable for the physical and simulated platforms to experience different frequencies of vibration while traveling at the same speed over similar terrain.

In order to determine a more precise definition of what *gross* platform behaviors must be modeled, it is desirable to focus the validation on the domains that one would expect the robot model to encounter. In our case, this includes cluttered, uneven terrains with the robot performing various driving maneuvers. This has led to focusing the validation tests on capabilities such as platform maneuverability (acceleration, maximum velocity, turning radius), and rough terrain handling (center of gravity, climbing ability).

Our validation approach then became one of comparing the performance of the physical system to the simulated system in various relevant scenarios. This led to the development of a test suite that allows objective comparisons to be made between the physical hardware and simulated models. However, it is often the case that the individual who designed or owns the platform that is to be modeled is not the same as the modeler. In addition, the robot owner may object to having to send their platforms to an outside test facility where potentially untrained individuals would perform the testing. Therefore, a critical design criterion for the validation test suite was ease of test construction and administration. It was desired that the entire physical test suite could be built out of readily available, low-cost materials in a matter of a few days and that the tests themselves could also be conducted in a short amount of time.

One such test suite has been under development at NIST and is in the process of becoming an international standard [26]. A sample test method from the test suite (the "step test") is shown in the left hand side of Fig. 6.1. The right hand side of

Fig. 6.1 Example test method from test suite. The figure on the *left* shows the physical method while the figure on the *right* depicts the simulated version

Fig. 6.1 shows the simulated version of this test. In the design of the simulated test methods, particular attention has been paid to validating that the test methods are accurately reproduced [29].

The tests that comprise the test suite may be decomposed (in order of increasing complexity) into the categories of static characteristics, hardware limits, hardware performance characterization, and interface performance characterization. The first two tests listed are the simplest to validate, and may usually be validated from design drawings of the physical platform. The static characteristics test verifies that the simulated model conforms to the same physical specification as the physical hardware. Robot dimensions, sensor placements, and actuator locations are verified. The true center of gravity (COG) of the robot is also established. This may be accomplished by either finding the balance point of the vehicle, or by using a set of scales to measure the force applied by each surface of the robot where ground contact is made. For the case of USARSim, the COG is an input parameter to the physics engine and no further tuning is required.

The hardware limits test verifies that the simulated model is constructed correctly. The range of motion of all moving parts is measured, and any motion constraints dictated by hard stoppages are verified. For example, a robot arm's range of motion may be reduced due to arm-body collisions. This test will verify that the simulated collision boxes are configured correctly so as to prevent the arm from traveling through the simulated robot body.

Hardware performance characterization is designed to verify the dynamic characteristics of the robotic system. As previous stated, it is not expected that a real-time physics engine will be capable of completely modeling the dynamics of a complex system such as a robot. However, it is important that the physics engine be "close enough" to modeling this system such that algorithms developed under simulation will be valid on the actual hardware. The hardware performance characterization consists of two separate tests. The difficulty of both of these tests is adjustable, and the tests are started from an "easy" configuration, with difficulty increasing until the platform is no longer able to accomplish the test (if possible).

Fig. 6.2 Dash-test in flat (*left*) and 30° (*right*) configurations. This test is used to measure the platform's acceleration, deceleration, and maximum velocity. The P3 robot's laser scanner (located in the near field for the flat configuration and at the top of the ramp for the 30° configuration) is used to measure the distance to the robot under test as it approaches. A flat target is attached to the robot under test in order to maximize the number of laser beams that hit the target and provide noise averaging

Figure 6.2 depicts the dash test that is designed to measure the acceleration profile, deceleration profile, and maximum velocity for a robot under the added stress of accelerating on an inclined plane. Once the maximum slope that a platform can climb is determined, the platform is tested at 100, 50, and 10% of the maximum slope as well as flat conditions. Certain platforms will flip over before reaching a slope that they lack the motor torque to climb. For these cases, a maximum slope is chosen based on platform and personnel safety.

Starting from zero velocity, the robot drives at maximum acceleration until full speed is reached. After a period of steady state velocity, the robot decelerates back down to zero velocity. During this test, a range sensor is used to observe the progress of a target fixed to the robot as the robot drives through the metric. In the case of Fig. 6.2, a SICK LMS 200 that is mounted on a spare platform was utilized.

By measuring the difference in distance provided by successive hits on the target by a laser beam, a velocity measure is computed. For this metric, the velocity computation is performed at 10 Hz for each beam that impacts the target. All of the measured velocities are averaged together to provide a velocity estimate for a given run at a particular time instant.

This test is performed 10 times for each slope setting. The results from the individual tests are then weighted based on the number of beams reporting velocities, and averaged together to form a velocity profile.

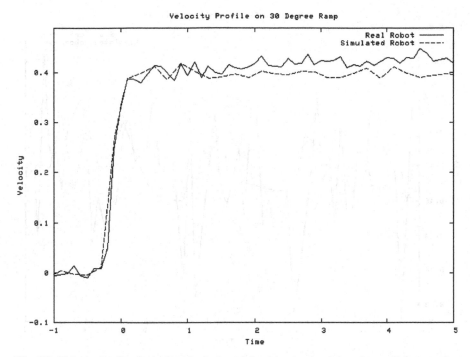

Fig. 6.3 Velocity profiles for both a physical model and its simulated counterpart while operating at a 30° slope. The plot depicts time(s) vs. velocity (m/s)

Figures 6.3 and 6.4 show the measured profile for a platform under test compared to a tuned model from simulation. The test for Fig. 6.3 was performed on the 30° slope ramp while the test shown in Fig. 6.4 was performed on a flat surface. In order to achieve this tuning, the maximum track velocity and motor torque settings of the simulated model are tuned. The exact technique used for this tuning procedure is simulation-system dependent. For USARSim, the physics engine has the ability to dynamically modify the model's parameters. The authors took advantage of this capability to dynamically modify parameters during a run until the average of actual velocities matched our desired average velocities. These modifications are currently carried out by hand; in the future it is desired that this tuning will be automated. Figure 6.5 depicts the velocity error between the simulated platform and the real platform as a percentage of the real platform's velocity. As may be seen in this figure, the model is able to consistently capture the velocity profile for the flat test area. However, there is an acceleration component that the real platform exhibits when traveling up a steep hill that is not captured by the simulation. It may be seen that there is increasing error with time as the real platform accelerates up the hill. This error has an upper bound of about 15%.

The second hardware performance test utilizes the metric shown in Fig. 6.1. This test measures the ability of a platform to climb steps of various heights. The test is started with a step of height 5 cm and the step height is increased by 5 cm increments

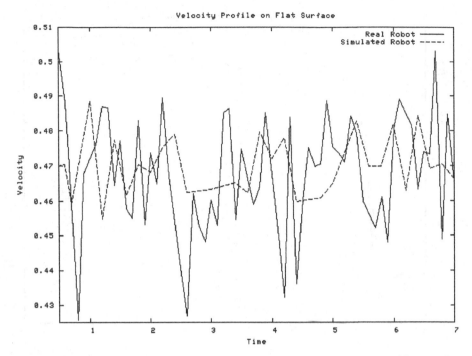

Fig. 6.4 Velocity profiles for both a physical model and its simulated counterpart while operating on flat ground. The plot depicts time(s) vs. velocity (m/s)

until the platform is unable to reach the landing. The tubes on the leading edge of the step are free-spinning and are designed to prevent the platform from gripping the step and pulling itself up. Video of the physical platform performing this test is collected, and the simulation is tuned so that the platform model achieves similar performance. The main simulation parameters that are adjusted during this process deal with tire properties. However, it may be necessary to tune the COG and motor torques as well.

The final area for platform validation concerns the control interface to the platform. It is desired that when presented with identical command streams, the physical and simulated platforms will exhibit similar responses. For hardware platforms that support a USARSim interface, this test is a simple matter of applying the same command stream to both the physical platform and the simulated platform. The platforms' trajectories may then be measured and compared. Such a test was first performed by Balaguer et al. [7]. Differences in behavior may usually be compensated for by scale factors on the command stream. When the interfaces do not support identical command streams, but a command API is available for the platform under test, a command translator may be constructed. The test is then able to proceed as described above. In the worst case, the command syntax is proprietary, and the only way to discern what input is being delivered to the physical platform is by measuring battery current to the motors using a system such as the one described in [36].

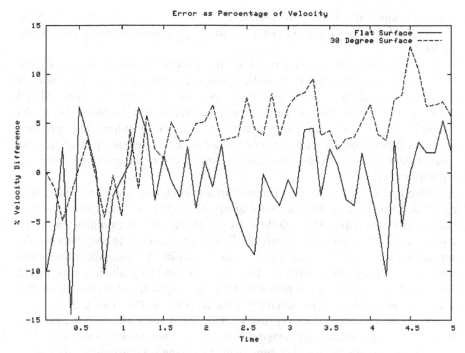

Fig. 6.5 This figure show the velocity error for the simulated system as a percentage of the desired velocity. Plots for both the 30° ramp and flat test methods are shown

Once all of the above tests have been performed and the platform model is complete, it is still necessary to validate the validation. In other words, does the validation produce behaviors that are consistent with real-robot performance across terrains other than the ones that were used in the validation process. This is accomplished by measuring the physical and simulated performance of the platform on a subset of the NIST test methods that were not used for the initial validation; for example, staircase climbing. If performance of the two platforms on these methods is similar, then one may consider the platform and its validation tests to have been successfully validated.

6.3 Sensor Validation

Sensor validation follows the same general-purpose methodology that is used to validate other components of the simulation system, e.g. robot platforms and actuators. The tenet is to implement the same experiment in simulation and in reality, and to numerically compare the results. The goal of the sensor validation stage is twofold. Not only is it important to determine the degree to which a simulated sensor delivers results that can be extrapolated to infer results of the real world system, but it is also pivotal to identify the sensors that fail to replicate phenomena observed with their real counterparts. In the past, a set of very different sensors have been modeled and validated [14, 13, 8], and we provide a short synopsis of the methodology and results.

The choice of the numeric metric to use for this comparison is in general non-trivial, and potentially amounts to a demanding research question on its own. In certain cases it will be possible to directly compare the outputs of simulated and real sensors, while in other scenarios it will be preferable to contrast the results of simple computational procedures that process the sensor output. The appropriate choice depends on the sensor under consideration, and no one catch-all formula exists, even though some indications can be extracted from the examples presented in the following.

Setting up a virtual-reality experiment implies various programming and modeling tasks. Clearly, an appropriate sensing model needs to be incorporated within the simulation. Additionally, a faithful model of the real environment has to be developed into the testing environment. This is an unavoidable step because the surrounding environment naturally influences the sensing process. According to our experience, and not surprisingly, the best way to develop a realistic sensor model is to embed, directly into the simulation engine, a computational method that replicates the physical phenomena occurring in the real device. This approach necessarily implies the inclusion of appropriate models of the noise affecting the process. This last step is particularly challenging because these noise sources are often difficult to characterize or add to the system. In the following we illustrate the methodology and findings we obtained while modeling and validating two different sensors, namely a laser range finder and a GPS sensor. Two reasons motivate the choice of these sensors as working examples to demonstrate our general-purpose validation method. First, these are among the most popular sensors used to solve the localization problem indoor and outdoor, respectively. Secondly, it is rather straightforward to implement their underlying working principles within the simulation environment by using ray-tracing primitives offered by the game engine.

6.3.1 Laser Range Finder

Laser range finders have application in a variety of robotic tasks, such as navigation and map building, just to name a few. These devices are essentially time of flight devices, i.e. they measure the time elapsed between the moment a beam is emitted and its reflection comes back to the sensor. By relating this time to the physical properties of the signal being sent and sensed, the distance between the sensor and the surface that reflected the beam can be inferred. These sensors usually emit a series of beams covering a wide angular range. The Sick PLS, probably the most popular sensor of this type, covers a range of 180° with beams spaced either 1° or 0.5° apart. Such a sensor is straightforwardly mimicked in USARSim by exploiting the ray-tracing primitive operation made available by the underlying game engine. Unfortunately, ray-tracing is an error-free function in the simulated environment that does not take into account the beam flight time or the surface of reflection—important error sources in real laser range finders. Therefore, appropriate noise needs to be added to the simulation in order to replicate real sensor behavior. Two noise sources mainly affect this sensor. The first are jitters observed in the measured distances due to errors while measuring time intervals. This error is fairly simple to model and imitate in the simulation environment. The current noise model consists of a piecewise linear mapping (true range vs. perceived range) that distorts the perceived range with a uniformly distributed random value added. In the future, a noise Application Programming Interface (API) is being developed for USARSim and several different noise models will be possible. The second error stems from unreturned beams, i.e. beams that are not reflected by the surface they hit. This second error is instead rather difficult to replicate in simulation because it would imply labeling all the surfaces in the simulated world as reflective or non-reflective, or assigning a reflective coefficient to each surface. While this can be achieved in principle, it implies an amount of work that is definitely not worth the effort and has therefore not been implemented in the simulator.

Since the sensor under consideration returns numerous values, it is possible to compare real and simulated data point by point, or to relate the result of a function computed on the sensed data. This latter approach is embraced in this case study. In particular, we will show the results of the well-known Hough algorithm for line detection. This example is particularly appealing because it illustrates the advantage of simulation from two points of view. Firstly, the algorithm requires hand tuning of a few parameters. This process is driven by the data being processed and may be particularly time consuming if being performed with a real device. Secondly, the algorithm is known to be fairly sensitive to noise in the data. By altering the amount of noise affecting the simulated sensor it is therefore possible to assess the robustness of the algorithm to different noise levels (i.e. width of the uniform jitter distribution). This later aspect is in our view one of the most appealing aspects of the simulation system we propose. Obviously, this assessment will be appropriate only if it will be possible to show a strong correlation between simulation and the real world system.

Figure 6.6 shows how the experiment is executed, i.e. the simulated and the real robot are positioned at the same place in the physical and simulated environment. Figure 6.7 shows a superimposition of the data collected by the two sensors while in the position depicted in Fig. 6.6. It is evident that the data returned by the two sensors are very close to each other. This visual evidence can be numerically assessed by comparing how the Hough transforms computed from the two datasets relate to each other. It should be stressed that this choice is not the only possible one, nor necessarily the best one, but was rather selected because the simulation environment was being used to perfect a robot control system where line detection via the Hough transform was a building block for the higher layers.

Without getting into a technical discussion of how the Hough transform is computed or used later on (the reader is referred to [14] for more details), we just mention that its final result is stored in a grid of non negative integer values. After computing the Hough transform for both datasets over a grid with 2,592 cells, the average difference between values stored in corresponding cells is 0.0328. In order

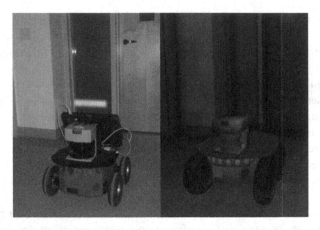

Fig. 6.6 Real and simulated robots observed while validating the range scanner simulation. The cross in Fig. 6.7 shows the position of the robot in the environment, where these pictures were taken

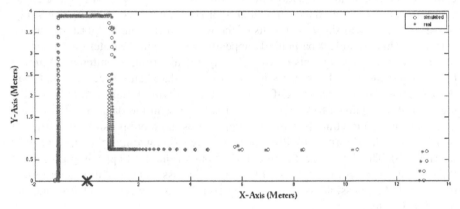

Fig. 6.7 Data collected by the simulated and real sensors. The location of the robot, from which the data was taken, is marked by a cross (at the 0, 0 coordinate). The cross also shows where the pictures in Fig. 6.6 were taken

to put this value into context, it should be mentioned that the highest value stored in the grids is 6, and that for more than 99% of the cells, the difference is smaller or equal than 1. These values support the claim that results obtained with the simulated sensor can be extrapolated to the real sensor as well.

6.3.2 Global Positioning System

Robots performing in outdoor environments rely more and more often on the availability of a Global Positioning System (GPS) sensor to obtain an estimate of their global position. The GPS system depends on the presence of a set of satellites orbiting around the earth according to known trajectories. In essence, a GPS sensor periodically receives signals from the orbiting satellites and uses this information to infer its position. A signal emitted by a satellite is received by a GPS receiver only if there is line of sight between the two. The GPS receiver then processes all information included in the received signals and computes its position. The more satellites that are within line of sight, the more accurate is the pose estimation produced by the GPS sensor (the reader is referred to [8] for a more detailed description of the computational aspects and additional results).

Based on this working principle, a GPS sensor simulator can be developed as follows. Every time a new reading from the GPS sensor is requested, ray tracing is performed between the simulated receiver and all of the satellites in order to determine which ones are visible from the receiver. The receiver's position is then computed through simple mathematical computations. Gaussian noise is superimposed to the computed position based on the number of detected satellites. This brief description reveals that the pose computed by the simulated sensor replicates the computational approach exploited by the real receiver. In particular, tracing between the receiver and the satellites entails two different kinds of knowledge. First, at any point in time it is necessary to know the position of the satellites. Secondly, while performing the tracing, it is necessary to take into account the presence of obstacles possibly obstructing the path between the receiver and a satellite.

Since the accuracy of the pose estimation is intimately related to the number of detected satellites, and this value is commonly made available by most GPS receivers, one good numeric value that can be used to compare the performance of the real and simulated device is the number of satellites detected. To do so, it is not only necessary to know the satellites' positions, but also to include into the simulated environment all the elements that may possibly preclude the detection of a satellite. In the real world experiment, a mobile robot equipped with a GPS receiver moves in the University of California, Merced campus between a set of multi-story buildings. Hence, in the simulated experiment, corresponding geometrical models of these buildings were inserted in the test environment. Given that the number of detected satellites influences the accuracy of the estimated pose, the other numerical parameter that can be numerically compared is the difference between the two position estimates. Obviously, this value should be as close to 0 as possible. For the GPS sensor, it is therefore possible to numerically compare the values produced by the sensor under consideration.

It should be mentioned that satellite occlusion is definitely not the only source of uncertainty for GPS sensors, but it is the easiest to faithfully model. Moreover, as evidenced in the experiments described in the following, while just including this source of error, a satisfactory performance is observed. In this context *satisfactory* means that the numerical values observed in the two experiments are close to each other thus providing evidence of the accuracy of the simulation.

The first experiment is illustrated in Fig. 6.8. Both the simulated and the real robots were driven though three paths in the campus quad. The picture shows the pose returned by the simulated and real GPS sensors. It is evident, by visual inspection, that the two paths are similar. However, this visual correspondence can be numerically evaluated as well. In fact, the average difference between the two is

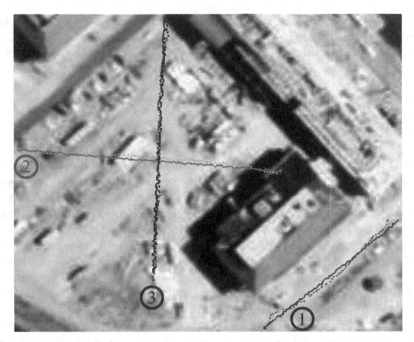

Fig. 6.8 Comparison between the traces produced by the real (dotted path) and simulated (solid path) GPS sensors for three different experiments. The traces have been superimposed to a satellite image of the University of California, Merced campus

2.37 for the first experiment, 1.49 m for the second, and 4.57 m for the third. Even though these numbers might seem high, the reader should keep in mind that the typical accuracy of civil GPS sensors is 3 m. The second experiment is illustrated in Fig. 6.9.

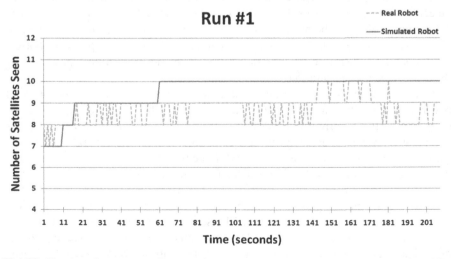

Fig. 6.9 Number of detected satellites as a function of time for the second experiment in Fig. 6.8

The chart shows the number of satellites seen by the real and simulated robot. In this case it is manifest that the simulated sensor almost always sees a greater number of satellites than the real one. This is easily explained by the fact that the simulated model does not include the numerous trees that are present in the real world. In any case, as evidenced in the first experiment, it turns out there is a good agreement between the data returned by the two sensors, therefore one can be confident that results obtained in simulation can be extrapolated to real world systems.

6.4 Algorithm Development

Robot simulation platforms provide versatile environments for the development of robotic algorithms. Developers can easily test algorithm execution using multiple robotic platforms in various synthetic environments and operational scenarios. This makes simulation-based algorithm development an indispensable research tool that allows extensive testing of robotic behavior in configurations and environments in which physical testing would otherwise be prohibitively expensive or even impossible to achieve.

Here, we focus simulation-based algorithm development analysis in the context of mobility and robot motion planning. Motion planning is an active research area and has been studied for decades. Depending on the approach, it has been referred to as motion planning, path planning or trajectory planning, but in essence it has always referred to the same class of problems. The seemingly trivial task of instructing a robot to move between two points on a plane via a collision-free path often becomes surprisingly difficult to generalize for any two points on any plane with any configuration space.

Early formulations of the problem appeared almost 40 years ago using graphs for shortest path searches (visibility graphs) between polygonal obstacles [28]. Another formulation of the motion planning problem, known as the *piano movers' problem* [11, 34, 35] was studied from a geometry perspective. Not unlike similar current problems of robots navigating past obstacles, the examined task was to successfully move a piano between house rooms while avoiding obstructions. The examination ignored any kinematic constraints such as how piano wheels-if they existed-would enable or constrain steering. It also disregarded speed and any temporal aspect such as elapsed time, related to the navigation of the piano from room to room.

Since this first formulation, many different motion planning approaches and techniques have been developed, such as grid sampling, configuration spaces [25], roadmaps [10], randomized potential fields [6], probabilistic roadmaps (PRMs) [17], rapidly-exploring Random Trees (RRTs) [21] and many others. Some of these techniques have crossed the boundaries of robotics research and have been applied in other domains and disciplines. For example, motion planning algorithms have been applied to biology, computer animation and medical surgery for protein folding pathways [22, 32, 39], virtual reality [31] and laparoscopy procedures [15], respectively.

Today the formulation of motion and mobility problems is significantly more complex. Task completion is extended with additional requirements such as maximizing speed, minimizing navigation distance (e.g. shortest-path) and energy consumption (i.e. battery power) or maximizing path accuracy. Compounding the complexity is: the operation in dynamic environments with uneven terrains and moving obstacles, the utilization of sensors and their imperfect readings and even the coordination and interaction of multiple robots in the same area.

The volume, breadth and depth of research and multi-disciplinary application of motion planning techniques emphasize the importance of the work in this area and yet it illustrates the inherent complexity of the problems to be solved. Even the original problem of moving a piano would be a discouraging proposition for anyone to physically validate using trial-and-error testing. Attempting to test, verify and validate new algorithms on complex robotic platforms operating in dynamic and even hostile environments, such as volcanoes, underwater, or on distant planets, is a daunting and prohibitively expensive, if not impossible task. This leads one to explore the utility of performing algorithm development in a safe, inexpensive simulated environment.

However, the effectiveness and appropriateness in simulation-based algorithm development is highly dependent on the characteristics of the simulation platform. The following are some desirable simulation platforms characteristics:

(a) *Generality*: to allow the testing and experimentation of different algorithms on different simulated robots with various sensor payloads and in different simulated operating environments.
(b) *Accuracy:* because simulated algorithm executions are expected to approximate reality. The narrower the gap between the simulation execution and the corresponding physical execution, the more accurate the simulation platform is.
(c) *Computational soundness:* which implies that executed algorithms and computations are deterministic. Non-deterministic simulations compromise the repeatability of the execution which is necessary to determine the stability of the simulation platform.

From a software engineering perspective, simulation platforms must also have features that support and augment the above characteristics:

(a) *Extensibility:* is related to (i) *framework extensibility* which includes *scenario* and *environment extensibility* so that new robot profiles, terrain information, environment artifacts, configuration spaces, sensor types and other components can be easily added or updated; (ii) *algorithm development extensibility* so that existing algorithms can be extended, new algorithms can be developed or easily replaced in either case without affecting the underlying framework.
(b) *Temporal control*: to allow real-time display and simulation speed and replay capability.

(c) *Realism:* in rendering the visual aspects of the simulation so robots do not seem to navigate over obstacles. Realism also applies in the rendering of sensor readings that may be inaccurate due to inherent noise in the environment.

(d) *Interface versatility and richness:* to enable easy control of the simulator and experiment design while providing a high degree of configurability and feedback during simulation executions.

(e) *Logging and meta-information collection mechanisms:* to provide a way to collect and analyze quantitative measurements from executed experiments.

(f) *Scalability:* to allow the inclusion of multiple robots or to allow framework extensibility as discussed earlier.

6.4.1 Criticisms and Advantages

Regardless of the particular characteristics and features of a simulation platform, the inherent complexities of robotic environments may introduce significant differences between the simulated and the real environment. The combination of extensive robot configuration parameters, diverse terrain options, high variability in sensor readings and error rates, coupled with operational scenarios that change in real-time, create very dynamic environments that are very hard to accurately replicate in a simulation. All of these negatively affect the reliability of simulation platforms and the algorithms developed using them. As mentioned earlier, further aggravating the situation is the lack of a formal methodology for simulation-based algorithm development.

Simulation-based algorithm development has advantages over the alternative, direct development on the physical platform and environment. Simulations allow for:

- Inexpensive and rapid setup of a virtual experimentation environment;
- Time execution acceleration, providing the developer insight on how an algorithm would execute after long time periods without having to execute them in real time;
- Consistent environment for experimentation;
- Test and experiment repetition;
- Easy examination of incremental development progress;
- Risk reduction for damage to hardware or the operating environment due to errors in execution.

Many simulation environments and frameworks have been developed and can be used to model different environments and simulate tasks. Some are described as general frameworks. Others specialize and focus on certain aspects of simulation. Specialized frameworks tend to provide greater accuracy and fidelity for the particular task they model, but lack extensibility which general frameworks tend to be better equipped to provide. For example, a non-comprehensive list of robotic simulation platforms and frameworks includes: Darwin2K [23], Gazebo [18], Player/Stage [16], SimRobot [20] and Webots [27].

In the context of a Virtual Manufacturing Automation Competition, detailed in Section 6.5, we embarked on the simulation-based development of two algorithms: one to address waypoint navigation and another to perform docking. Both of the algorithms were developed, exercised, tested and verified using a simulation infrastructure comprised of USARSim [2, 12], a Gamebots [1] variant of the Unreal Engine [40] and the Mobility Open Architecture Simulation and Tools (MOAST) framework [4].

The USARSim infrastructure supports algorithm experimentation. The MOAST framework allows the separation of algorithm development from the different system levels (echelons) and configurations which control different aspects of the experiment.

The operational scenario of the navigation algorithm was based on waypoints for terrestrial navigation. Waypoints are longitude and latitude-based coordinate locations used to construct routes. In the context of robotic navigation, waypoints can serve as the guides for a robot to follow a specific path. While following waypoints is trivial, idiosyncrasies of the robotic platform and the operating environment may render this type of navigation inaccurate and ineffectual under some circumstances. An example of such a case is with Ackerman-steered vehicles. The limitation of such vehicles is their inherent inability to perform sharp turns to reach waypoints in very close proximity, specifically, waypoints located between the vehicle and its instantaneous center of rotation (ICR). This case presents two navigation challenges. First, it may cause the vehicle to deviate considerably from the prescribed path. Second, it may cause the vehicle to enter an endless spiral path attempting to reach the practically unreachable waypoint. To address these challenges we developed an algorithm that visits all waypoints on an arbitrary projected path, minimizing path deviation, avoiding spiral movements and continuing navigation on the projected path. Using an arbitrary navigation path consisting of segments of circular arcs, the algorithms were executed using a simulated Ackerman-steered, automated guided vehicle (AGV) to navigate the path. The navigation path or the "world," along with an arc file, were provided as input. The simulated vehicle generated the waypoints to be followed to complete the path from the initial location to a goal waypoint. The simulation environment provided a number of configurable parameters for the vehicle that could be adjusted to modify the navigation performance.

Table 6.1 Unit loader default navigation parameters

Vehicle (unit loader) parameters	
Maximum translation speed	5
Maximum translation acceleration	50
Maximum rotational speed	600
Maximum rotational acceleration	6,000
V cutoff angle	210
W cutoff angle	10
Control point	0.3 0 0 0 0 0

Table 6.1 is an example of the default, configurable navigation parameters. In this table, the maximum translation speed (acceleration) represents how fast the vehicle can move (accelerate) when it is directly facing its desired heading, the maximum rotational speed (acceleration) is how fast the vehicle can change its heading, and the control point is the location of the instantaneous center of rotation. The V and W cutoff angles affect the relationship between the vehicle forward and angular velocities as the vehicle heading deviates from its goal heading. More information on these parameters may be found in the MOAST documentation.

For the experiment, a Unit Loader AGV was used, and all parameters but velocity were kept at their default settings. In one experiment, the focus was on path navigation accuracy and speed of path completion. To examine the range of the results, six simulation executions were performed with different vehicle velocity parameters. By examining the results and plotting the velocity with respect to the time it took to complete the path, it is shown that as velocity increases, the time required to complete the path decreased (Fig. 6.10).

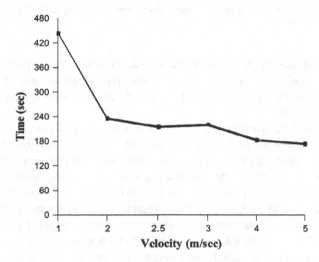

Fig. 6.10 Plot of the time over variable velocity of multiple executions of the same navigation path

Intuitively, this is an expected result; however, it does not reveal any information about the accuracy of the followed path. In fact, it hides the effect that the higher velocities have on accuracy in following the path. Figure 6.11 reveals that while the time for path completion is reduced with higher speeds, the amount of corrective actions, in this case, the number of times the vehicle must reverse to continue reaching all of the path waypoints also increases dramatically.

For velocities that do not exceed 2.0 m/s, the number of reverse corrective actions is the same and negligible. However, even a 0.5 m/s velocity increase causes the number of corrective actions to increase by a factor of seven. Doubling the velocity from 2.0 m/s to 4.0 m/s results in a sixteen-fold increase of the number of corrective actions. The steep increases in the number of corrective actions indicate that

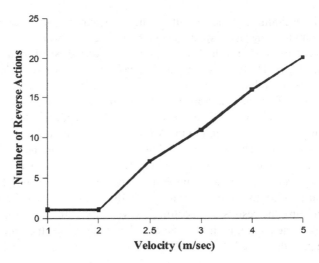

Fig. 6.11 Plot of the number of reverse actions over variable velocity of multiple executions of the same navigation path

the vehicle must maneuver, and therefore deviate from the original path to compensate for poor orientation from waypoints. These measurements and analysis were possible only because of the ease in tuning different parameters, afforded by the simulation platform.

The simulation environment allowed us to develop a successful navigation algorithm and to analyze the performance of the algorithms. To examine the range of the results, six simulation executions were performed with different vehicle velocity parameters. By examining the results and plotting the velocity with respect to the time it took to complete the path, we were able to identify additional elements about our algorithm. As expected, when velocity increased, the time required to complete the path decreased. At the same time, increased velocity affected path accuracy. While the time for path completion reduced with higher speeds, the number of corrective actions, where the vehicle must reverse and change orientation to continue reaching all of the path waypoints, also dramatically increased. We were also able to identify velocity ranges where the number of corrective actions was negligible or very small, along with velocity ranges that caused a seven-fold, and on other occasions, sixteen-fold increase in the number of corrective actions.

Clearly, the simulation-based algorithm development gave us an advantage by being able to execute the algorithms multiple times and being able to collect quantifiable, measurable data related to the performance of the robotic platform. Attempting to perform the same development on an actual platform would have taken a tremendous amount of time and would have risked significant damage to the robot during the early stages of development.

The operational scenario of the second algorithm was based on properly docking a vehicle to a conveyor belt docking station. Autonomous robot docking requires accurate path following and accurate alignment with the target location, typically a docking station.

Using the simulator, we developed a novel, partially heuristic algorithm that allows accurate docking for Ackerman-steered vehicles. Multiple experiments were performed to better understand and analyze the heuristic element of the technique. The algorithm was exercised in gradually smaller rooms to evaluate the versatility of the algorithm under more constrained environments. The results underscored the impact of the vehicle's steering characteristics in docking precision and may prove valuable in attempting to remove the heuristic element in future algorithms.

Again the simulation-based algorithm development gave us an advantage by being able to execute the algorithms multiple times and in variable-sized rooms. Attempting to perform the same development on an actual platform would have taken a tremendous amount of time, would have been difficult to replicate the extensive testing in various sized rooms, and would have risked significant damage to the robot during the early stages of development. Preliminary validation based on visual inspection of the algorithm on a physical platform (a NIST-modified ATRV platform) demonstrated that the docking objective was achieved.

6.5 Competitions

The framework under which the above algorithm development took place was the IEEE Virtual Manufacturing Automation Competition (VMAC) [42]. Both this competition and the RoboCup Rescue Virtual Robots Competition [30] utilize the USARSim simulator as part of their infrastructure.

The RoboCup Rescue Virtual Robots Competition falls under the umbrella of the RoboCup Rescue competitions. The RoboCup Rescue competitions provide a benchmark for evaluating robot platforms for their usability in disaster mitigation and are experiencing ever-increasing popularity. The league vision is the ability to deploy teams of robots that cooperatively explore a devastated area and locate victims. Farsighted goals include the capability to identity hazards, provide structural support and more. RoboCup Rescue is structured in two leagues, the Rescue Robot League and the Rescue Simulation League. The Rescue Robot League fosters the development of high-mobility platforms with adequate sensing capabilities, e.g. to identify human bodies under harsh conditions. The Rescue Simulation League promotes research in planning, learning, and information exchange in an inherently distributed rescue effort. The Rescue Simulation League contains three competitions: the Virtual Robot Competition, the Agent Competition, and the Infrastructure Competition. The Virtual Robots competition simulates, compared to the Rescue Agents competition, small teams of agents with realistic capabilities operating on a city block-sized scenario.

The Virtual Robot competition, first held during the RoboCup competitions in 2006, provides a realistic environment for simulating conditions after a real disaster, such as an earthquake, a major fire, or a car wreck on a highway. Robots are simulated at the sensor and actuator level, making a transparent migration of code between real robots and their simulated counterparts possible. The simulation environment allows evaluation of the performance of large robot teams and their interactions. For example, in the real robot competition there are usually only one or two robots deployed, while in the Virtual Robot competition, teams of up to twelve robots are deployed. Furthermore, the simulator provides accurate ground truth data allowing an objective evaluation of the robots' capabilities in terms of localization, exploration and navigation, e.g. avoidance of bumping. More information on the virtual rescue competition may be found in Balakirsky et al. [3].

The VMAC competition focuses on Automated Guided Vehicles (AGVs). These vehicles represent an integral component of today's manufacturing processes. Major corporations use them on factory floors for jobs as diverse as intra-factory transport of goods between conveyors and assembly sections, parts and frame movements, and truck trailer loading/unloading.

The competition design was based on the successful RoboCup Rescue Virtual Robots Competitions. Since all code used in these competitions is open source, participants are able to learn from their competitors and concentrate their research in their particular areas of expertise. It was envisioned that researchers from multi-agent cooperation, mapping, communications networks, and sensory processing backgrounds would all be interested in participating.

The initial competition design was formulated by using the System, Component, and Operation Relevant Evaluations (SCORE) framework [33]. This framework specifies that an overall system scenario be defined, and then basic elemental skills that allow for the successful completion of the scenario be extracted. Systems are then evaluated on both their ability in the elemental tasks as well as the overall scenario.

From the outset, the competition was to be based on real-world scenarios. Based on NIST's industry outreach effort, the scenario chosen was a factory setting that had significant clutter, maze-like passageways of various widths, and dynamic obstacles. The objective was to have several Ackerman-steered AGVs pick-up packages at a central loading station, and deliver these packages to one of several unloading stations. The package destinations were encoded in a Radio Frequency IDentification (RFID) Tag on each package.

Utilizing the SCORE framework, this scenario was decomposed into elemental tasks that included traffic management, route planning, accurate path following, and docking with loading/unloading stations. While the baseline code provided to the teams was capable of performing the objectives, it was far from optimal.

For the first running of the competition, a decision was made to only compete two of the basic elemental tasks: accurate path following, and docking. One team's experiences with the virtual development cycle for this task are outlined in the previous section. More information on the VMAC may be found on the VMAC webpage [42] or in an overview by Balakirsky et al. [5].

6.6 Conclusion

Robot simulators are useful tools for developing algorithms to control robot behavior. However, the execution of robotic algorithms is not confined to the inner workings of a CPU but have a physical manifestation in real environments with physical robots traversing real terrains and avoiding real obstacles. This means that robotic algorithms are assessed both theoretically and practically.

Simulation platforms allow the development of algorithms in a safe environment where execution errors are benign and hardware reliability is not an issue. Such platforms allow great flexibility in designing complex environments and testing algorithms repeatedly under multiple scenarios.

At the same time, these platforms have limitations. Algorithm development using a simulation assumes that the information about the environment is accurate. Yet the complexity of the operating environment can be daunting and certain variables such as terrain characteristics or sensor sensitivity may be omitted or ignored. In addition, all of the elements of a simulation—the robot, the terrain and the sensor payload—are subject to simplifying assumptions. These assumptions may reduce the realism of the simulation enough to render the developed algorithm insufficient for deployment in a real world environment. Still, none of these concerns are enough to offset the tremendous benefits of simulation platforms in terms of cost savings, risk reduction by testing on the real platform, having a consistent experimentation environment and having the ability to repeatedly test and collect measurable, quantifiable data.

References

1. Adobbati, R., Marshall, A.N., Scholer, A., Tejada, S., Kaminka, G.A., Schaffer, S., Sollitto, C. "GameBots: A 3D virtual world test bed for multiagent research," Proceedings of the Second International Workshop on Infrastructure for Agents, MAS, and Scalable MAS, Montreal, Canada, 2001.
2. Balakirsky, S., Scrapper, C., Carpin, S., Lewis, M., "USARSim: providing a framework for multi-robot performance evaluation," Proceedings of PerMIS, Gaithersburg, MD, 2006.
3. Balakirsky, S., Scrapper, C., Carpin, S., Lewis, M., "USARSim: A RoboCup Virtual Urban Search and Rescue Competition," Proceedings of the 2007 SPIE Unmanned Systems Technology IX, Defense and Security Symposium, 2007.
4. Balakirsky, S., Proctor, F., Scrapper, C., and Kramer, T., "An Integrated Control and Simulation Environment for Mobile Robot Software Development," Proceedings of the ASME Computers and Information in Engineering Conference, New York, NY, 2008.
5. Balakirsky, S., Madhavan, R., and Scrapper, C., "NIST/IEEE Virtual Manufacturing Automation Competition: From Earliest Beginnings to Future Directions," Proceedings of PerMIS, Gaithersburg, MD, 2008.
6. Barraquand, J., Kavraki, L., Latombe, J., Motwani, R., Li, T., and Raghavan, P. 1997. "A random sampling scheme for path planning," International Journal on Robotics Research 16(6), Dec. 1997, 759–774.
7. Balaguer, B., Carpin, S., Balakirsky, S., "Towards Quantitative Comparisons of Robot Algorithms: Experiences with SLAM in Simulation and Real World Systems," Workshop on Performance Evaluation and Benchmarking for Intelligent Robots and Systems at IEEE/RSJ IROS, 2007.
8. Balaguer, B., Carpin, S. "Where Am I? A Simulated GPS Sensor for Outdoor Robotic Applications," Proceedings of the First International Conference on Simulation, Modeling and Programming for Autonomous Robots, Springer-Verlag, Berlin, 2008, 222–233.
9. Brooks, R., Matarić, M., "Real Robots, Real Learning Problems," Robot Learning, Kluwer Academic Press, Hingham, MA, 1993, 193–213.

10. Canny, J. F., The Complexity of Robot Motion Planning, MIT Press, 1988.
11. Canny, J., "On the "Piano Movers" series by Schwartz, Sharir, and Ariel-Sheffi". In the Robotics Review 1, O. Khatib, J. J. Craig, and T. Lozano-Pérez, Eds. MIT Press, Cambridge, MA, 1989, pp. 33–40.
12. Carpin, S., Lewis, M., Wang, J., Balakirsky, S., Scrapper, C., "USARSim: a robot simulator for research and education," Proceedings of the IEEE 2007 International Conference on Robotics and Automation, IEEE, Roma, 2007, pp. 1400–1405.
13. Carpin, S., Stoyanov, T., Nevatia, Y., Lewis, M., Wang, J., "Quantitative Assessments of USARSim Accuracy," Proceedings of PerMIS 2006.
14. Carpin, S., Wang, J., Lewis, M., Birk, A., Jacoff, A., "High fidelity tools for rescue robotics: results and perspectives," Robocup 2005: Robot Soccer World Cup IX, LNAI, Vol. 4020, Springer, Berlin, 2006, pp. 301–311.
15. Faraz al, A., Payandeh, S., "Kinematic modelling and trajectory planning for a tele-laparoscopic manipulating system," Robotica, 18(4), 2000, 347–360, Cambridge University Press, Cambridge, MA.
16. Gerkey, B., Vaughan, R., Howard, A. "The player/stage project: Tools for multi-robot and distributed sensor systems," Proceedings of the International Conference on Advanced Robotics (ICAR), North-Holland Publishing Co., Amsterdam, 2003, pp. 317–323.
17. Kavraki, L. E.; Svestka, P.; Latombe, J.-C.; Overmars, M. H., "Probabilistic roadmaps for path planning in high-dimensional configuration spaces," IEEE Transactions on Robotics and Automation 12(4), 1996, 566–580.
18. Koenig, N., Howard, A. "Design and use paradigms for Gazebo, an open-source multi-robot simulator," IEEE/RSJ International Conference on Intelligent Robots and Systems, Sage Publications, Inc., Thousand Oaks, CA, 2004, pp. 2149–2154.
19. Kyriacou, T., Nehmzow, U., Iglesias, R., Billings, S. A. "Accurate robot simulation through system identification", Robotics and Autonomous Systems, 56(12), 2008, 1082–1093.
20. Laue, T., Spiess, K., Röfer, T. (2006). "SimRobot – A General Physical Robot Simulator and Its Application in RoboCup," RoboCup 2005: Robot Soccer World Cup, IX(4020), 2006, 173–183.
21. LaValle, S.M., Kuffner, J.J., "Rapidly-exploring random trees: Progress and prospects," Proceedings Workshop on the Algorithmic Foundations of Robotics, 2000.
22. Latombe, J.-C. "Motion planning: A journey of robots, molecules, digital actors, and other artifacts," International Journal on Robotic Research, 18(11), 1999, 1119–1128.
23. Leger, C., "Darwin2k, An Evolutionary Approach to Automated Design for Robotics," Kluwer Academic Publishers, Hingham, MA, 2000, 0-7923-7929-2.
24. Lewis, M., Jacobson, J. "Game Engines in Research," Communications of the Association for Computing Machinery (CACM), NY:ACM 45(1), 2002.
25. Lozano-Perez, T. "Spatial Planning: A Configuration Space Approach," IEEE Transactions on Computers, C-32(2), February 1983, pp. 108–120. Also, IEEE Tutorial on Robotics, IEEE Computer Society, 1986, pp. 26–38. Also, AI Memo 605, December 1980.
26. Messina, E., "Performance Standards for Urban Search & Rescue Robots: Enabling Deployment of New Tools for Responders," Defense Standardization Program Office Journal, July/December 2007, pp. 43–48.
27. Michel, O., "Webots: a Powerful Realistic Mobile Robots Simulator," Proceeding of the Second International Workshop on RoboCup. LNAI. Springer-Verlag, Berlin, 1998.
28. Nilsson, N., "A mobile automaton: An application of artificial intelligence techniques," Proceedings of International Joint Conference on Artificial Intelligence, Washington, DC, (IJCAI-69), 1969.
29. Pepper, C., Balakirsky, S., and Scrapper, C., "Robot Simulation Physics Validation," Proceedings of the Performance Metrics for Intelligent Systems (PerMIS) Workshop, 2007.
30. RoboCup Rescue Virtual League. http://www.robocuprescue.org/wiki/index.php?title= Virtualrobots, accessed 01/15/09.

31. Sheng, X., "Motion planning for computer animation and virtual reality applications," Computer Animation, IEEE Computer Society, Washington, DC, 1995, p. 56.
32. Song, G., Amato, N. M., "Using motion planning to study protein folding pathways," Proceedings of the Fifth Annual international Conference on Computational Biology, ACM, New York, NY, 2001, pp. 287–296.
33. Schlenoff, C., Steves, M., Weiss, B., Shneier, M., Virts, A., "Applying SCORE to field-based performance evaluations of soldier worn sensor technologies," Journal of Field Robotics, 24(8–9), 2006, 671–698.
34. Schwartz, J. T., Sharir, M., "On the piano movers problem: I. The case of a rigid polygonal body moving amidst polygonal barriers", Communications on Pure and Applied Mathematics, 36, 1983, 345–398.
35. Schwartz, J. T., Sharir, M., "On the Piano movers Problem II: General techniques for computing topological properties of algebraic manifolds," Advances in Applied Mathematics, 4, 1983, 298–351.
36. Southwest Research Institute, Applied Physics Division, Micrologger Introduction.
37. Sycara, K., Pannu, A. S., "The RETSINA multiagent system (video session): towards integrating planning, execution and information gathering," Proceedings of the Second International Conference on Autonomous Agents, ACM, New York, NY, 1998.
38. Taylor, B., Balakirsky, S., Messina, E., Quinn, R., "Design and Validation of a Whegs Robot in USARSim," Proceedings of the Performance Metrics for Intelligent Systems (PerMIS) Workshop, 2007.
39. Shawna, T., Song, G., Amato, N. M., "Protein Folding by Motion Planning," Physical Biology, 2, 2005, S148–S155.
40. UNR, Unreal engine, http://www.epicgames.com, accessed June 1, 2008.
41. USARSim Homepage, http://www.sourceforge.net/projects/usarsim, accessed 01/15/ 2009.
42. VMAC Home page. http://vmac.hood.edu, accessed 01/15/2009.

Chapter 7
Cognitive Systems Platforms using Open Source

Patrick Courtney, Olivier Michel, Angelo Cangelosi, Vadim Tikhanoff, Giorgio Metta, Lorenzo Natale, Francesco Nori, and Serge Kernbach

Abstract This chapter reports to the development of the tools and methodologies that are in development within the EU, with an emphasis on the Open Source approaches with a view to performance analysis and comparison, and to provide an overview of cooperative research and especially on the use of Open platforms.

7.1 Introduction

A significant research programme under the banner of "cognitive systems" is now underway within the European Union with some €400 M committed in the period to 2010. This programme focuses on developing the technology and the necessary scientific understanding to provide new levels of autonomy and decision making into computer-based systems. Active research approaches in the area range broadly, from traditional rule-based AI, through to connectionist, dynamical and emergent systems and include embodied systems combining computing and robotic systems. A major practical motivation for the development of cognitive systems is to overcome the problems faced by traditional computer systems in dealing robustly with the uncertainties and changing demands that characterise the real world. Potential applications cited span a very wide range and have included care-giver robots, and easier-to-use interfaces.

This chapter reports to the development of the tools and methodologies that are in development within the EU. One major approach that has emerged across the projects has been in the use of Open Source platforms in order to share experiences and run larger scale experiments. This clearly also has a role to play in facilitating

P. Courtney (✉)
Perkinelmer, Beaconsfield, UK
e-mail: patrick.courtney@perkinelmer.com

R. Madhavan et al. (eds.), *Performance Evaluation and Benchmarking of Intelligent Systems*, DOI 10.1007/978-1-4419-0492-8_7,
© Springer Science+Business Media, LLC 2009

performance analysis and comparison. Here we present 3 significant research efforts in cognitive robotics with varying degrees of development with respect to benchmarking[1]:

7.1.1 The Rats Life Benchmark

Rat's Life is a complete cognitive robotics benchmark that was carefully designed to be easily reproducible in a research lab with limited resources. It relies on two e-puck robots, some LEGO bricks and the Webots robot simulation software. This benchmark is a survival game where two robots compete against each other for resources in a previously unseen maze. Like the rats in cognitive animal experimentation, the e-puck robots look for feeders which allow them to live longer than their opponent. Once a feeder is reached by a robot, the robot draws energy from it and the feeder becomes unavailable for a while. Hence, the robot has to further explore the maze, searching for other feeders.

7.1.2 The iCub Platform

The iCub is a new Open Source humanoid robot developed for research in embodied cognition. At around the size of a 3.5 year old child, it can crawl on all fours and sit up to manipulate objects. Its hands have been designed to support sophisticate manipulation skills. The iCub is distributed as Open Source following the GPL/FDL licenses. The entire design is available for download from the project homepage and repository. In the following, we concentrate on the description of the hardware and software systems, as well as an iCub simulator, each of which may be used in comparative studies.

7.1.3 The Swarm Platform of the Replicator and SYMBRION Projects

Cooperation and competition among stand-alone swarm agents can increase the collective fitness of the whole system. An interesting form of collective system is demonstrated by some bacteria and fungi, which can build symbiotic organisms. Symbiotic communities can enable new functional capabilities which allow all members to survive better in their environment. Here we present an overview of two large European projects dealing with new collective robotic systems which utilize principles derived from natural symbiosis. We also describe some of the typical hardware, software and methodological challenges arising, as well as prototypes

---

[1] For a full list of past and current EU funded projects in Cognitive Systems and Robotics research, see http://cordis.europa.eu/fp7/ict/programme/challenge2_en.html

and on-going experiments. The use of multiple robotic agents lends itself to the replication of experiments and thus benchmarking of behavior and functionality.

7.2 The Rat's Life Benchmark: Competing Cognitive Robots

7.2.1 Motivation

Most scientific publications in the area of robotics research face tremendous challenges: comparing the achieved result with other similar research results and hence convincing the reader of the quality of the research work. These challenges are very difficult because roboticists lack common tools allowing them to evaluate the absolute performance of their systems or compare their results with others. As a result, such publications often fail at providing verifiable results, either because the studied system is unique and difficult to replicate or they do not provide enough experimental details so that the reader could replicate the system accurately.

This fact is unfortunately impairing the credibility of robotics research. A number of robotics researchers proposed to develop series of benchmarks to provide a means of evaluation and comparison of robotics research results [1–7].[2]

7.2.2 Existing Robot Competitions and Benchmarks

Several popular robot competitions are organized on a regular basis, usually once a year. The Robocup soccer [8] is a robot soccer tournament with several categories (small size league, middle size league, standard platform league, simulation league, etc.). The Robocup Rescue is based on the Urban Search And Rescue (USAR) benchmark developed by the NIST [9] where robots have to search and rescue the victims of a disaster in a urban environment. MicroMouse[3] involves wheeled robots solving a maze. The AAAI Robot Competition[4] proposes different scenarios each year during the AAAI conference, but often lack clear performance metrics. The DARPA Grand Challenge and Urban Challenge[5] and the European Land-Robot Trial[6] focus on unmanned ground and sometimes aerial vehicles racing against each other.

---

[2]   EURON Benchmarking. www.euron.org/activities/benchmarks/index.html
Robot Benchmark EURON website www.robot.uji.es/EURON/en/euron07.htm
NIST ISD website http://www.isd.mel.nist.gov/
Euron GEM SIG website http://www.heronrobots.com/EuronGEMSig/
RAWSEEDS web site http://www.rawseeds.org/
RoSta website http://www.robot-standards.eu/
[3]MicroMouse contest. http://www.micromouseinfo.com
[4]AAAI Robot Competition and exhibition. http://palantir.cs.colby.edu/aaai07
[5]DARPA Grand Challenge. http://www.darpa.mil/grandchallenge
[6]European Land-Robot Trial ELROB. http://www.elrob.org/

Such competitions are useful as they can provide elements of comparison between different research results. However one of the major problem is that the rules often change across the different editions of the same competition. Hence it is difficult to compare the progress achieved over time. Also these competitions are very specific to particular problems, like Robocup is focused mostly on robot soccer and has arguably a limited interest for cognitive robotics [10].

Among all the benchmarks we reviewed which are mostly robot competitions, none of them provides both stable rules with advanced cognitive robotics challenges and an easy setup. In this Chapter, we propose a new robotics benchmark called "Rat's Life" that addresses a number of cognitive robotics challenges while being cheap and very easy to setup for any research lab. The aim of this benchmark is to foster advanced robotics and AI research.

7.2.3 Rat's Life Benchmark: Standard Components

7.2.3.1 The e-puck mobile robot

The e-puck mini mobile robot was originally developed at the Swiss Federal Institute of Technology in Lausanne (EPFL) for teaching purposes by the designers of the successful Khepera robot. The e-puck hardware and software is fully Open Source, providing low level access to every electronic device and offering unlimited extension possibilities. The robot is already equipped with a large number of sensors and actuators (Fig. 7.1). It is well supported by the Webots simulation software with simulation models, remote control and cross-compilation facilities. The robot is commercially available from Cyberbotics[7] for about EUR 570.

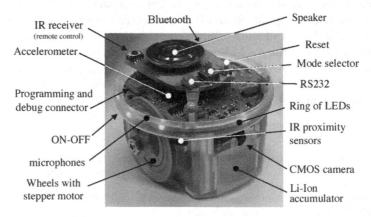

Fig. 7.1 The e-puck robot

---

[7]Cyberbotics Ltd. http://www.cyberbotics.com

7.2.3.2 LEGO bricks

The LEGO bricks are used to create an environment for the e-puck robot. This environment is actually a maze which contains "feeder" devices (see next sections) as well as visual landmarks made up of patterns of colored LEGO brick in the walls of the maze (see Fig. 7.2).

Fig. 7.2 The Rat's Life maze: LEGO bricks, e-puck robots and feeder device

7.2.3.3 The Webots robot simulation software

Webots [11] is a commercial software package for fast prototyping and simulation of mobile robots. It was originally developed at EPFL from 1996 and has been continuously developed, documented and supported since 1998 by Cyberbotics Ltd. Over 500 universities and industrial research centers worldwide are using this software for research and educational purposes. Webots has already been used to organize robot programming contests (ALife contest and Roboka contest). Although Webots is a commercial software, a demo version is available from Cyberbotics's web site and includes the complete Rat's Life benchmark usable for free.

7.2.4 Rat's Life Benchmark Description

This section does not claim to be a technical reference for the Rat's Life benchmark. Such a technical reference is available on the Rat's Life web site.[8]

7.2.4.1 Software-only Benchmark

The Rat's Life benchmark defines precisely all of the hardware necessary to run the benchmark (including the robots and their environment). Hence the users of

---

[8] Rat's Life contest. http://www.ratslife.org

the benchmarks don't have to develop any hardware. Instead, they can focus on robot control software development only. This is similar to the Robocup standard league where the robot platforms (Aibo robots) and the environment is fully defined and the competitors are limited to develop control software only. This has the disadvantage of preventing hardware research and is constraining the contest to the defined hardware only. However, it has the great advantage of letting the users focus on the most challenging part of cognitive robotics, i.e., the control software.

7.2.4.2 Configuration of the Maze

For each evaluation, the maze is randomly chosen among a series of 10 different configurations of the maze. In each configuration, the walls, landmarks and feeder are placed at different locations to form a different maze. Each configuration also has 10 different possible initial positions and orientations for the two robots, one of which is chosen at random. This makes 100 possible initial configurations. This random configuration of the maze prevents the robots from having prior knowledge of the maze, and forces them to discover their environment by exploring it. This yields to much more interesting robot behaviors. A possible configuration is depicted in Fig. 7.3.

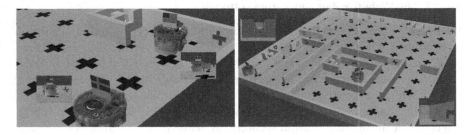

Fig. 7.3 Closeup of the Rat's Life simulated robots in Webots (*left*) and general overview (*right*)

7.2.4.3 Virtual Ecosystem

The Rat's Life benchmark is a competition where two e-puck robots compete against each other for resources in a LEGO maze. Resources are actually a simulation of energy sources implemented as four feeder devices. These feeder devices are depicted on Fig. 7.4. They are made up of LEGO NXT distance sensors which are controlled by a LEGO NXT control brick. They display a red light when they are full of virtual energy. The e-puck robots can see this colored light through their camera and have to move forward to enter the detection area of the distance sensor. Once the sensor detects the robot, it turns its light off to simulate the fact that the feeder is now empty. Then, the robot is credited an amount of virtual energy corresponding

Fig. 7.4 A full feeder facing an e-puck robot (*left*) and an empty one (*right*)

to the virtual energy that was stored in the feeder. This virtual energy will be consumed as the robot is functioning and could be interpreted as the metabolism of the rat robot. The feeder will remain empty (i.e., off) for a while. Hence the robot has to find another feeder with a red light on to get more energy before its energy level reaches 0. When a robot runs out of virtual energy (i.e., its energy level reaches 0), the other robot wins.

7.2.4.4 Robotics and AI Challenges

Solving this benchmark in an efficient way requires the following cognitive capabilities:

- Recognize a feeder (especially a full one) from a camera image.
- Navigate to the feeder and dock to it to grab energy.
- Navigate randomly in the maze while avoiding getting stuck.
- Remember the path to a previously found feeder and get back to it.
- Optimize energy management.
- Try to prevent the other robot from getting energy.

7.2.5 Evolution of the Competition over Time

Observing the evolution of the competition over days was very interesting and we decided to store all the simulation movies in a data base to be able to analyze this evolution afterwards. The movie database contains more than 2,500 movies (totaling more than 50 GB of data) and is freely available online.[9] During the contest, several major performance breakthroughs could be observed simply by analyzing the behavior of the robots in the simulation movies. One could identify five major breakthroughs which happened chronologically one after the other, bringing each time an improved performance:

---

[9] http://www.cyberbotics.com/ratslife/movies

Random Walkers. The random walkers came actually from the very first version of the sample source code included with the contest software development kit, made available to all the competitors. This simple control algorithm similar to Braitenberg vehicles [12] let the robots move randomly while avoiding the obstacles. By chance some of them met a feeder from time to time, but this behavior is very inefficient are rely mostly on luck. Also, this very first version was not very efficient at navigating and often caused the robot to get stuck in some unexpected situations, like facing a corner.

Vision-Enabled Random Walkers. The so called vision-enabled random walkers are an improved version of the original random walker making an extensive use of vision to recognize the feeders and adjust the trajectory of the robot to reach the feeder instead of simply moving randomly. This results in slightly more efficient robots who won't pass in front of a feeder without getting energy from it. A vision-enabled random walker is included in the sample code currently distributed to the competitors. This sample version has however been largely improved by different competitors over time.

Right Hand Explorers. One of the problems with the random walkers is that a Braitenberg vehicle behavior is not very efficient at exploring extensively a maze and hence at finding the feeders. Maze exploration algorithms exist and are much more efficient. The right hand algorithm is one of the simplest and best known maze exploration algorithms. It consists in simply following the first wall found on the right hand side of the robot (this also works with the left hand side of course). Using this algorithm combined with some vision to reach efficiently the feeders, a significant performance breakthrough was reached. The first right hand explorer appeared on February 22nd, with a robot named Tony (which reached rank #1 of the hall of fame on February 22th very rapidly) and was rapidly copied by many other competitors as this behavior is both easy to understand and to re-program.

Energy-Aware Robots. Getting energy from the feeder as soon as you find the feeder is nice, but there is an even better strategy: Once a robot finds a feeder, it can simply stop and sit in front of the feeder, thus preventing the other robot from reaching this feeder. In the meantime the robot sitting in front of the feeder should watch its energy level and decide to move to the feeder once its energy level reached a very low value, just enough to make that move to the feeder and refuel. During this waiting time, the other robot may be struggling to find a feeder and possibly loose the game if it runs out of energy. This kind of energy-aware robots appeared on February 28th, with a robot named Ratchou (which reached rank #1 thanks to this breakthrough). Similarly to the right hand explorer, it was rapidly copied by other competitors as it was easy to understand and to re-program.

SLAMers. SLAM stands for Self Localization And Mapping. Compared to other techniques mentioned above, it involves a much more complicated algorithm and requires efficient image processing. SLAMer robots actually seems to use the right hand algorithm on a first stage to explore extensively the maze, but they build dynamically a map of this maze while exploring it and eventually don't use the right hand algorithm at all. Their internal representation of the environment contains the walls, the feeders and likely the landmarks. This map is then used by the robot to get back to previously found feeders. It turned out to be very efficient and clearly outperformed the simpler reactive controllers. The first SLAMer robot is Ratatouille who implemented a first version of visual SLAM-based navigation on April 6th and reach rank #1. This first version was however probably not well tuned (or somehow buggy) and it happened to lose in rare occasions against lucky and efficient right-hand explorers. However, the author of Ratatouille continued to improve the performance of his SLAMer robot and finally sat steadily on the very top of the hall of fame for more than 2 months. The other competitors, including Tony among others, tried hard to implement such an efficient SLAM-based navigation controller, they were not very successful until June 5th. At this point a competitor with a robot controller named Gollum developed a pretty efficient SLAMer robot able to challenge Ratatouille. Golum reached rank #2 on June 5th and had a fierce and very interesting match against Ratatouille, but was not successful.

Super-SLAMers. The author of Ratatouille actually never stopped from April 6th to improve his SLAM-based robot controller. A major improvement was probably the estimation of the status of the feeders, combined with an estimation of the time needed to travel the maze to reach the feeder. From the most recent simulation movies, Ratatouille seems to be able to anticipate that a mapped feeder will become available again: when the feeder is still red, Ratatouille starts to navigate towards this feeder and about 1 s before it reaches the feeder, the feeder becomes green again. This makes Ratatouille the most efficient robot controller currently on the Rat's Life benchmark. At this point, it is difficult to imagine a better behavior than the one exhibited by Ratatouille.

7.2.6 Discussion

Thanks to the Rat's Life benchmark, it becomes possible to evaluate the performance of various approaches to robot control for navigation in an unknown environment, including various SLAM and bio-inspired models. The performance evaluation allow us to make a ranking between the different control programs submitted, but also to compare the progresses achieved over a short period of time of research on this problem. However, this period of time could be extended and we could, for example, compare the top 5 controller programs developed in 2008 to the top 5 controller programs developed in 2012 to evaluate how much the state of the art progressed.

The control program resulting from the best robot controllers could be adapted to real world robotics applications in the areas of surveillance, mobile manipulators, UAV, cleaning, toys, etc. Also, interesting scientific comparisons with biological intelligence could be drawn by opposing the best robot controllers to a real rat (or a rat-controlled robot) in a similar problem. Similarly, we could also pit the best robot controllers against a human (possibly a child) remote controlling the robot with a joystick and with limited sensory information coming only from the robot sensors (mainly the camera).

We hope that this initiative is a step towards a more general usage of benchmarks in robotics research. By its modest requirements, simplicity, but nevertheless interesting challenges it proposes, the Rat's Life benchmark has the potential to become a successful reference benchmark in cognitive robotics and hence open the doors to more complex and advanced series of cognitive robotics benchmarks.

7.3 The Open Source Humanoid Robot Platform iCub

The RobotCub project has the two-fold goal of: (i) creating an Open hardware/software humanoid robotic platform for research in embodied cognition, and (ii) advancing our understanding of natural and artificial cognitive systems by exploiting this platform in the study of the development of cognitive capabilities.

The RobotCub stance on cognition posits that manipulation plays a fundamental role in the development of cognitive capability [13–16]. As many of these basic skills are not ready-made at birth, but developed during ontogenesis [17], RobotCub aims at testing and developing this paradigm through the creation of a child-like humanoid robot: the iCub. The small, compact size and high number of degrees of freedom combined with the Open Source approach distinguishes RobotCub from other humanoid robotics projects developed worldwide. The iCub is also an attempt at standardization and community-building, since several copies of the robot are being built and distributed to research laboratories in Europe and the USA. Together with Open Source licensing, this allows leveraging on the work of others, and possibly benchmarking. Software for the iCub developed at one site can be effectively tested and improved by other scientists.

On one hand this makes software development harder; it is expected that software modules will be tested in different environments and by researchers who will try to layer additional behaviors on top of them. Minimal standards and quality are required to guarantee interoperability. On the other hand, this initial difficulty can be the seed of more reliable and effective software for cognitive systems; systems that have to run multiple times instead of once.

7.3.1 The iCub

The iCub has been designed to allow manipulation and mobility. For this reason 38° of freedom (DOF) have been allocated to the upper part of the body (including the waist). The hands, in particular, have 9 DOF each with three independent fingers and the fourth and fifth to be used for additional stability and support (only one DOF). They are tendon driven, with most of the motors located in the forearm. The legs have 6 DOF each and are strong enough to allow bipedal locomotion.

From the sensory point of view, the iCub is equipped with digital cameras, gyroscopes and accelerometers, microphones, and force/torque sensors. A distributed sensorized skin is under development using capacitive sensor technology.

Each joint is instrumented with positional sensors, in most cases using absolute position encoders. A set of DSP-based control cards, designed to fit the small size of the iCub, takes care of the low-level control loop in real-time. The DSPs talk to each other via CAN bus. Four CAN bus lines connect the various segments of the robot.

All sensory and motor-state information is transferred to an embedded Pentium based PC104 card that handles acquisition, synchronization and reformatting of the various data streams. Time consuming computation is typically carried out externally on a cluster of machines. The communication with the robot occurs via a Gbit Ethernet connection.

The overall weight of the iCub is 22 kg. The umbilical cord contains both the Ethernet cable and power to the robot. At this stage there is no plan for making the iCub fully autonomous in terms of power supply and computation (e.g. by including batteries and/or additional processing power on board).

The mechanics and electronics were optimized for size, starting from an evaluation and estimation of torques in the most demanding situations (e.g. crawling). Motors and gears were appropriately sized according to the requirements of a set of typical tasks. The kinematics was also defined following similar criteria. The controller cards were designed to fit the available space. Figure 7.5 shows the prototype of the iCub.

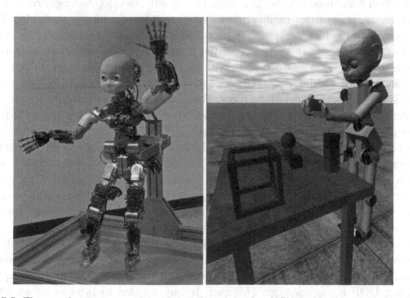

Fig. 7.5 The complete prototype of the iCub (*left*) and simulated iCub looking at and manipulating an object in its environment (*right*)

Several copies of the iCub are now available in the EU. These have been developed as part of the RobotCub competitive call, which awarded seven iCubs free of charge to the six best project proposals. Many more copies of the iCub are also being constructed for other EU funded projects.

7.3.2 Mechanics

The kinematic specifications of the body of the iCub, the definition of the number of DOF, their actual locations as well as the actual size of the limbs and torso were based on ergonomic data and human X-ray images.

The possibility of achieving certain motor tasks is favored by a suitable kinematics and, in particular, this translates into the determination of the range of movement and the number of controllable joints (where clearly replicating the human body in detail is impossible with current technology). Kinematics is also influenced by the overall size of the robot which was imposed a priori. The size is that of a 3.5 years old child (approximately 100 cm tall). This size can be achieved with current technology. The Sony QRIO is an example of a robot of an even smaller size although with less degrees of freedom. In particular, our task specifications, especially manipulation, require at least the same kinematics of QRIO with the addition of the hands and moving eyes. Also, we considered the workspace and dexterity of the arms and thus a 3° of freedom shoulder was included. This was elaborated into a proper list of joints, ranges, and sensory requirements.

Considering dynamics, the most demanding requirements appear in the interaction with the environment. Impact forces, for instance, have to be considered for locomotion behaviors, but also and more importantly, developing cognitive behaviors such as manipulation might require exploring the environment erratically. As a consequence, it is likely that high impact forces need to be sustained by the robot mechanical structure. This requires strong joints, gearboxes, and more in general powerful actuators and appropriate elasticity (for absorbing impacts). In order to evaluate the range of the required forces and stiffness, various behaviors were simulated in a dynamical model of the robot. These simulations provided the initial data for the design of the robot. The simulations were run using Webots and were later cross-checked by conventional static analysis.

At a more general level, we evaluated the available technology, compared to the experience within the project Consortium and the targeted size of the robot: it was decided that electric motors were the most suitable technology for the iCub, given also that it had to be ready according to the very tight schedule of the overall project. Other technologies (e.g. hydraulic, pneumatic) were left for a "technology watch" activity and were not considered further for the design of the iCub.

From the kinematic and dynamic analysis, the total number of degrees of freedom for the upper body was set to 38 (7 for each arm, 9 for each hand, and 6 for the head). For the legs the simulations indicated that for crawling, sitting and squatting a 5 DOF leg is adequate. However, it was decided to incorporate an additional DOF at the ankle to support standing and walking. Therefore each leg has 6 DOF: these include 3 DOF at the hip, 1 DOF at the knee and 2 DOF at the ankle (flexion/extension and abduction/adduction). The foot twist rotation was not implemented. Crawling simulation analysis also showed that for effective crawling a 2 DOF waist/torso is adequate. However, to support manipulation a 3 DOF waist was incorporated. A 3 DOF waist provides increased range and flexibility of motion for the upper body resulting in a larger workspace for manipulation (e.g. when sitting).

The neck has a total of 3 DOF and provides full head movement. The eyes have further 3 DOF to support both tracking and vergence behaviors.

The actuation solution adopted for the iCub is based on a combination of a harmonic drive reduction system (CSD series, 100:1 ratio for all the major joints) and a brushless frameless motor (BLM) from the Kollmorgen frameless RBE series. The harmonic drive gears provide zero backlash, high reduction ratios on small space with low weight while the brushless motors exhibit the desired properties of robustness, high power density, and high torque and speed bandwidths (especially when compared with conventional DC brushed motors). The use of frameless motors permits integration of the motor and gears in an endoskeletal structure that minimizes size, weight and dimensions. Smaller motors (brushed-DC type) were used for the hands and head joints.

7.3.3 The Software: YARP

The iCub software was developed on top of Yet Another Robot Platform (YARP) [18]. RobotCub supported a major overhaul of the YARP libraries to adapt to a more demanding collaborative environment. Better engineered software and interface definitions are now available in YARP.

YARP is a set of libraries that support modularity by abstracting two common difficulties in robotics: namely, modularity in algorithms and in interfacing with the hardware. Robotics is perhaps one of the most demanding application environments for software recycling where hardware changes often, different specialized OSs are typically encountered in a context with a strong demand for efficiency. The YARP libraries assume that an appropriate real-time layer is in charge of the low-level control of the robot and instead takes care of defining a soft real-time communication layer and hardware interface that is suited for cluster computation.

YARP takes care also of providing independence from the operating system and the development environment. The main tools in this respect are ACE [19] and CMake.[10] The former is an OS-independent communication library that hides the quirks of interprocess communication across different OSs. CMake is a cross-platform make-like description language and tool to generate appropriate platform specific project files.

YARP abstractions are defined in terms of protocols. The main YARP protocol addresses inter-process communication issues. The abstraction is implemented by the port C++ class. Ports follow the observer pattern by decoupling producers and consumers. They can deliver messages of any size, across a network using a number of underlying protocols (including shared memory when possible). In doing so, ports decouple as much as possible (as function of a certain number of user-defined parameters) the behavior of the two sides of the communication channels. Ports can be commanded at run time to connect and disconnect.

---

[10]http://www.cmake.org

The second abstraction of YARP is about hardware devices. The YARP approach is to define interfaces for classes of devices to wrap native code APIs (often provided by the hardware manufactures). Change in hardware will likely require only a change in the API calls (and linking against the appropriate library). This easily encapsulates hardware dependencies but leaves dependencies in the source code. The latter can be removed by providing a "factory" for creating objects at run time (on demand).

The combination of the port and device abstractions leads to remotable device drivers which can be accesses across a network: e.g. a grabber can send images to a multitude of listeners for parallel processing.

Overall, YARP's philosophy is to be lightweight and to be "gentle" with existing approaches and libraries. This naturally excludes hard real-time issues that have to be necessarily addressed elsewhere, likely at the OS level.

7.3.4 Research with the iCub

One of the goals of the RobotCub project is to further our understanding of cognitive systems through the ontogenic development of a humanoid robot. That is, it is a program of enquiry into emergent embodied cognitive systems whereby a humanoid robot, equipped with a rich set of innate action and perception capabilities, can develop over time an increasing range of cognitive abilities by recruiting ever more complex actions and thereby achieving an increasing degree of prospection (and hence, adaptability and robustness) in dealing with the world around it.

Cognitive development involves several stages, from coordination of eye-gaze, head attitude, and hand placement when reaching, through to more complex—and revealing—exploratory use of action. This is typically achieved by dexterous manipulation of the environment to learn the affordances of objects in the context of one's own developing capabilities. Our ultimate goal is to create a humanoid robot—the iCub—that can communicate through gestures simple expressions of its understanding of the environment, an understanding that is achieved through rich manipulation-based exploration, imitation, and social interaction.

This program of research is carried out both by investigating natural cognition and by implementing some of these skills on the iCub. In particular, there is solid evidence that sustains the hypothesis of a significant involvement of the motor system in supporting processes traditionally considered as "high level" or cognitive, such as action understanding, mental imagery of actions, objects perception and discrimination. The *biologically plausible stance* guiding the RobotCub project requires a deeper study of these processes not only because of their scientific value but also because their implementation on the iCub can provide better understanding of the corresponding neural processes.

A typical example of how sensorimotor integration is used by the brain in practical tasks is provided by a population of neurons in the monkey ventral premotor cortex (mirror neurons) that discharge both when the monkey performs a grasping action and when it observes the same action performed by other individuals [20].

A further example of the involvement of the motor system in cognitive functions is given by spatial attention. The view that we favor is that attention derives from the same circuits that are responsible for the control of eye movements. This view was originally proposed for visuospatial attention by Rizzolatti and Camarda [21]. Similar properties apply to the selection of visual stimuli even if not necessarily connected to eye movements. Craighero and colleagues [22] have shown that this is the case for other actions like for example grasping.

From the robotics side, RobotCub has started investigating several elements of these sensorimotor processes.

One first example is concerned with the basis of motor control for the iCub. In this case, we hypothesized the existence two basic types of motor primitives: i.e. discrete (aperiodic and finite) and rhythmic (periodic) movements [23]. We model these motor primitives as solutions of a dynamical system with a globally attractive fixed point and an oscillator, respectively. Such an approach allows us to use the stability properties of dynamical systems to ensure a robust control of the movements. This controller allows the simultaneous execution of rhythmic and point to point movements, as required for example in locomotion. Separately, we developed a task-space based controller for the iCub that can take into account joint limits together with any number of constraints in generating appropriate kinematic inverses.

Finally, we would like to mention the implementation on the iCub of exploratory procedures for objects. Building on the existing sensorimotor behaviors, the iCub can touch and elicit movement from objects. Following the object behavior, the association of sensory and motor cues allows the autonomous acquisition of object affordances [24]. Object affordances are a powerful tool for a cognitive system, allowing the interpretation of scenes (because of the behavior of objects) as well as the imitation of other people's actions (as a consequence of the behavior of objects) [25].

Examples of these experiments are illustrated in Fig. 7.6.

Fig. 7.6 iCub experimenting with affordances (*left*) and exercising the reaching-rhythmic controller (*right*)

The Open Source approach used in the production and dissemination of the iCub platform supports the process of standardization and platform benchmarking in cognitive science research [4]. Currently, more than 10 iCub platforms are being produced to be used in various labs in the EU and USA. Moreover, the distribution of the Open Source iCub simulator (see next section) will further permit the use of the same robotic platform in many more labs, further contributing to the use of iCub as a benchmark platform.

7.4 The iCub Simulator

Computer simulations play an important role in robotics research. Despite the fact that the use of a simulation might not provide a full model of the complexity present in the real environment and might not assure a fully reliable transferability of the controller from the simulation environment to the real one, robotic simulations are of great interest for cognitive scientists [26].

The iCub simulator has been designed to reproduce, as accurately as possible, the physics and the dynamics of the robot and its environment with the constraint of running approximately in real-time. The simulated iCub robot is composed of multiple rigid bodies connected via joint structures. It has been constructed collecting data directly from the robot design specifications in order to achieve an exact replication (e.g. height, mass, Degrees of Freedom) of the first iCub prototype developed by the RobotCub Consortium. The environment parameters on gravity, objects mass, friction and joints are based on known environment conditions. The iCub simulator presented here has been created using Open Source libraries in order to make it possible to distribute the simulator freely to any researcher without requesting the purchase of restricted or expensive proprietary licenses.

7.4.1 Physics Engine

The iCub simulator uses ODE (Open Dynamic Engine)[11] for simulating rigid bodies and the collision detection algorithms to compute the physical interaction with objects. The same physics library was used for the Gazebo project and the Webots commercial package. ODE is a widely used physics engine in the Open Source community, whether for research, authoring tools, gaming etc. It consists of a high performance library for simulating rigid body dynamics using a simple C/C++ API. ODE was selected as the preferred Open Source library for the iCub simulator because of the availability of many advanced joint types, rigid bodies (with many parameters such as mass, friction, sensors . . .), terrains and meshes for complex object creation.

---

[11] Open Dynamics Engine http://opende.sourceforge.net/.

7.4.2 Rendering Engine

Although ODE is a good and reliable physics engine, computing all the physical interaction of a complex system can take a good deal of processing power. Since ODE uses a simple rendering engine based on OpenGL, it has limitations for the rendering of complex environments comprising many objects and bodies. This can significantly affect the simulation speed of complex robotic simulation experiments. It was therefore decided to use OpenGL directly combined with SDL,[12] an Open Source cross-platform multimedia library. This makes it possible to render the scene with more ease and to carry out computationally-efficient simulation experiments.

7.4.3 YARP Protocol for Simulated iCub

As the aim was to create a replica of the physical iCub robot, the same software infrastructure and inter-process communication will have to be used as those used to control the physical robot. As the physical iCub is based on YARP as its software architecture, the simulator and the actual robot share the same interface either when viewed via the device API or across network and are interchangeable from a user perspective. The simulator, like the real robot, can be controlled directly via sockets and a simple text-mode protocol; use of the YARP library is not a requirement. This can provide a starting point for integrating the simulator with existing controllers in esoteric languages or complicated environments. The user code can send and receive information to both the simulated robot itself (motors/sensors/cameras) and the world (manipulate the world). Network wrappers allow device remotization. The Network Wrapper exports the YARP interface so that it can be accessed remotely by another machine (Fig. 7.7).

7.4.4 iCub Body Model

The iCub simulator has been created using the data from the physical robot in order to have a replica of it. As for the physical iCub, the total height is around 100 cm, weighs approximately 22 kg and has a total of 53° of freedom (DoF). These include 12 controlled DoFs for the legs, 3 controlled DoFs for the torso, 32 for the arms and six for the head.

The robot body model consists of multiple rigid bodies attached through a number of different joints. All the sensors were implemented in the simulation on the actual body, such as touch sensors and force/torque sensors. As many factors impact on the torque values during manipulations, the simulator might not guarantee to be perfectly correct. However the simulated robot torque parameters and their verification in static or motion are a good basis and can be proven to be reliable [27].

---

[12] SDL – Simple DirectMedia Layer http://www.libsdl.org

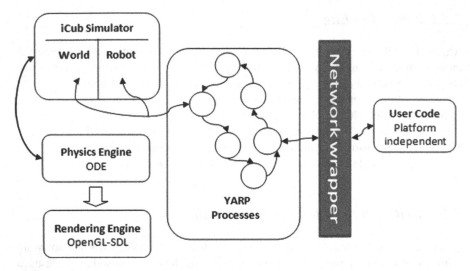

Fig. 7.7 This figure shows the architecture of the simulator with YARP support

All the commands sent to and from the robot are based on YARP instructions. For the vision we use cameras located at the eyes of the robot which in turn can be sent to any workstation using YARP in order to develop vision analysis algorithms.

The system has full interaction with the world/environment. The objects within this world can be dynamically created, modified and queried by simple instructions which are exactly those that YARP uses in order to control the real robot.

7.4.5 Simulator Testing and Further Developments

The current version of the iCub simulator has been used for preliminary testing by partners in the RobotCub and ITALK projects. The ITALK project [28] aims at the development of cognitive robotic agents, based among others on the iCub humanoid platform, that learn to handle and manipulate objects and tools autonomously, to cooperate and communicate with other robots and humans, and to adapt their abilities to changing internal, environmental, and social conditions. The main theoretical hypothesis behind the project is that the parallel development of action, conceptualisation and social interaction permits the bootstrapping of language capabilities, which on their part enhance cognitive development. This is possible through the integration and transfer of knowledge and cognitive processes involved in sensorimotor learning and the construction of action categories, imitation and other forms of social learning, the acquisition of grounded conceptual representations and the development of the grammatical structure of language.

In addition to being used for experiments on the development of controllers for the iCub robot, some groups have used the simulator to create a mental model [29] used by the robot to represent the current state of the environment.

Future plans on the simulator development will mostly involve the design of functionalities to model and interact with the physical environment. For example, this will allow the users to modify the objects in the world where the iCub resides, in order to allow different types of experiments. Finally, further work will focus on the systematic testing and replication of simulation studies with the physical robot. The latest version of the iCub simulation is available Open Source in the RobotCub/iCub repository.[13]

7.5 Symbiotic Robot Organisms: Replicator and SYMBRION Projects

7.5.1 Introduction

Nature shows several interesting examples for cooperation of individuals. Most prominent examples of cooperation are found in social insects [30], where specialized reproductive schemes (in most cases just a few out of thousands of colony members are able to reproduce) and the close relationships of colony members favoured the emergence of highly cooperative behaviours [31]. However, also non-eusocial forms of cooperative communities evolved, like the collective hunting in predatory mammals [32] (e.g., lions, whales, . . .) or the reciprocal trophallactic altruism in vampire bats. Such cooperative behaviours are mostly explained by reciprocal advantages due to the cooperative behaviours and/or by the close relationship among the community members. In contrast, cooperation sometimes arises also among individuals that are not just very distant in a gene pool, sometimes they do not even share the same gene pool: Cooperative behaviours between members of different species is called "Symbiosis". A non-exhaustive list of prominent examples includes the pollination of plants by flying insects (or birds), the cooperation between ants and aphids. Also lichens, which are a close integration of fungi and algae and the cooperation between plant roots and fungi represent symbiotic interactions.

A common pattern in all these above-mentioned forms of cooperation is that single individuals perform behaviours, which—at first sight—are more supportive for the collective of the group than for themselves. However, as these behaviours have emerged through natural selection, we can assume that these cooperative behaviours have their ultimate reasoning in a sometimes delayed and often non-obvious individual egoistic advantage.

From symbiotic forms of organization emerge new functional capabilities which allow aggregated organisms to achieve better fitness in the environment. When the need of aggregation is over, symbiotic organisms can disaggregate and exists further as stand-alone agents, thus an adaptive and dynamical form of cooperation is often advantageous.

Lately, technical systems mimic natural collective systems in improving functionality of artificial swarm agents. Collective, networked or swarm robotics are

---

[13] http://www.robotcub.org/

scientific domains, dealing with cooperation in robotics [33]. Current research in these domains is mostly concentrated on cooperation and competition among stand-alone robots to increase their common fitness [34]. However, robots can build a principally new kind of collective systems, when to allow them to aggregate into a multi-robot organism-like-forms. This *robot organism* can perform such activities that cannot be achieved by other kind of robotic systems and so to achieve better functional fitness.

To demonstrate this idea, we consider a collective energy foraging scenario for the Jasmine micro-robots [35]. Swarm robots can autonomously find an energy source and recharge. The clever collective strategy can essentially improve the efficiency of energy foraging, but nevertheless a functional fitness of a swarm is limited. For instance, if the recharging station is separated from a working area by a small barrier, robots can never reach the energy source. However, if robots aggregate into more complex high-level organism which can pass the barrier, they will reach the docking station. In this way a cooperative organization of robotic system allows an essential increase of functional capabilities for the whole group. The large integrated project REPLICATOR[14] deals with such issues as reconfigurability of sensors and actuators, adaptive control and learning strategies as well as working in real environments.

The cooperative (swarm-based or symbiotic) organization of the robotic system provides essential plasticity of used hardware and software platforms. The robot organism will be capable of continuously changing its own structure and functionality. Such an evolve-ability opens many questions about principles and aspects of long- and short-term artificial evolution and controllability of artificial evolutionary processes. The large integrated project SYMBRION[15] is focused on evolve-ability, dependability and artificial evolution for such robot organisms based on bio-inspired and computational paradigms. Both projects are Open Science and Open Source.

Both projects, consortia and the European commission are closely cooperating to achieve the targeted goals. It is expected that results of both projects create new technology for making artificial robotic organisms self-configured, self-healing, self-optimizing and self-protecting from a hardware and software point of view. This leads not only to extremely adaptive, evolve-able and scalable robotic systems, but also enables the robot organisms to reprogram themselves without human supervision, to develop their own cognitive structures and, finally, to allow new functionalities to emerge.

7.5.2 New Paradigm in Collective Robotic Systems

Collective intelligence is often associated with macroscopic capabilities of coordination among robots, collective decision making, labor division and tasks allocation in the group [36]. The main idea behind this is that robots are achieving better

---

[14] http://www.replicatores.eu
[15] http://www.symbrion.eu

performance when working collectively and so are capable of performing such activities which are not possible for individual robots. The background of collective intelligence is related to the capability of swarm agents to interact jointly in one medium. There are three different cases of such interactions:

1. In the first case agents communicate through a digital channel, capable for semantic messages exchange. Due to information exchange, agents build different types of common knowledge [37]. This common knowledge in fact underlies collective intelligence.
2. The second case appears when macroscopic capabilities are defined by environmental feedback. The system builds a closed macroscopic feedback-loop, which works in a collective way as a distributed control mechanisms. In this case there is no need of complex communication, agents interact only by kinetic means. This case of interaction is often denoted as a spatial reasoning, or spatial computing.
3. The third case of interactions we encounter in nature, when some bacteria and fungi (e.g. *dictyostelium discoideum*) can aggregate into a multi-cellular organism when this provides better chances of survival [38]. In this way, they interact not only through information exchange or spatial interactions, they build the closest physical, chemical and mechanical interconnections, though the agents still remain independent from each other. The first two cases of interactions are objects of extensive research in many domains: robotics, multi-agents systems, bio-inspired and adaptive community and so on. However, the practical research in the last case represents essential technological difficulties and therefore is not investigated enough. Despite the similarities between a robot swarm and multi-robot organism, such as a large number of robots, focus on collective/emergent behavior, a transition between them is a quite difficult step due to mechanical, electrical and, primarily, conceptual issue. Now, we believe that research around the third case of interactions is concentrated on three important questions:

- *Reconfigurability*, adaptability and learn-ability of the symbiotic systems. These issues include flexible and multifunctional sensors and actuators, distributed computation, scalability, modelling, control and other issues, which are closely related to the reconfigurable robotic research. The REPLICATOR project is focused on these points.
- *Evolve-ability* of the symbiotic systems, which includes principles and aspects of long- and short-term artificial evolution and adaptivity as well as exploring and analogies to biological systems. The SYMBRION project is focused on these points.
- *Embodiment* of evolutionary systems for different environments and medias as well as investigation of information properties of such systems. These points are covered by other research initiatives and projects.

In this way, the next step in further research within the collective robotic community can consist in investigation of multi-robot organisms or, in other words, a transition from robot swarm to multi-robot organisms. All further sections are devoted to demonstrate diverse aspects of such a transition.

7.5.3 Example: Energy Foraging Scenario

In this section, we demonstrate the advantages of symbiotic organization of autonomous robotic systems. We choose for this purpose an example of energy homeostasis, because it is applicable to both living and robotic organisms and so we can draw several analogies between them.

The distinctive property of any living organism is energy homeostasis and, closely connected, foraging behavior and strategies [32]. The robots, equipped with on-board recharging electronics, can also possess its own energy homeostasis. In this way, when swarm robots get *hungry*, they can collectively look for energy resources and execute different strategies in a cooperative energy foraging [39]. In critical cases robots can even decide to perform individual foraging, competing with other robots for resources.

The need of energy is a perfect example of natural fitness. If robots that are performing individual strategies find enough energy, they can survive in the environment. In turn, this means that these strategies were sufficient enough to balance these robots energetic budgets. Simultaneously, other energetically die if their behavioral strategy was poor. Based on such energy foraging, many of evolutionary approaches for different robotic species can be developed, compared and tested.

However, if there are many robots foraging in the environment, several undesired effects can emerge: (1) the docking station can become a *bottleneck* resource that essentially decreases the swarm efficiency; (2) robots with a high-energy level can occupy the docking station and block low-energetic robots. These robots can energetically die (and so decrease the swarm efficiency); (3) many robots can create a *crowd* around a docking station and essentially hinder a docking approach. This can increase the total recharging time and makes worse the energetic balance of the whole swarm.

Robots, in pursuing their energetic homeostasis, have only two possible decisions to make: (1) to execute a current collective task or (2) to move for recharging. In balancing these two behaviours, a cooperative strategy may find the right timing and the right combination between these individual decisions of all robots. Lately, several strategies of energy foraging for a robot swarm up to 70 swarm agents are implemented, see Fig. 7.8. These cover different bio-inspired approaches [40, 41] and hand-coded strategies [42].

In one of these experiments, a few robots died close to the docking station and blocked the recharge area (we *simulated* this in the Fig. 7.8b). Robots that were in front of this barrier (away from the docking station) finally also died. This is the limit of functional fitness of swarm robots. There is no strategy that allows swarm robots to overpass the barrier. Only when swarm robots would collectively emerge new functionality, like *pass the barrier*, they would solve the *barrier problem*.

Fig. 7.8 (**a**) Docking of a few robots for recharging. Shown is the two-line approach: the first line—recharging robots, the second line—robots waiting for recharging; (**b**) The "barrier problem"—robots separated form docking stations by a barrier; (**c**) Possible solution to the "barrier problem": swarm robots form a symbiotic multi-robot organism & collective pass the barrier

Thus, an ideal solution for the *barrier problem* can be the aggregation of many single robots into one cooperative multi-robot organism. This way, they can reach the docking stations by "growing legs" and stepping over the barrier. In that case, the robots are helping each other in a cooperative manner, (see Fig. 7.8c).

Obviously, such a robotic behaviour is extremely challenging from many viewpoints: Cooperative (symbiotic) robot systems have many similarities with known robotic research as e.g. mechanical self-assembling [43] or reconfigurable robotics [44]. However, the symbiotic form, show in Fig. 7.8 essentially differs from this robotic research, namely: (1) Robots should be capable for autonomous aggregation and disaggregation; (2) Robots in the disaggregate state should possess individual locomotion; (3) There is no central control neither for disaggregated state (swarm) nor for the aggregate state (organism); (4) Stand-alone robots should profit from the aggregation into organism.

The swarm-based approaches, which is underlying the aggregation processes, differs primarily from aggregated systems which are studied in the field of *reconfigurable robotics*. In the following we consider on-going work with aggregated (symbiotic) robot organisms.

7.5.4 Hardware and Software Challenges

The main feature of a modular robot consists in being composed of several independent modules, with limited complexity and capabilities, which are able to connect to each other in different configurations, in order to form a robot with greater capabilities. As a consequence, the overall functionalities and capabilities of a robotic modular organism are deeply related to the hardware structure and functions of its basic composing modules. At the current stage of development of the projects (both projects started in 2008), the development of the hardware represents one of the hardest issues. In general, the concept of hardware design is as follows:

1. Independence for separate robots, this includes capabilities for communication, computation and sensing as a stand-alone robot, as well as individual locomotion and energy management.

2. Large computational power of the organism, required for performing on-line and on-board evolutionary approaches.
3. Heterogeneity of individual robots, which allows their later specialization within the organism.
4. Rich sensing and communication capabilities of the organism. The more robots are joined in the organism, the more functional diversity the organism can demonstrate.
5. Possible higher independency from human in term of energy, support and maintenance.

The consortia considered many state-of-the-art reconfigurable solutions, such as superBot [45], M-Tran [46], Poly-Bot [47], molecube [48], HYDRA/ATRON [49] and others, even visited some of these labs for exchange of experience. Currently, we follow three different developmental lines, which will be later fused into one or two first prototypes.

The electronic design is also a huge challenge due to strong restrictions of the size of the robot and the complexity of the mechanical design. Each stand-alone robot is equipped with several 32 bit ARM-based microcontrollers and one main microprocessor with a large external memory. The breakdown in microcontrollers and microprocessor was deliberately intended to separate computational tasks within the single cell. The microcontrollers perform basic functionality (e.g. sensor pre-processing, running artificial immune system) and keep the robot alive. The microprocessor is mainly responsible for bio-inspired approaches like the genetic algorithms, sensor fusion, ANNs etc. and is more powerful in comparison to the microcontrollers. Due to higher computational power results in higher energy consumption, the microprocessor is able to run at different power-down modes when computational power isn't needed. One of the biggest challenges during the electronic design is the development of all modules in a tiny size as well as finding solutions for shared resources like memory, power and communication capabilities.

Beside the hardware challenges, the project is faced with many software and controller challenges. Because robots can either run independently, as a swarm, as an organism, or even as a swarm of organisms, the interaction has to be managed in an organized and efficient way. The whole software is divided into several layers. On the bottom layers, such as BIOS and real time operational system, there are different processes, which are able to communicate with each other and to control a low-level behavior of all components. To cope with the additional difficulties a swarm or an organism causes, a middleware-like system is necessary. On top of this abstraction layer high-level control mechanisms, distributed applications and genetic framework are integrated.

7.5.5 Towards Evolve-Ability and Benchmarking of Robot Organisms

Within the projects, the creation of evolvable or otherwise adaptive software and hardware is the main focus. Achievement of evolve-ability for the robot organism is planned in two complementary ways, which we call bio-inspired (or bio-mimicking) and engineering-based approaches. Comparison between both approaches allows benchmarking the behavior and functionality of evolved organisms. It is also planned to use four benchmark tests on reconfigurability of the platform, sensing and sensor-fusion capabilities, behavioral and learning tasks as well as on evolving approaches. Evaluation can be performed qualitatively in terms of success in solving corresponding tasks and quantitatively by comparing running time and consumed energy.

7.5.5.1 Bio-inspired/Bio-mimicking Approach

Any bio-inspired approach is based on analogies to living organisms and is carried out by the biological partners in our consortia. Our bio-inspired control algorithms use neither any global point of information nor any form of complex knowledge. Our algorithms are stable to a wide range of environmental conditions and are extremely robust. Therefore, the bio-inspired strategies in projects are going to draw advantage from the well-known robustness/simplicity as well as from the plasticity/adaptability derived from natural systems. Our goal is to create stable, robust and adaptable robotic organisms. Here we will investigate a variety of concepts, such as:

Genome. All robotic organisms will carry one or several Genomes. A Genome is a collection of genes, which carry information about controller structure and controller dynamics. A gene can be a simple part of a blueprint, which "depicts" a part of the final controller. But a gene can also work as a rule, which is used to "construct" parts of the final controller. In the latter case, there can be interferences between different genes, thus competition or cooperation can arise also on the genetic level. A self-organized process can be established which will be able to create a flexible, but robust controller structure.

Controller. We will investigate several controller types, ranging from rules-based controllers, to Evolvable Artificial Neural Networks (EANN) and Artificial Immune Networks (AIS) to hormone-based controllers and to even hand-coded controllers that execute hand-optimized (modular) parts of the whole organism's behavioural repertoire.

Sexuality/Reproduction. We plan to enhance and to speed-up the dynamics of artificial evolution by implementing virtual-reproduction of robots. A separate process will allow removing controllers from the least fit robots and to re-initialize them with mixtures (interbreeds) of the controllers of more fit robots. We will also investigate the advantages of sexual reproduction in such scenarios.

Embryology. To allow well-ordered controllers to emerge from the information stored in the Genome, we will mimic embryological processes, driven by a virtual hormone system.

7.5.5.2 Engineering-Based Approach

The engineering-based approach is complementary to the bio-inspired one and focuses in such issues as learning, distributed decision making, navigation and so on. Generally, consortium focuses on three following approaches (these approaches are closely connected so that finally it will be a kind of hybrid framework):

On-line learning. On-line learning is based on the behavior level and uses automatically generated feedback. The feedback comes from internal, external and virtual sensors. Some direct feedback can be sensed through vision-based subsystem, by using FRID-based identification or localization technologies, by using smart laser scanner, sound, light, humidity, temperature, internal energy sensor and other sensors. It is intended to use middleware and sensor-fusion approach to generate complex non-direct feedbacks through virtual sensors. Since off-line mechanisms can hardly be applied to real robots, the challenge of the proposed approach is to perform non-supervised learning without any off-line mechanisms (or at least with a minimum of them). This can be achieved by combining evolving computation with rewards/feedback/fitness calculated on-line. Therefore the whole approach can be named "on-line learning".

Evolutionary computation. High computational power of the system allows running on-line and on-board such well-known approaches as genetic programming (GP) (e.g. [50]), Genetic Algorithms (GA) (e.g. [51]). To avoid the problems posed by a huge search space, we intend to integrate limitations, originating from hardware platform. Another set of problems we are aware of are the fitness functions required for these algorithms. These fitness functions are very difficult to calculate based only on local sensor data. Moreover these functions are evaluated extremely delayed because the organism mostly assesses their fitness after accomplishing the task.

Approaches from the domain of Distributed Artificial Intelligence (DAI). On-line learning as well as GA/GP include diverse aspects of DAI such as a distributed knowledge management, semantic information processing, navigation and actuation in the environment, planning, sensor fusion and others. Development and implementation of these approaches is an important step towards evolve-ability of the robot organisms.

7.5.6 Discussion

We provided an overview of two large European projects, dealing with a new paradigm in collective systems, where the swarm robots get capable of self-assembling into a single symbiotic multi-robot organism. We introduced an energy foraging scenario for both robot species and demonstrated that a transition between collective and symbiotic robot forms represents a very hard problem. It involves not only hardware and software issues, but also very basic questions being also open not only in biological but also in engineering sense. We demonstrated the main hardware and software challenges and the road-map how to achieve the evolve-ability of the robot organisms.

7.6 Other Projects and Future Work

We have described three specific approaches to cognitive robotics, each with a significant Open Source aspect. The clearest of these is the Rat's Life benchmark which is applied to competitively evaluate the performance of various approaches to robot control for navigation in an unknown environment.

In the iCub project, Open Source forms the basis of a platform for experimental work in humanoid robotics that can be replicated across multiple labs. The motivation is primarily community building and leveraging the work of others, although as a consequence, performance evaluation is also made possible. The relatively large number of iCub platforms being produced is an encouraging indicator of the willingness of research labs to cooperate at this level, with the distribution of the Open Source iCub simulator, an additional mechanism for sharing.

Within the field of swarm robotics, the creation of evolvable or otherwise adaptive software and hardware is the main focus. Here the Open Source aspect is perhaps more subtle, as it lies both in the creation and operation of multiple identical agents, as well as in the execution of the experiments which allow different approaches to be compared and replicated.

As these Open Source cognitive paradigms start to gain a foothold and the benefits within the research community spread, so we can expect the Open Source approach to be applied in a wide range of robotics research. Not surprisingly, a number of the other projects within the current EU programme also have an element of open source as listed in Table 7.1:

Table 7.1 List of current projects with Open Source element

Project	Open Source component
RoSta	architectures and standards for robotics
PHRIENDS	plans to deliver open-source code at the end of the project
DEXMART	plans to use a open source controller
SEARISE	implementation of software based on open-source modules
GRASP	open source platform for benchmarking in grasping and dextrous manipulation. Open source development in detecting the unexpected and learning from it
SF	use of multi-level neuronal simulation environment (iqr) released under GNU general public Licence
SCOVIS	Use of Open CV tool
POETICON	open resource for grounding action (movement) semantics

Further details of these projects are available on the website.[16]

Acknowledgments The Rat's Life benchmark was supported by the European Commission ICEA project,[17] while the work on iCub was supported by the European Commission FP6 Project RobotCub and FP7 Project ITALK[18] within the Cognitive Systems and Robotics unit. The authors would like to thank the RobotCub Consortium. Paul Fitzpatrick is gratefully acknowledged for the continuous support to YARP. The REPLICATOR and SYMBRION projects are funded by European Commission within the 7th framework program. The authors also acknowledge the support of the FP6 euCognition[19] Coordinated Action project funded under the same Cognitive Systems and Robotics unit.

References

1. J. Baltes, 2000, A benchmark suite for mobile robots, IROS 2000 conference proceeding, 2:1101–1106, ISBN: 0-7803-6348-5.
2. A. Serri, 2004, A Lego robot for experimental benchmarking of robust exploration algorithms, Circuits and Systems, iii – 3:163–166, ISBN: 0-7803-8346-X.
3. M. Eaton, J.J. Collins, and L. Sheehan, 2001, Toward a benchmarking framework for research into bio-inspired hardware-software artefacts, Artificial Life and Robotics, Springer Japan, 5(1):40–45, ISSN: 1433-5298.
4. R. Dillmann, 2004, Benchmarks for Robotics Research, EURON, April.
5. J.S. Albus, 2002, Metrics and performance measures for intelligent unmanned ground vehicles. In Proceeding of the performance Metrics for Intelligent System Workshop, NIST, USA.
6. A.P. del Pôbil, 2006, Why do we need benchmarks in robotics research?, International Conference on Intelligent Robot and Systems, Beijing, China.
7. F. Bonsignorio, J. Hallam, and A.P. del Pôbil, 2007, Good experimental methodologies in robotics: State of the art and perspectives, Workshop on Performance Evaluation and Benchmarking for Intelligent Robots and Systems, IEEE/RSJ International Conference on Intelligent Robots and Systems, San Diego, USA.
8. H. Kitano, M. Asada, Y. Kuniyoshi et al., 1995, RoboCup: the robot world cup initiative, IJCAI-95 workshop on entertainment and AI/ALife.
9. A. Jacoff, B. Weiss, and E. Messina, 2003, Evolution of a Performance Metric for Urban Search and Rescue Robots, Proceedings of the 2003 Performance Metrics for Intelligent Systems (PerMIS) Workshop, Gaithersburg, MD, September 16–18.
10. J.J. Collins, M. Eaton, M. Mansfield, D. Haskett, and S. O'Sullivan, 2004, Developing a benchmarking framework for map building paradigms, Proceedings of the 9th International Symposium on Artificial Life and Robotics, January, Oita, Japan, pp.614–617.
11. O. Michel, 2004, Webots: Professional Mobile Robot Simulation, Journal of Advanced Robotics Systems, 1(1):39–42.
12. V. Braitenberg, 1984, Vehicles: Experiments in Synthetic Psychology. Cambridge, MA: MIT Press.
13. L. Fadiga, L. Craighero, and E. Olivier, 2005, Human motor cortex excitability during the perception of others' action, Current Biology, 14:331–333.

---

[17] http://ICEA European Project (IST 027819) ICEA stands for Integrating Cognition, Emotion and Autonomy and is focused on brain-inspired cognitive architectures, robotics and embodied cognition, bringing together cognitive scientists, neuroscientists, psychologists, computational modelers, roboticists and control engineers. It aims at developing a cognitive systems architecture integrating cognitive, emotional and bioregulatory (self-maintenance) processes, based on the architecture and physiology of the mammalian brain. http://www.iceaproject.eu

[18] robotcub.org and italkproject.org

[19] http://cordis.europa.eu/ist/cognition/index.html

14. L. Fadiga, L. Craighero, G. Buccino, and G. Rizzolatti, 2002, Speech listening specifically modulates the excitability of tongue muscles: a TMS study, European Journal of Neuroscience, 15:399–402.

15. G. Rizzolatti and L. Fadiga, 1998, Grasping objects and grasping action meanings: the dual role of monkey rostroventral premotor cortex (area F5), in Sensory Guidance of Movement, Novartis Foundation Symposium, G. R. Bock and J. A. Goode, Eds. Chichester: John Wiley and Sons, New York, NY, pp. 81–103.

16. D. Vernon, G. Metta, and G. Sandini, 2007, A Survey of Cognition and Cognitive Architectures: Implications for the Autonomous Development of Mental Capabilities in Computational Systems, IEEE Transactions on Evolutionary Computation, special issue on AMD, vol. 11.

17. C. von Hofsten, 2003, On the development of perception and action. In J. Valsiner and K. J. Connolly, (Eds.) Handbook of Developmental Psychology. Sage, London, pp. 114–140.

18. P. Fitzpatrick, G. Metta, and L. Natale, 2008, Towards Long-Lived Robot Genes, Journal of Robotics and Autonomous Systems, Special Issue on Humanoid Technologies, 56(1):29–45.

19. S. D. Huston, J. C. E. Johnson, and U. Syyid, 2003, The ACE Programmer's Guide, Addison-Wesley, Boston, MA.

20. V. Gallese, L. Fadiga, L. Fogassi, and G. Rizzolatti, 1996, Action recognition in the premotor cortex. Brain, 119:593–609.

21. G. Rizzolatti and R. Camarda, 1987, Neural circuits for spatial attention and unilateral neglect. In M. Jeannerod (Ed.), Neurophysiological and neuropsychological aspects of spatial neglect. North Holland, Amsterdam, pp. 289–313.

22. L. Craighero, M. Nascimben, and L.Fadiga, 2004, Eye Position Affects Orienting of Visuospatial Attention. Current Biology, 14: 331–333.

23. S. Degallier, L. Righetti, L. Natale, F. Nori, G. Metta, and A. Ijspeert, 2008, A modular bio-inspired architecture for movement generation for the infant-like robot iCub. In Proceedings of 2nd IEEE RAS/EMBS International Conference on Biomedical Robotics and Biomechatronics (BioRob), Scottsdale, AZ.

24. L. Montesano, M. Lopes, A. Bernardino, and J. Santos-Victor, 2008, Learning object affordances: From sensory motor maps to imitation, IEEE Transactions on Robotics, 24(1):15–26.

25. G. Metta and P. Fitzpatrick, 2003, Early integration of vision and manipulation. Adaptive Behavior, 11(2):109–128.

26. T. Ziemke, 2003, On the role of robot simulations in embodied cognitive science, AISB Journal, 1(4):389–399.

27. N. Nava,V, Tikhanoff, G. Metta, and G Sandini, 2008, Kinematic and Dynamic Simulations for The Design of RoboCub Upper-Body Structure ESDA.

28. A. Cangelosi, T. Belpaeme, G. Sandini, G. Metta, L. Fadiga, G. Sagerer, K. Rohlfing, B. Wrede, S. Nolfi, D. Parisi, C. Nehaniv, K. Dautenhahn, J. Saunders, K. Fischer, J. Tani, and D. Roy, 2008, The ITALK project: Integration and transfer of action and language knowledge. In: Proceedings of Third ACM/IEEE International Conference on Human Robot Interaction (HRI 2008), Amsterdam, 12–15 March.

29. P.F. Dominey, 2007, Sharing Intentional Plans for Imitation and Cooperation: Integrating Clues from Child Developments and Neurophysiology into Robotics, Proceedings of AISB 2007 Workshop on Imitation.

30. E. Bonabeau, M. Dorigo, and G. Theraulaz, 1999, Swarm intelligence: from natural to artificial systems. Oxford University Press, New York, NY.

31. S. Camazine, J-L. Deneubourg, N.R. Franks, J. Sneyd, G. Theraulaz, and E. Bonabeau, 2003, Self-Organization in Biological Systems. Princeton University Press, Princeton, NJ.

32. D.W. Stephens and J.R. Krebs, 1987, Foraging Theory. Princeton University Press, Princeton, NJ.

33. E. Sahin, 2004, Swarm Robotics: From sources of inspiration to domains of application. Springer-Verlag, Heidelberg.

34. S. Kernbach, R. Thenius, O. Kernbach, and T. Schmickl, 2009, Re-embodiment of honeybee aggregation behavior in artificial micro-robotic system. Adaptive Behavior, 17(3):237–259.

35. S. Kornienko, O. Kornienko, and P. Levi, 2005, IR-based communication and perception in microrobotic swarms. In Proceedings of IROS 2005, Edmonton, Canada.
36. G. Weiss, 1999, Multiagent systems. A modern approach to distributed artificial intelligence. MIT Press, Cambridge, MA.
37. J.Y. Halpern and Y. Moses, 1990, Knowledge and common knowledge in a distributed environment. Journal of ACM, 37(3):549–587.
38. H. Haken, 1983, Synergetics: An introduction, third edition. Springer-Verlag, New York, NY.
39. S. Kornienko, O. Kornienko, A. Nagarathinam, and P. Levi, 2007, From real robot swarm to evolutionary multi-robot organism. In Proceedings of CEC2007, Singapore.
40. D. Häbe, 2007, Bio-inspired approach towards collective decision making in robotic swarms. Master Thesis, University of Stuttgart, Germany.
41. T. Kancheva, 2007, Adaptive role dynamics in energy foraging behavior of a real microrobotic swarm. Master Thesis, University of Stuttgart, Germany.
42. A. Attarzadeh, 2006, Development of advanced power management for autonomous microrobots. Master Thesis, University of Stuttgart, Germany.
43. A. Ishiguro and T. Maegawa, 2006, Self-assembly through the interplay between control and mechanical systems. In Proceedings of IEEE/RSJ06 International Conference on Intelligent Robots and Systems. Beijing, China, pp. 631–638.
44. S. Murata, K. Kakomura, and H. Kurokawa, 2006, Docking experiments of a modular robot by visual feedback. In Proceedings of IEEE/RSJ06 International Conference on Intelligent Robots and Systems. Beijing, China, pp. 625–630.
45. W.-M. Shen, M. Krivokon, H. Chiu, J. Everist, M. Rubenstein, and J. Venkatesh, 2006, Multimode locomotion for reconfigurable robots. Autonomous Robots, 20(2):165–177.
46. H. Kurokawa, K. Tomita, A. Kamimura, S. Kokaji, T. Hasuo, and S. Murata, 2008, Distributed self-reconfiguration of m-tran iii modular robotic system. International Journal of Robotics Research, 27(3–4):373–386.
47. A. Golovinsky, M. Yim, Y. Zhang, C. Eldershaw, and D. Duff, 2004, Polybot and polykinetic/spl trade/system: a modular robotic platform for education. In IEEE ICRA, 1381–1386.
48. V. Zykov, E. Mytilinaios, B. Adams, and H. Lipson, 2005, Self-reproducing machines. Nature, 435(7039):163–164.
49. D.J. Christensen, E.H. Ostergaard, and H.H. Lund, 2004, Metamodule control for the atron self-reconfigurable robotic system. In Proceedings of IAS-8. Amsterdam, pp.685–692.
50. J. Koza, 1992, Genetic programming: on the programming of computers by means of natural selection. MIT Press, Cambridge, MA.
51. M. Srinivas and L.M. Patnaik, 1994, Genetic algorithms: A survey. Computer, 27(6):17–26.

Chapter 8
Assessing Coordination Demand in Cooperating Robots

Michael Lewis and Jijun Wang

Abstract Controlling multiple robots substantially increases the complexity of the operator's task because attention must constantly be shifted among robots in order to maintain situation awareness (SA) and exert control. In the simplest case an operator controls multiple independent robots interacting with each as needed. Control performance at such tasks can be characterized by the average demand of each robot on human attention. In this Chapter, we present several approaches to measuring, coordination demand, CD, the *added* difficulty posed by having to coordinate as well as operate multiple robots. Our initial experiment compares "equivalent" conditions with and without coordination. Two subsequent experiments attempt to manipulate and measure coordination demand directly using an extension of the Neglect Tolerance model.

8.1 Introduction

Borrowing concepts and notation from computational complexity, control of robots by issuing waypoints, could be considered $O(n)$ because demand increases linearly with the number, N, of robots to be serviced. Another form of control such as designating a search region by drawing a box on a GUI (Graphical User Interface), being independent of the number of robots, would be $O(1)$. From this perspective the most complex tasks faced in controlling teams are likely to be those that involve choosing and coordinating subgroups of robots. Simply choosing a subteam of m out of n robots to perform a task (the iterated role assignment problem), for example,

M. Lewis (✉)
School of Information Sciences, University of Pittsburgh, 135 N. Bellefield Ave., Pittsburgh, PA 15260, USA
e-mail: ml@sis.pitt.edu

R. Madhavan et al. (eds.), *Performance Evaluation and Benchmarking of Intelligent Systems*, DOI 10.1007/978-1-4419-0492-8_8,
© Springer Science+Business Media, LLC 2009

has been shown to be O(mn) [7]. The three experiments presented in this Chapter develop methods to assess the operator effort required to coordinate robots in tasks representative of expected application areas.

8.1.1 Coordination Demand

Despite the apparent analogy between command complexity and the workload imposed by a command task there is no guarantee that human operators will experience difficulty in the same way. The performance of human-robot teams is complex and multifaceted reflecting the capabilities of the robots, the operator(s), and the quality of their interactions. Recent efforts to define common metrics for human-robot interaction [15] have favored sets of metric classes to measure the effectiveness of the system's constituents and their interactions as well as the system's overall performance. In this Chapter, we present new measures of the demand coordination places on operators of multirobot systems and three experiments evaluating our approach and the usefulness of these measures.

Controlling multiple robots substantially increases the complexity of the operator's task because attention must constantly be shifted among robots in order to maintain situation awareness (SA) and exert control. In the simplest case an operator controls multiple independent robots interacting with each as needed. A search task in which each robot searches its own region would be of this category although minimal coordination might be required to avoid overlaps and prevent gaps in coverage. Control performance at such tasks can be characterized by the average demand of each robot on human attention [5]. Under these conditions increasing robot autonomy should allow robots to be neglected for longer periods of time making it possible for a single operator to control more robots.

For more strongly cooperative tasks and larger teams, individual autonomy alone is unlikely to suffice. The round-robin control strategy used for controlling individual robots would force an operator to plan and predict actions needed for multiple joint activities and be highly susceptible to errors in prediction, synchronization or execution. Estimating the cost of this coordination, however, proves a difficult problem. Established methods of estimating multirobot system, MRS, control difficulty, neglect tolerance, and fan-out [5] are predicated on the independence of robots and tasks. In neglect tolerance, the period following the end of human intervention but preceding a decline in performance below a threshold is considered time during which the operator is free to perform other tasks. If the operator services other robots over this period, the measure provides an estimate of the number of robots that might be controlled. Fan-out, when measured empirically, works from the opposite direction, adding robots and measuring performance until a plateau without further improvement is reached. Both approaches presume that operating an additional robot imposes an additive demand. These measures are particularly attractive because they are based on readily observable aspects of behavior: the time an operator is engaged controlling the robot, interaction time (IT), and the time an operator is not engaged in controlling the robot, neglect time (NT).

8.2 Coordination Demand

To separate coordination demand (CD) from the demands of interacting with independent robots we have extended Crandall's [5] neglect tolerance model by introducing the notion of occupied time (OT) as illustrated in Fig. 8.1.

Fig. 8.1 Extended neglect tolerance model

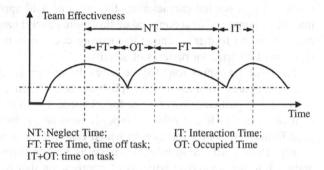

NT: Neglect Time;   IT: Interaction Time;
FT: Free Time, time off task; OT: Occupied Time
IT+OT: time on task

   The neglect tolerance model describes an operator's interaction with multiple robots as a sequence of control episodes in which an operator interacts with a robot for period IT raising its performance above some upper threshold after which the robot is neglected for the period NT until its performance deteriorates below a lower threshold when the operator must again interact with it. To accommodate dependent tasks we introduce occupied time, OT, to describe the time spent controlling other robots in order to synchronize their actions with those of the target robot. The episode depicted in Fig. 8.1 starts just after the first robot is serviced. The ensuing free time (FT) preceding the interaction with a second dependent robot, the OT for robot-1 (that would contribute to IT for robot-2), and the FT following interaction with robot-2 but preceding the next interaction with robot-1 together constitute the neglect time for robot-1. Coordination demand, CD, is then defined as:

$$CD = 1 - \frac{\sum FT}{NT} = \frac{\sum OT}{NT} \tag{8.1}$$

where, CD for a robot is the ratio between the time required to control cooperating robots and the time still available after controlling the target robot, i.e.; the portion of a robot's free time that must be devoted to controlling cooperating robots. Note that OT_n associated with $robot_n$ is less than or equal to NT_n because OT_n covers only that portion of NT_n needed for synchronization. A related measure total attention demand, TAD, includes IT for the target robot to measure the portion of time devoted to the entire task.

Most MRS research has investigated homogeneous robot teams where additional robots provide redundant (independent) capabilities. Differences in capabilities such as mobility or payload, however, may lead to more advantageous opportunities for cooperation among heterogeneous robots. These differences among robots in roles and other characteristics affecting IT, NT, and OT introduce additional complexity to assessing CD. Where tight cooperation is required as in box-pushing, task requirements dictate both the choice of robots and the interdependence of their actions. In the more general case requirements for cooperation can be relaxed allowing the operator to choose the subteams of robots to be operated in a cooperative manner as well as the next robot to be operated. This general case of heterogeneous robots cooperating as needed characterizes the types of field applications our research is intended to support. To accommodate this more general case, the Neglect Tolerance model must be further extended to measure coordination between different robot types and for particular patterns of activity.

The resulting expression [18] measures the way in which the team's capabilities or resources are combined to accomplish the task without reference to the operation or neglect of particular robots. So, for example, it would not distinguish between a situation in which one robot of type, X, was never operated while another was used frequently from a situation in which both robots of type, X, were used more evenly. The incorporation of action patterns further extends the generality of the approach to accommodate patterns of cooperation that occur in episodes such as dependencies between loading and transporting robots. When an empty transporter arrives, its brief IT would lead to extended OTs as the loaders do their work. When the transporter has been filled the dependency would be reversed.

8.2.1 Experimental Plan

One approach to investigating coordination demand has been to design experiments that allow comparisons between "equivalent" conditions with and without coordination demands. If performance is poorer in the condition with coordination demand we infer that CD has consumed cognitive resources used in the other condition to improve performance. The first experiment and one comparison within the third experiment follow this approach. The first experiment compares search performance between a team of autonomously coordinating robots, manually (waypoint) controlled robots, and mixed initiative teams with autonomously coordinated robots that followed operator entered waypoints but avoided overlapping searches. The impact of coordination demand was observable through the difference in performance between the manually controlled teams and the mixed initiative ones. The fully automated teams provided a control ensuring that the benefits in the mixed initiative condition were not due solely to the superior performance of the automation.

While experiment 1 examines coordination demand indirectly by comparing performance between conditions in which it was filled either manually or through automation, experiments 2 & 3 attempt to manipulate and measure coordination demand directly. In experiment 2 robots perform a box pushing task in which CD is varied by control mode and robot heterogeneity. By making the actions of each robot entirely dependent on the other, this choice of task eliminates the problem of distinguishing between interactions intended to control a target robot and those needed to coordinate with another. The third experiment attempts to manipulate coordination demand in a loosely coordinated task by varying the proximity needed to perform a joint task in two conditions and by automating coordination within subteams in the third. Because robots must cooperate in pairs and interaction for control needs to be distinguished from interaction for coordination for this task, CD is computed between robot types rather than directly between robots (Equation 8.1) as done in experiment 2.

All three experiments used paid participants from the University of Pittsburgh and lasted approximately one and a half hours. All used repeated measures designs and followed a standard sequence starting with collection of demographic data. Standard instructions for the experiment were presented followed by a 10 min training session during which the participant was allowed to practice using the MrCS. Participants then began their first trial followed by a second with a short break in between. Experiments 2 and 3 included a third trial with break.

8.3 Simulation Environment

The reported experiments were performed using the USARSim robotic simulation with 2–6 simulated robots performing Urban Search and Rescue (USAR), experiments 1 & 3, or box pushing (experiment 2) tasks. USARSim is a high-fidelity simulation of USAR robots and environments developed as a research tool for the study of Human Robot Interaction (HRI) and multi-robot coordination. Validation studies showing agreement for a variety of feature extraction techniques between USARSim images and camera video are reported in [3], showing close agreement in detection of walls and associated Hough transforms for a simulated Hokuyo laser range finder [2] and close agreement in behavior between USARSim models and the robots being modeled [4, 10, 12, 16, 20].

8.3.1 MrCS—The Multirobot Control System

A multirobot control system (MrCS) was developed to conduct these experiments. The system was designed to be scalable to allow control of different numbers of robots, reconfigurable to accommodate different human-robot interfaces, and reusable to facilitate testing different control algorithms. The user interface of MrCS is shown in Fig. 8.2. The interface is reconfigurable to allow the user to resize the components or change the layout. Shown in the figure is a configuration that used in the RoboCup 2006 competition in which a single operator controls six robots. On the upper and center portions of the left-hand side are the robot list and team

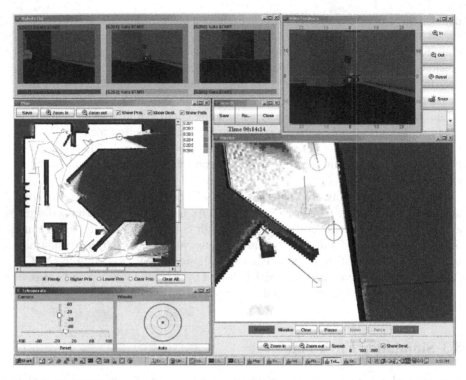

Fig. 8.2 MrCS Interface

map panels, which show the operator an overview of the team. The destination of each of robot is displayed on the map to help the user keep track of current plans. Using this display, the operator is also able to control regional priorities by drawing rectangles on the map. On the center and lower portions of the right-hand side are the camera view and mission control panels, which allow the operator to maintain situation awareness of an individual robot and to edit its exploration plan. On the mission panel, the map and all nearby robots and their destinations are represented to provide partial team awareness so that the operator can switch between contexts while moving control from one robot to another. The lower portion of the left-hand side is a teleoperation panel that allows the operator to teleoperate a robot.

8.4 Experiment 1

Fourteen participants were asked to control 3 P2DX robots simulated in USARsim to search for victims in a damaged building. Each robot was equipped with a pan-tilt camera with 45° Field of View (FOV) and a front laser scanner with 180° FOV and resolution of 1°. When a victim was identified, the participant marked its location on NIST Reference Test Arena, Yellow Arena [9]. Two similar testing arenas were built using the same elements with different layouts. In each arena, 14 victims were evenly distributed in the world. We added mirrors, blinds, curtains, semitransparent boards, and wire grid to add difficulty in situation perception. Bricks, pipes, a ramp, chairs, and other debris were put in the arena to challenge mobility and SA in robot control.

Presentation of mixed initiative and manual conditions were counterbalanced. Under mixed initiative, the robots analyzed their laser range data to find possible exploration paths. They cooperated with one another to choose execution paths that avoided duplicating efforts. While the robots autonomously explored the world, the operator was free to intervene with any individual robot by issuing new waypoints, teleoperating, or panning/tilting its camera. The robot returned back to auto mode once the operator's command was completed or stopped. While under manual control robots could not autonomously generate paths and there was no cooperation among robots. The operator controlled a robot by giving it a series of waypoints, directly teleoperating it, or panning/tilting its camera. As a control for the effects of autonomy on performance we conducted "full autonomy" testing as well. Because MrCS doesn't support victim recognition, based on our observation of the participants' victim identification behaviors, we defined detection to have occurred for victims that appeared on camera for at least 2 s and occupied at least 1/9 of the thumbnail view. Because of the high fidelity of the simulation, and the randomness of paths picked through the cooperation algorithms, robots explored different regions on every test. Additional variations in performance occurred due to mishaps such as a robot getting stuck in a corner or bumping into an obstacle causing its camera to point to the ceiling so no victims could be found. Sixteen trials were conducted in each area to collect data comparable to that obtained from human participants.

8.4.1 Results

All 14 participants found at least 5 of a possible 14 (36%) victims in each of the arenas [19]. These data indicate that participants exploring less than 90% of the area consistently discovered 5 to 8 victims while those covering greater than 90% discovered between half (7) and all (14) of the victims. Within participant comparisons found wider regions were explored in mixed-initiative mode, $t(13) = 3.50$, $p < 0.004$, as well as a marginal advantage for mixed-initiative mode, $t(13) = 1.85$, $p = 0.088$, in number of victims found. Comparing with "full autonomy", under mixed-initiative conditions two-tailed t-tests found no difference ($p = 0.58$) in the area explored.

No difference was found between area explored in autonomous or mixed initiative searches, however, autonomously coordinating robots explored significantly, $t(44) = 4.27$, $p < 0.001$, more regions than under the manual control condition (see Fig. 8.3). Participants found more victims under both mixed-initiative and manual control conditions than under full autonomy with $t(44) = 6.66$, $p < 0.001$, and $t(44) = 4.14$, $p < 0.001$, respectively (see Fig. 8.8). The median number of victims found under full autonomy was five. Comparing the mixed-initiative with the

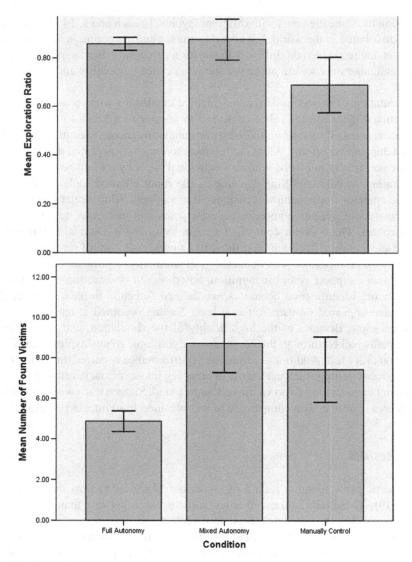

Fig. 8.3 Area explored and victims found for the three conditions

manual control, most participants (79%) rated team autonomy as providing either significant or minor help.

8.4.1.1 Human Interactions

Participants intervened to control the robots by switching focus to an individual robot and then issuing commands. Measuring the distribution of attention among robots as the standard deviation of the total time spent with each robot, a measure that captures unevenness in control, no difference (p = 0.232) was found between mixed initiative and manual control modes. However, we found that under mixed initiative, the same participant switched robots significantly more often than under manual mode (p = 0.027). Across participants, the frequency of shifting control among robots explained a significant proportion of the variance in number of victims found for both mixed initiative, $R^2 = .54$, F (1, 11) = 12.98, p = 0.004, and manual, $R^2 = 0.37$, F (1, 11) = 6.37, p < 0.03, modes.

In this experiment, cooperation was limited to deconfliction of plans so that robots did not re-explore the same regions or interfere with one another. The experiment found that even this limited degree of autonomous cooperation helped in the control of multiple robots. The results showed that cooperative autonomy among robots helped the operators explore more areas and find more victims. The fully autonomous control condition demonstrates that this improvement was not due solely to autonomous task performance as found in [14] but rather resulted from mixed initiative cooperation with the robotic team.

8.5 Experiment 2

Finding a metric for coordination demand (CD) is difficult because there is no widely accepted standard. In this experiment, we investigated CD (as defined in Section 8.2) by comparing performance across three conditions selected to differ substantially in their coordination demands. When an operator teleoperates the robots one by one to push the box forward, he must continuously interact with one of the robots because neglecting both would immediately stop the box.

Because the task allows no free time (FT) we expect CD to be 1. However, when the user is able to issue waypoints to both robots, the operator may have FT before she must coordinate these robots again because the robots can be instructed to move simultaneously. In this case CD should be less than 1. Intermediate levels of CD should be found in comparing control of homogeneous robots with heterogeneous robots. Higher CD should be found in the heterogeneous group since the unbalanced pushes from the robots would require more frequent coordination. In the present experiment, we compared computed CDs between these three conditions.

Figure 8.4 shows our experiment setting simulated in USARSim [10]. The controlled robots were either two Pioneer P2AT robots or one Pioneer P2AT and one less capable three wheeled Pioneer P2DX robot. Each robot was equipped with a GPS, a laser scanner, and a RFID reader. On the box, we mounted two RFID tags to enable the robots to sense the box's position and orientation. When a robot pushes the box, both the box and robot's orientation and speed will change. Furthermore, because of irregularities in initial conditions and accuracy of the physical simulation the robot and box are unlikely to move precisely as the operator expected. In addition, delays in receiving sensor data and executing commands were modeled presenting participants with a problem very similar to coordinating physical robots.

We introduced a simple matching task as a secondary task to allow us to estimate the FT available to the operator. Participants were asked to perform this secondary task as possible when they were not occupied controlling a robot. Every operator action and periodic timestamped samples of the box's moving speed were recorded for computing CD.

Fig. 8.4 Box-pushing robots

A within subject design was used to control for individual differences in operators' control skills and ability to use the interface. To avoid having abnormal control
behavior, such as a robot bypassing the box bias the CD comparison, we added
safeguards to the control system to stop the robot when it tilted the box.

8.5.1 Procedure

Fourteen participants performed three testing sessions in counterbalanced order. In
two of the sessions, the participants controlled two P2AT robots using teleoperation alone or a mixture of teleoperation and waypoint control. In the third session,
the participants were asked to control heterogeneous robots (one P2AT and one
P2DX) using a mixture of teleoperation and waypoint control. The participants were
allowed 8 min to push the box to the destination in each session.

8.5.2 Results

Figure 8.5 shows samples of time distributions of robot control commands recorded
in the experiment. As we expected no free time was recorded for robots in the teleoperation condition and the longest free times were found in controlling homogeneous robots with waypoints. The box speed shown on Fig. 8.5 is the moving speed
along the hallway that reflects the interaction effectiveness (IE) of the control mode.
The IE curves in this picture show the delay effect and the frequent bumping that
occurred in controlling heterogeneous robots revealing the poorest cooperation performance.

None of the 14 participants were able to perform the secondary task while teleoperating the robots. Hence, we uniformly find TAD = 1 and CD = 1 for both robots
under this condition. Within participants comparison found that under waypoint
control the team attention demand in heterogeneous robots is significantly higher

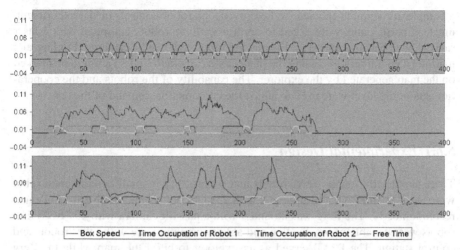

Fig. 8.5 Sample of time distribution curves for teleoperation (*upper*) and waypoint control (*middle*) for homogeneous robots, and waypoint control (*bottom*) for heterogeneous robots

than the demand in controlling homogeneous robots, t(13) = 2.213, p = 0.045 (Fig. 8.5). No significant differences were found between the homogeneous P2AT robots in terms of the individual cooperation demand (P = 0.2). Since the robots are identical, we compared the average CD of the left and right robots with the CDs measured under heterogeneous condition. Two-tailed t-test shows that when a participant controlled a P2AT robot, lower CD was required in homogeneous condition than in the heterogeneous condition, t(13) = −2.365. p = 0.034. The CD required in controlling the P2DX under heterogeneous demand (CD) condition is marginally higher than the CD required in controlling homogenous P2ATs, t(13) = −1.868, p = 0.084 (Fig. 8.5). Surprisingly, no significant difference was found in CDs between controlling P2AT and P2DX under heterogeneous condition (p = 0.79). This can be explained by the three observed robot control strategies: (1) the participant always issued new waypoints to both robots when adjusting the box's movement, therefore similar CDs were found between the robots; (2) the participant tried to give short paths to the faster robot (P2DX) to balance the different speeds of the two robots, thus we found higher CD in P2AT; (3) the participant gave the same length paths to both robots and the slower robot needed more interactions because it trended to lag behind the faster robot, so lower CD for the P2AT was found for the participant. Among the 14 participants, 5 of them (36%) showed higher CD for the P2DX contrary to our expectations.

8.6 Experiment 3

To test the usefulness of the extended CD measurement for a weakly cooperative MRS, we conducted an experiment assessing coordination demand using an Urban Search And Rescue (USAR) task requiring high human involvement and of a complexity suitable to exercise heterogeneous robot control. In the experiment, nineteen participants were asked to control *explorer* robots equipped with a laser range finder but no camera and *inspector* robots with only cameras. Finding and marking a victim required using the inspector's camera to find a victim to be marked on the map generated by the explorer. The capability of the robots and the cooperation autonomy level were used to adjust the coordination demand of the task.

8.6.1 Experimental Design

Three simulated Pioneer P2AT robots and 3 Zergs [1], a small experimental robot were used. Each P2AT was equipped with a front laser scanner with 180° FOV and resolution of 1°. The Zerg was mounted with a pan-tilt camera with 45° FOV. The robots were capable of localization and able to communicate with other robots and control station. The P2AT served as an explorer to build the map while the Zerg could be used as an inspector to find victims using its camera. To accomplish the task the participant must coordinate these two types robot to ensure that when an inspector robot finds a victim, it is within a region mapped by an explorer robot so the position can be marked.

Three conditions were designed to vary the coordination demand on the operator. Under condition 1, the explorer had 20 m detection range allowing inspector robots considerable latitude in their search. Under condition 2, scanner range was reduced to 5 m requiring closer proximity to keep the inspector within mapped areas. Under condition 3, explorer and inspector robots were paired as subteams in which the explorer robot with a sensor range of 5 m followed its inspector robot to map areas being searched. We hypothesized that CDs for explorer and inspector robots would be more evenly distributed under condition-2 (short range sensor) because explorers would need to move more frequently in response to inspectors' searches than in condition-1 in which CD should be more asymmetric with explorers exerting greater demand on inspectors. We also hypothesized that lower CD would lead to higher team performance. Three equivalent damaged buildings were constructed from the same elements using different layouts. Each environment was a maze like building with obstacles, such as chairs, desks, cabinets, and bricks with 10 evenly distributed victims. A fourth environment was constructed for training. Figure 8.6 shows the simulated robots and environment. A within subjects design with counterbalanced presentation was used to compare the cooperative performance across the three conditions.

8.6.2 Results

Overall performance was measured by the number of victims found, the explored areas, and the participants' self-assessments. To examine cooperative behavior in finer detail, CDs were computed from logged data for each type robot under the

Fig. 8.6 Scout and Explorer robots

three conditions. We compared the measured CDs between condition 1 (20 m sensing range) and condition 2 (5 m sensing range), as well as condition 2 and condition 3 (subteam). To further analyze the cooperation behaviors, we evaluated the total attention demand in robot control and control action pattern as well. Finally, we introduce control episodes showing how CDs can be used to identify and diagnose abnormal control behaviors.

8.6.2.1 Overall Performance

Examination of data showed two participants failed to perform the task satisfactory. One commented during debriefing that she thought she was supposed to mark inspector robots rather than victims. After removing these participants a paired t-test shows that in condition-1 (20 m range scanner) participants explored more regions, $t(16) = 3.097$, $p = 0.007$, as well as found more victims, $t(16) = 3.364$, p $= 0.004$, than under condition-2 (short range scanner). In condition-3 (automated subteam) participants found marginally more victims, $t(16) = 1.944$, $p = 0.07$, than in condition-2 (controlled cooperation) but no difference was found for the extent of regions explored. In the posttest survey, 12 of the 19 (63%) participants reported they were able to control the robots although they had problems in handling some interface components, 6 of the 19 (32%) participants thought they used the interface very well, and only one participant reported it being hard to handle all the components on the user interface but still maintained she was able to control the robots. Most participants (74%) thought it was easier to coordinate inspectors with explorers with long range scanner. 12 of the 19 (63%) participants rated auto-cooperation between inspector and explorer (the subteam condition) as improving their performance, and 5 (26%) participants though auto-cooperation made no difference. Only 2 (11%) participants judged team autonomy to make things worse.

8.6.2.2 Coordination Effort

During the experiment we logged all the control operations with timestamps. From the log file CDs were computed for each type robot. Figure 8.7 shows a typical (IT,FT) distribution under condition 1 (20 m sensing range) in the experiment with a calculated CD for the explorer of 0.185, a CD for the inspector of 0.06. The low CDs reflect that in trying to control 6 robots the participant ignored some robots while attending to others. The CD for explorers is roughly twice the CD for inspectors. After the participant controlled an explorer, he needed to control an inspector multiple times or multiple inspectors since the explorer has a long detection range and large FOV. In contrast, after controlling an inspector, the participant needed less effort to coordinate explorers.

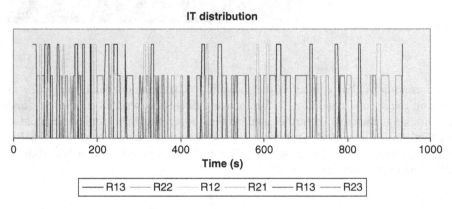

Fig. 8.7 Typical (IT,FT) distribution (higher line indicates the interactions of explorers)

Figure 8.8 shows the mean of measured CDs. We predicted that when the explorer has a longer detection range, operators would need to control the inspectors more frequently to cover the mapped area. Therefore a longer detection range should lead to higher CD for explorers. This was confirmed by a two tailed t-test that found higher coordination demand, t(18) = 2.476, p = 0.023, when participants controlled explorers with large (20 m) sensing range.

We did not find a corresponding difference, t(18) = 0.149, p = 0.884, between long and short detection range conditions for the CD for inspectors. This may have occurred because under these two conditions the inspectors have exactly the same capabilities and the difference in explorer detection range was not large enough to impact inspectors' CD for explorers. Under the subteam condition, the automatic cooperation within a subteam decreased or eliminated the coordination requirement when a participant controlled an inspector. Within participant comparisons shows that the measured CD of inspectors under this condition is significantly lower than the CD under condition 2 (independent control with 5 m detection range), t(18) = 6.957, p < 0.001. Because the explorer always tries to automatically follow an inspector, we do not report CD of explorers in this condition.

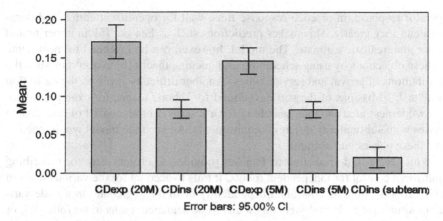

Fig. 8.8 CDs for each robot type

As auxiliary parameters, we evaluated the total attention demand, i.e. the occupation rate of total interaction time in the whole control period, and the action pattern, the ratio of control times between inspector and explorer, as well. Paired t-test shows that under long sensing conditions, participants interacted with robots more times than under short sensing which implies that more robot interactions occurred. The mean action patterns under long and short range scanner conditions are 2.31 and 1.9 respectively. This means that with 20 and 5 m scanning ranges, participants controlled inspectors 2.31 and 1.9 times respectively after an explorer interaction. Within participant comparisons shows that the ratio is significantly larger under long sensing condition than under short range scanner condition, $t(18) = 2.193$, $p = 0.042$.

8.6.2.3 Analyzing Performance

As an example of applying CDs to analyze coordination behavior, the performance over explorer CD and total attention demand under the 20 m sensing range condition reveals three abnormal cases A, B, and C low on both CD and TAD. Associating these cases with recorded map snapshots, we observed that in case A, one robot was entangled by a desk and stuck for a long time, for case B, two robots were controlled in the first 5 min and afterwards ignored, and in case C, the participant ignored two inspectors throughout the entire trial.

8.7 Conclusions

The Neglect Tolerance model for describing human-machine interaction as a series of periodic interventions needed to restore the system to proper functioning is conceptually very general and appears to apply across the board to a wide variety of human interactions with automated systems. While the model was presented in terms of regular cyclic interaction with fixed duration intervals and immediate operator response, in practice response must wait for operator attention and durations can vary greatly. This makes predictions such as Fan-out [5] an upper bound rather than realistic estimate. The model, however, can be extended to meet some of these objections by using scheduling and queuing theory [6] to accommodate the distributions of arrival and service times. Another difficulty involves the definition of a "task". Missions of the sort envisioned for robotic teams, for example, flying to a wilderness area and then searching for a lost hiker often consist of discernable phases with substantially different requirements. An accurate model would need to take these phases into account.

While NT based prediction of Fan-out provides an elegant way for describing multirobot control for independent robots, it fails to account for the various ways in which robots might be controlled dependently or in the aggregate. In a wide variety of situations such as designating areas to be searched, paths to be followed, or targets to be sought, aggregated commands are both more direct and simpler for the human commander than single robot alternatives. In addition most current work in coordinating robot teams whether it be biologically inspired control laws for swarms [11], optimizing path planners [7], or BDI agents [13, 14] involve commands to an aggregate.

Our work to develop measures of coordination demand are part of a wider effort to extend the Neglect Tolerance model beyond independently controlled robots to aggregate commands and varying degrees of dependent multirobot control. We envision control over robotic teams that involves commands of all these different types with some issued to the entire team, such as a search area, others involving a single robot, such as verifying a reported victim, and still others to subteams such as our explorer/inspector dyads. We have started by extending the Neglect Tolerance model to allow us to evaluate coordination demand in applications where an operator must coordinate multiple robots to perform dependent tasks. Results from the first experiment that required tight coordination conformed closely to our hypotheses with the teleoperation condition producing $CD = 1$ as predicted and heterogeneous teams exerting greater demand than homogenous ones. The CD measure proved useful in identifying abnormal control behavior as well [18]. As most target applications such as construction or search and rescue require weaker cooperation among heterogeneous platforms the second experiment extended NT methodology to such conditions. Results in this more complex domain were mixed, although the predicted effects on CD for sensor range and autonomous coordination were found.

These experiments have demonstrated the utility of measuring the process of human-robot interaction as well as outcomes to diagnosing operator performance and identifying aspects of the task, particularly for multiple robots, that might benefit from automation. While our approach to measuring CD was supported in both experiments the second experiment suggests the need for more sophisticated measures that can take into account strategies and patterns of actions as well as their durations.

Although our CD measures are objective and based solely on measurement of durations they cannot discriminate interactions performed in service of coordination from those with other purposes. Consequently the measurements are only useful under conditions where this attribution can be made a priori. Another difficulty with the current model is its assumptions that humans will control to a fixed threshold, or equivalently (from our measurement perspective) that operators will immediately recognize a drop below that threshold. There is ample evidence from scaling experiments [8,17] that operators relax their thresholds to accommodate additional robots while observation of completely neglected robots and substantial variation in neglect times suggests that operators do not routinely control to a fixed threshold. These problems might be partially addressed by developing a self-reflection capability for the robots. If robots could recognize their approach to the threshold or CD actions requiring intervention they could alarm, capturing the operator's attention and lessening the monitoring problem. This could help bring performance in line with the Neglect Tolerance model and perhaps improve it by using self-reflection to drive adjustable autonomy.

Acknowledgments This work was supported in part by the Air Force Office of Scientific Research under Grant FA9550-07-1-0039.

References

1. Balakirsky, S., Carpin, S., Kleiner, A., Lewis, M., Visser, A., Wang, J., and Zipara, V. (2007). Toward hetereogeneous robot teams for disaster mitigation: Results and performance metrics from RoboCup Rescue, Journal of Field Robotics, 24(11–12): 943–967.
2. Carpin, S., Wang, J., Lewis, M., Birk, A., and Jacoff, A. (2005). High fidelity tools for rescue robotics: Results and perspectives, Robocup 2005 Symposium.
3. Carpin, S., Stoyanov, T., Nevatia, Y., Lewis, M., and Wang, J. (2006a). Quantitative assessments of USARSim accuracy. Proceedings of PerMIS 2006.
4. Carpin, S., Lewis, M., Wang, J., Balakirsky, S., and Scrapper, C. (2006b). Bridging the gap between simulation and reality in urban search and rescue. Robocup 2006: Robot Soccer World Cup X, Lecture Notes in Artificial Intelligence, Springer, Berlin.
5. Crandall, J., Goodrich, M., Olsen, D., and Nielsen, C. (2005). Validating human-robot interaction schemes in multitasking environments. IEEE Transactions on Systems, Man, and Cybernetics, Part A, 35(4): 438–449.
6. Cummings, M., Nehme, C., and Crandall, J. (2007). Predicting Operator Capacity for Supervisory Control of Multiple UAVs. In J. S. Chahl, L. C. Jain, A. Mizutani, and M. Sato-Ilic (Eds.) Innovations in Intelligent Machines vol. 70, Studies in Computational Intelligence, Springer, Berlin.
7. Gerkey, B. and Mataric, M. (2004). A formal framework for the study of task allocation in multi-robot systems. International Journal of Robotics Research, 23(9): 939–954.
8. Humphrey, C., C. Henk, C., G. Sewell, G., B. Williams, B., and Adams, J. (2007). Assessing the Scalability of a Multiple Robot Interface. Proceedings of the 2nd ACM/IEEE International Conference on Human-Robotic Interaction, ACM, New York, NY.
9. Jacoff, A., Messina, E., and Evans, J. (2001, September). Experiences in deploying test arenas for autonomous mobile robots. In Proceedings of the 2001 Performance Metrics for Intelligent Systems (PerMIS) Workshop, Mexico City, Mexico.
10. Lewis, M., Hughes, S., Wang, J., Koes, M., and Carpin, S. (2005). Validating USARsim for use in HRI research. Proceedings of the 49th Annual Meeting of the Human Factors and Ergonomics Society, Orlando, FL, pp. 457–461.
11. Loizou, S.,Tanner, H., Kumar, V., and Kyriakopoulos, K. (2003). Closed Loop Navigation for Mobile Agents in Dynamic Environments. IEEE/RSJ International Conference on Intelligent Robots and Systems, Beijing, China.

12. Pepper, C., Balakirsky, S., and Scrapper, C. (2007). Robot Simulation Physics Validation, Proceedings of PerMIS'07.
13. Scerri, P., et al. (2004). Coordinating large groups of wide area search munitions. In D. Grundel, R. Murphey, and P. Pandalos (Ed.) Recent Developments in Cooperative Control and Optimization, World Scientific, Singapore, pp. 451–480.
14. Schurr, N., Marecki, J., Tambe, M., Scerri, P., Kasinadhuni, N., and Lewis, J. (2005). The future of disaster response: Humans working with multiagent teams using DEFACTO. In AAAI Spring Symposium on AI Technologies for Homeland Security.
15. Steinfeld, A., Fong, T., Kaber, D., Lewis, M., Scholtz, J., Schultz, A., and Goodrich, M. (2006, March) Common Metrics for Human-Robot Interaction, 2006 Human-Robot Interaction Conference, ACM, New York, NY.
16. Taylor, B., Balakirsky, S., Messina, E., and Quinn, R. (2007). Design and Validation of a Whegs Robot in USARSim, Proceedings of PerMIS'07, ACM, New York, NY.
17. Velagapudi, P., Scerri, P., Sycara, K., Wang, H., Lewis, M., and Wang, J. (2008). Scaling effects in multi-robot control. 2008 International Conference on Intelligent Robots and Systems (IROS08), Nice, France.
18. Wang, J. (2007). Human Control of Cooperating Robots, dissertation, University of Pittsburgh, http://etd.library.pitt.edu/ETD/available/etd-01072008-135804/unrestricted/Wang_EDC_2007-final2.pdf (accessed 7/22/2008).
19. Wang, J. and Lewis, M. (2007). Human control of cooperating robot teams. 2007 Human-Robot Interaction Conference, ACM, New York, NY.
20. Zaratti, M., Fratarcangeli, M., and Iocchi, L. (2006). A 3D Simulator of Multiple Legged Robots based on USARSim. Robocup 2006: Robot Soccer World Cup X, Springer, LNAI.

Chapter 9
Measurements to Support Performance Evaluation of Wireless Communications in Tunnels for Urban Search and Rescue Robots

Kate A. Remley, George Hough, Galen Koepke, and Dennis Camell

Abstract We describe general methods for evaluating the over-the-air performance in various radio propagation environments of wireless devices used for control and telemetry of urban search and rescue robots. These methods are based on identification and evaluation of performance metrics that can be used to assess impairments to the wireless link. The type and level of each impairment are derived from measurement data in a given environment, here a subterranean tunnel. We illustrate how parameters can be extracted from the measurement data to determine specific values for the performance metrics and discuss how these values can be used to develop standardized test methods for assessing, verifying, or predicting robot performance.

9.1 Performance Requirements for Urban Search and Rescue Robot Communications

Robots have been employed with great success in a wide variety of settings where precise, repetitive, or dangerous tasks need to be carried out. For example, they are commonly found on the production floor of heavy manufacturing facilities where they weld, assemble, and even deliver parts. A relatively new use of robots is in the urban search and rescue (US&R) environment. The majority of robots utilized for potentially dangerous tasks such as explosive ordinance disposal and search and rescue may be considered as extensions of one's eyes, ears, nose, and hands. In this manner, robots have the potential to provide enormous utility for responders that perform vital search and rescue missions at sites of disasters. Robotic sensing devices can access dangerous areas more efficiently in many instances, and can

K.A. Remley (✉)
NIST Electromagnetics Division, 325 Broadway, Boulder, CO, USA
e-mail: kate.remley@nist.gov

Work of the United Stated government, not protected by US copyright.

R. Madhavan et al. (eds.), *Performance Evaluation and Benchmarking of Intelligent Systems*, DOI 10.1007/978-1-4419-0492-8_9,
© Springer Science+Business Media, LLC 2009

provide information on trapped or missing people while minimizing the danger to which responders expose themselves at such events.

Wireless telemetry and control of US&R robots is desirable in many situations where, for example, a tether may become tangled, broken, or damaged in debris or other objects in the environment. Evaluation of the performance of the wireless telecommunication devices used for US&R robots typically follows the same fundamental procedures that are used to evaluate wireless devices for civilian applications. Performance evaluation of any wireless device is complicated by the fact that every environment presents a different set of parameters that may impact the wireless device differently. The geometry of features in the environment, the material from which they are made, other radio traffic, and even the movement of the radio within the environment are all factors that may impact the behavior of wireless devices. Evaluating a wireless device in one environment may not adequately represent its performance in another environment. This is not an insurmountable problem. The wireless industry has identified a number of characteristics common to many environments, and from these, they have developed models of representative environments. Devices are designed to perform to a specified level of service within a given type of environment. Wireless device operation is then verified in a test environment whose physical characteristics mimic the representative environment before the device is released for sale. Such "over-the-air" performance evaluation is the focus of this work.

The same general procedure used for civilian devices can be used to predict the performance of wireless communication devices used for US&R robotic applications. For both commercial and US&R applications, the ability to predict performance in real time can also enable real-time modification of system parameters to overcome signal impairments. For example, many wireless devices reduce their transmitted data rate to compensate for a harsh propagation environment. Some robotic systems also may automatically deploy repeaters when signal strength becomes weak. Both of these are examples of intelligent systems.

Because there are some key differences between commercial use of wireless devices and the use of search and rescue robots, existing specifications for wireless device performance are not entirely sufficient for US&R robot applications. Some specifications must either be modified or newly developed. One key difference between the two applications is the need for a high level of reliability in US&R applications. For commercial applications, such as wireless local area networking (WLAN) or cellular telephone communications, if the application is interrupted the user may be inconvenienced, but the session can be reconnected with little more than time lost. Obviously, for US&R use, lives may depend on the reliability of the wireless device, so a higher standard for reliability of service must be applied.

A second difference between many types of commercial wireless applications and US&R robotic applications is that the latter typically involve "point-to-point" radio communications, where the robot and controller interact directly with each other without the use of a base station or other hub that rebroadcasts the signal. Many existing applications, both for commercial wireless and for trunked public safety radio, use the "cellular" model, where a base station serves a number of nodes.

There are some fundamental differences between cellular and point-to-point communication channels. First, for point-to-point communications used in an emergency response event, the transmitter and receiver are often physically closer together than for cellular systems. Second, a cellular base station usually has a much higher elevation than for an emergency response point-to-point scenario. For the robotics application, the robot itself and often the controller, are both located relatively near to the ground, often at a height of one meter or less.

A third difference between commercial wireless applications and US&R applications is that many responder applications involve multiple challenging environmental impairments. For example, cellular telephones are typically designed to operate in outdoor environments, where long delay spreads may result in multipath that can cause intersymbol interference. Emergency response communications often must overcome outdoor multipath as well, but then the responder may enter a large structure, causing significant attenuation in addition to multipath. As another example, many wireless local area networks were designed to operate in home or office environments, where multipath may be overcome by deploying nodes in close proximity to receivers. US&R robots, as well as most emergency response equipment, need to operate reliably within large building structures, in highly reflective industrial environments, and within subterranean tunnels, to name but a few examples. These environments can be much more challenging in terms of reduced received signal level and the amount of both self-interference (multipath) and interference from external sources.

Also, fewer channel models exist for many public-safety environments compared to the well standardized commercial sector. In particular, there is a lack of open-literature data on radio-signal characteristics in responder environments. In the following, we develop methodologies for acquiring and evaluating such data, a key step in development of both performance evaluation metrics and standards for US&R wireless communications.

One final difference between civilian wireless devices and those used by the emergency-response community is the size of the market. Vast resources have been spent on the multi-billion dollar cellular and WLAN communication industries. The public-safety market share is a small fraction of that. The use of wireless technology, other than for voice communications, is relatively new in the emergency-response sector. As a result, few standards exist for specifying the performance of wireless devices for responder applications in responder environments. To mitigate this, the Department of Homeland Security (DHS) Office of Standards is providing resources for development of technically sound performance metrics and standards that cover the use of wireless communications for US&R robots, as well as for other wireless devices used by the response community.

In 2004, the DHS Science and Technology (S&T) Directorate initiated an effort to support the National Institute of Standards and Technology (NIST) in the development of comprehensive standards for testing and certification of effective robotic technologies for US&R applications [1–3]. By assisting in the process of creating such standards, DHS seeks to provide guidance to local, state, and federal homeland security organizations regarding the purchase, deployment, and use of robotic systems for US&R applications. The NIST/US&R Responder informal advisory board was created, and was able to define over 100 initial performance requirements, and generate 13 deployment categories. The performance requirements were grouped into categories such as human-system interaction, mobility, logistics, sensing, power, and communications. For each requirement, the responders defined how they would measure performance [2].

NIST has since organized a standards effort through ASTM[1] International Standard E54.08 on Homeland Security Standards. In this effort, industry representatives and US&R responders have endeavored to slice the problem into manageable categories. The head of each working group is responsible for producing one or more standard test method that objectively measures a robot's performance in a particular area. Ultimately, the response organization will be able to determine which robots best suit its requirements. Robot researchers and manufacturers will benefit from the definition of test methods and operational criteria, enabling them to provide innovative solutions to meet the universal requirements.

In the area of wireless communications, the performance requirements specified by the responders included:

1. *Expandable Bandwidth:* Will support additional operational components, without loss of data transmission rate, sufficient to allow each component to perform its function.
2. *Range—Beyond Line of Sight:* Must be able to ingress specified number of meters in worst-case collapse. Worst case is a reinforced steel structure.
3. *Security:* System must be shielded from jamming interference and encrypted (to prevent civilians, reporters, and terrorists from listening in).
4. *Range—Line of Sight*
5. *Data Logging—Status and Notes:* Ability to pick up and leave notes.

Items (2) and (4) were designated as critical in the initial standards development effort. Predicting the range of a given robot depends on the technical specifications of the robot's radio link, as well as the radio environment in which the robot is deployed. The technical specifications of the robot's radio are determined by factors such as FCC regulations on output power, frequency of operation, and occupied bandwidth. Additionally, most robot manufacturers rely on the use of existing transmission formats to take advantage of the significant amount of work done on efficient and standardized wireless data transfer by the commercial sector. Consequently, there is little leeway in changing the technical specifications for the radios used in these robots. However, to study the effect of the environment on the range of US&R robots, data needs to be acquired on the radio environments in which robots are likely to be deployed. The Electromagnetics Division at NIST has conducted a multi-year study to acquire open-literature data in several representative environments [4, 5]. After discussing the steps used in evaluating the wireless link

---

[1] Formerly the American Society for Testing and Materials

for US&R robots, this chapter will focus on the methods used for acquiring data to evaluate the expected received-signal characteristics for a given environment. We then discuss how the data can be interpreted to develop models and predict performance of US&R robots in a representative responder environment.

9.2 Performance Evaluation Procedures

We describe a commonly used procedure for evaluating the performance of wireless devices and highlight areas where the US&R robot performance evaluation may differ from commercial device evaluation. Procedures are described to identify and extract the key characteristics, or "signal impairments," that will affect the performance of a wireless device in a given radio propagation environment. Knowledge of these impairments can then be used to classify representative environments for the development of models that can help to predict device performance (such as propagation models or data throughput models), and/or to develop test methods that place the wireless device in a sufficient number of operating states that it can be expected to operate satisfactorily in the field for a given environment. We can summarize this procedure in a few steps:

1. Develop an understanding of how signal impairments impact the performance of a specific wireless device or class of wireless devices.
2. Develop performance metrics that can be used to quantify this impact on performance.
3. Conduct measurements and/or simulations to determine the type and level of signal impairments expected in a given propagation environment.
4. Develop models of, for example, the environment and/or of the system performance, or gather sufficient measurement data in order to predict device performance in the presence of representative impairments.
5. Evaluate device performance when subjected to representative impairments by determining whether the signal impairments cause the device to exceed specified values of performance metrics. This can be done either through measurement verification or, at least for preliminary verification, by looking at the output of the model.

We go through the above procedure step by step. The evaluator first determines what impairments to the received signal affect the performance of the wireless device. Some examples are: a low received signal level, the amount and duration of self or external radio interference, excessive multipath, or the movement of the transmitter relative to the receiver.

Performance metrics are identified that summarize the degradation of device performance when the transmitted signal experiences impairments such as those described above. For example, bit error rate (BER), frame error rate (FER), and block error rate (BLER) are common wireless device performance metrics that indicate a receiver's inability to accurately decode an impaired signal. For US&R robot applications, the control channel can easily be evaluated by a go/no-go performance metric. For the video and telemetry links, performance metrics are currently being developed [6, 7].

Based on a representative environment similar to that where the device will be deployed, the evaluator next ascertains what environmental factors create signal impairments, either through measurements or by modeling the environment. Measuring and extracting the type and level of signal impairments in a tunnel environment will be the focus of this chapter. While every tunnel environment will be different, it is possible to identify physical characteristics of tunnels that affect the electrical performance of radio signals. With knowledge of these characteristics, the evaluator can develop or use a model to predict the performance of a robot in other types of tunnel environments.

To benchmark the performance of the wireless device, the evaluator often will set limits on acceptable values for various performance metrics. For example, a performance benchmark for a US&R robot application may be "if the control channel is expected to work 95% of the time in a tunnel environment having x, y, and z characteristics, the robot is deemed appropriate for use in this type of tunnel environment." Predictive benchmarking can often be carried out using appropriate models of a given environment. Often, a final measurement verification step is carried out to ensure that the predicted performance approximates the true performance in a representative environment before the models and/or predictions are deemed satisfactory.

An example of this procedure can be illustrated well by the current state-of-the-art in cell phone performance verification. Three key signal impairments that affect the performance of cellular telephones are (a) reduction in received signal level, (b) the existence of multipath (reflected) signals, and (c) the period required for the multipath to decay below a certain amplitude level. For cell phone systems, engineers have developed channel correction algorithms (also called channel equalization) to minimize the effects of (b) as long as the reflections decay within a certain period (c) and the signal attenuation (a) is not beyond the error-correcting capability of the code used.

Cell phone standards bodies have developed propagation channel classifications to describe common environments in terms of the signal impairments (a)–(c). For example, the PB3 model [8] specifies the signal level and amount of multipath likely to be experienced by a pedestrian in an urban environment walking at an average speed of 3 km/h. Extensive data collection has taken place to determine the values of the various signal impairments expected in each of these environments. While not every pedestrian will experience the conditions specified by the PB3 model, standards organizations have determined that this model provides a sufficiently representative description of the signal impairments in these types of environments.

Cell phone system engineers design new cell phones to withstand the signal impairments specified by the PB3 model over a certain percentage of time. In the verification stage, each new model of cell phone that is produced is tested in a facility that simulates the impairments specified by the model. To be accepted for use, the model must meet or exceed the value of the performance metric specified for each relevant application. For example, for a wideband code-division-multiple-access (W-CDMA) signal, 1.2% BER is specified by the Cellular Telecommunications Industry Association's "over the air" standard.

There are many types of US&R robots being tested for the ASTM standard, including aerial, ground, and aquatic robots. For now the tests target mainly ground robots. Because ground robots move slowly, effects of distortion due to movement (Doppler spread, narrowband fading, and/or wideband fading) are not expected to be an issue. As a result, the main signal impairments expected to degrade the wireless links used by US&R ground robots are also those listed in (a)–(c) above. For applications where the robot goes into or behind a structure, the reduction in received signal strength (a) can be significant. As a second example of the performance evaluation technique described above, we discuss the proposed standardized test method for non-line-of-sight wireless communications. We discuss this test method to illustrate a typical simplified test that captures key performance metrics while providing both portability and repeatability.

The non-line-of-sight test method is intended to simulate the condition where a robot moves behind a building and only a few diffracted signal paths exist between the robot and the operator. The metric that we use for testing the control of the robot is whether the operator can maneuver the robot in a figure-eight pattern. Both measurements and models have been used to develop this test, with the goal of providing a reduction in the signal level comparable to what may be experienced in the field.

In the non-line-of-sight test method, the robot moves 500 m down range from the operator in an open area, such as an airstrip, where reflections are minimal. The robot then turns 90°, moving behind a large obstruction such as a row of large metallic shipping containers stacked two or three high. Once the robot is in a non-line-of-sight condition, the received signal is weak and propagation is ideally limited to a few diffracted paths. The use of a metallic structure enhances repeatability by minimizing the effects of various construction materials, and the use of an open environment enhances repeatability by minimizing multipath reflections. Propagation models of this test environment have helped to optimize the test method in terms of size, shape, and location of the shipping containers for various frequencies of operation and even modulation types. The work presented in the following sections will help with the development of a similarly representative test method for the tunnel environment.

For the tunnel environment discussed below, we will see that once the robot turns a corner in the tunnel, the received signal level drops significantly. The reduction in signal strength is not equal across frequency, however. Received signals at both the lowest and highest frequencies of the test band experience more signal-level reduction than signals at frequencies in the middle of the band.

As we will show, the level (b) and duration (c) of multipath in the tunnel changes, depending on whether or not a line-of-sight path exists. When a line-of-sight exists, the operator may receive the direct-path signal plus one or more strong reflections. The reflected signals can cause deep nulls or sharp increases in the received signal depending on whether the reflections add constructively or destructively with the main signal. Again, we will see that this effect is frequency dependent. Once the robot has rounded a corner from the operator, all of the received signals arrive via reflected paths. A direct signal plus strong reflections is replaced by multiple weaker reflections and the received signal takes on a much noisier appearance.

Well-developed channel propagation models exist for tunnel environments. Thus, for many wireless applications, including the US&R robot application, it is possible to extract the level of signal impairment in one tunnel and derive the expected level of impairment for another tunnel. The key to using this predictive method of performance evaluation effectively is extraction of signal impairments that are expected for the application at hand. In our case, data must be gathered under conditions that represent the US&R robot operating conditions mentioned in the introduction, including point-to-point communications and low-to-the-ground operation. We will demonstrate a method for acquisition and extraction of data and its use in models that predict the performance of one type of US&R robot. We can verify the model by measuring the performance of the robot under the same conditions as those where the model parameters were extracted. We then describe how this model can be used to predict the performance of robots in a tunnel more representative of the type encountered in US&R operations. Throughout, the goal of the work is to present a framework for performance evaluation, rather than to conduct a comprehensive evaluation of a specific robot in a specific environment.

9.3 Measurement of Signal Impairments in a Tunnel Environment

Researchers from the Electromagnetics Division of the National Institute of Standards and Technology and the Fire Department of New York (FDNY) conducted field tests to quantify the expected type and level of signal impairments in a representative tunnel environment. Tests were conducted at the Black Diamond Mines Regional Park near Antioch, California on March 19–21, 2007. The goal of these field tests was to investigate propagation channel characteristics that affect the reliability of wireless telemetry and control of US&R robots in tunnels and other weak-signal environments. In this section, we describe measurement methods used to study parameters relevant to robot performance.

We used both time- and frequency-domain techniques to study the signal attenuation (loss) and reflections (multipath) that may impair successful wireless communications in tunnels. We also directly evaluated the performance of both the video and control links for a robot inside one of the mine tunnels. In this section, we summarize the data we collected and interpret the key findings from the study, which is described in its entirety in [5]. In the next section, we use the measured data to model both path loss and channel capacity in tunnels. The models are verified by comparison to our tests of the robot in the tunnel. Models such as these can be used to predict robot performance in tunnels having characteristics different from the ones we measured, such as subways and utility tunnels, as shown in Section 9.5.

Recently, the wireless field has seen a renewed interest in studies of signal propagation in both mine and subway tunnels, following a good deal of study on mine communications in the 1970s. A seminal work on mine tunnel propagation by Emslie et al. [9], studied radio wave propagation in small underground coal tunnels (4.3 m wide × 2.1 m high) for frequencies ranging from 200 MHz to 4 GHz. Emslie's model for propagation in tunnels is still used today. Recently, Rak and Pechak [10] applied Emslie's work to small cave galleries for speleological applications, confirming Emslie's findings that once a few wavelengths separate the transmitter and receiver, the tunnel acts as a waveguide that strongly attenuates signals below the waveguide's cutoff frequency. Because the walls of the tunnel are not perfectly conducting, signals operating above the cut-off frequency also experience significant loss. In a recent paper, Dudley et al. [11] performed a detailed assessment of operating frequency in a variety of tunnels. They found that as frequency increases, the lossy waveguide effect decreases.

Other work on propagation measurement and modeling in mine tunnels was reported in [12], whose group conducted narrowband and wideband measurements centered at 2.4 GHz. A model that describes tunnel propagation as a cascade of impedances was reported in [13, 14]. Studies of radio wave propagation in subway tunnels at 945 MHz and 1.853.4 GHz were presented in [15]. A ray tracing model was implemented to study the effects of the tunnel geometry and materials on propagation.

Our measurements, covering a much wider frequency range than [11–15], and implementation of the model of [10] confirm the lossy waveguide effect in the tunnels we studied. This effect can have a significant impact on the choice of frequency for critical applications such as US&R operations, where typically infrastructure such as a repeater network is not available and lives may be at stake.

Another factor in tunnel communications is multipath caused by reflections off the walls, floor, and ceiling of the tunnel. This was clearly seen in the work of Dudley, et al. [11] and was studied carefully over a 200 MHz bandwidth in [12]. Multipath can have a pronounced effect on successful transmission of wideband data. Some types of multipath interference may affect certain frequencies in a wideband signal while simultaneously having little impact on other frequencies. This frequency selectivity can make decoding signals difficult for the demodulator in a receiver.

We studied the frequency selectivity of the multipath in the tunnel environment by measuring the received signal power across a wide frequency range. We studied the decay time of the multipath by measuring the root-mean-square (RMS) delay spread, a common figure of merit that describes the period needed for reflected signals to decay below a threshold value. We compare our measured results to a model of channel capacity based on a modified form of Shannon's theory of channel capacity [16]. This theorem provides a basis for predicting the success of wireless communication in multipath environments.

We will first describe the tunnel environment where we made the measurements. We next discuss the types of measurements we made in sufficient detail so that other organizations could reproduce them in other environments. We give selected results of the measurements and a brief interpretation of the results. Finally, we report on tests made of the control and video channels of a robot in this tunnel environment.

9.3.1 The Test Environment

The Black Diamond Mines are part of an old silica mine complex that was used early in the 1900s to extract pure silica sand for glass production. As such, the walls of the mine shafts are rough and consist of sandy material. Two tunnels were studied, the Hazel-Atlas North (here called the "Hazel-Atlas" tunnel) and Hazel-Atlas South (here called the "Greathouse" tunnel). The tunnels are located beneath a mountain and are joined together several hundred meters inside, as shown in Fig. 9.1. Several chambers and tunnels intersect with the main tunnels, some of which reach the surface to provide air shafts. These airshafts can create alternative paths for RF signals traveling within the mine complex.

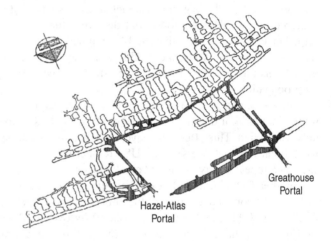

Fig. 9.1 Overview of the Black Diamond mine tunnel complex The *dark-shaded* areas are accessible. The distance between the two portals is around 400 m

Greathouse
Portal

Hazel-Atlas
Portal

The cross-sectional dimensions of the Hazel-Atlas tunnel varied from approximately 1.9 m (6′, 3″) × 1.9 m to as much as 2.6 m (8′, 5″) × 2.4 m (8′, 0″). The dimensions of the Greathouse tunnel were somewhat bigger, up to approximately 3 m square in places. The Hazel-Atlas tunnel contained railroad tracks spaced 61 cm (24") apart. Both tunnels consisted of a straight section followed by a 90° turn around a corner, as shown in Fig. 9.1.

Figure 9.2 shows photographs of the Hazel-Atlas tunnel. Figure 9.2(a) shows the portal (entrance) of the Hazel-Atlas mine and Fig. 9.2(b) shows a location deep inside the tunnel The photographs show the rough, uneven walls in the tunnels, some with wooden shoring, and railroad tracks. Figure 9.3 shows photographs of the Greathouse tunnel. Figure 9.3(a) shows the receive antenna located at the junction of the large chamber where it meets the Greathouse tunnel just inside and to the left of the portal. This was our reference location. Figure 9.3(b) shows a wooden walkway deep inside the Greathouse tunnel.

Fig. 9.2 (**a**) Portal into the Hazel-Atlas mine tunnel. (**b**) Wood shoring approximately 150 m into the tunnel. The robot tested can be seen on the cart between the railroad tracks

Fig. 9.3 (**a**) Receive antenna site in the Greathouse mine tunnel. (**b**) Walkway at the top of the stairs in the Greathouse tunnel

9.3.2 Measurements

9.3.2.1 Narrowband Received Power

We measured the received power from a transmitter placed at various locations inside the tunnels. We collected single-frequency (unmodulated) received-power data at frequencies near public-safety bands (approximately 50, 160, and 450 MHz). Gathering information at these frequencies helps to provide a choice of optimal frequency for the US&R community for this environment, both for robot communications and for other types of radio communication. These data provide insight into the lossy waveguide effect mentioned in the Introduction.

The handheld transmitters used were radios similar to those of first responders, except they were placed in ruggedized cases and were modified to transmit continuously. Each radio transmitted a signal of approximately 1 watt through an omnidirectional "rubber duck" antenna mounted on the case. During the tests, the radio antennas were approximately 0.75 m from the floor, a height similar to that of the robot we studied.

We carried the radio transmitters from the receive antenna location to locations deep within the tunnels while continuously recording the received signal. From the Hazel-Atlas tunnel portal, we carried the transmitter approximately 100 m down a straight tunnel, then turned a corner and proceeded another 100 m, as shown in Fig. 9.4(a). For the Greathouse tunnel, we went deeper into the mountain, as shown in Fig. 9.4(b). We carried the transmitter approximately 100 m down the tunnel, turned left and took an approximately 60 m hairpin path in order to continue

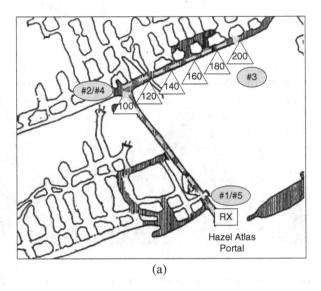

(a)

Fig. 9.4 (a) Close-up view of the Hazel-Atlas tunnel. (b) Close-up view of the Greathouse tunnel. The *dashed lines* show the paths along which we took measurements, including the 90° turns at 100 m in both tunnels. The *triangles* indicate the distance in meters, the ovals correspond to locations shown in Figs. 9.5 and 9.6, and the receiving equipment is labeled RX

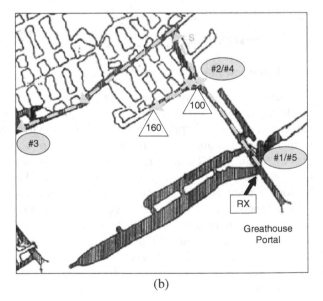

(b)

further into the tunnel. After the hairpin, we climbed several stairs (marked "S" in Fig. 9.4(b)), turned left and continued approximately 120 m almost to the junction with the Hazel-Atlas tunnel. We then returned by the same route.

The receiving equipment was located just outside the portal for the Hazel-Atlas measurements, and at the junction of the Greathouse chamber and the main tunnel for the Greathouse measurements. Omnidirectional discone receiving antennas were mounted on tripods, as shown in Fig. 9.2(a). We used a narrowband communications receiver to convert the received signal to audio frequencies, where it was digitized by a computer sound card and recorded on a computer. This instrument, when combined with NIST-developed post-processing techniques [5, 17], provides a high-dynamic-range measurement system that is affordable for most public-safety organizations. Part of the intent of this project was to demonstrate a user-friendly system that could be utilized by US&R organizations to assess their own unique propagation environments. A rough estimate for the uncertainty in this measurement, based on repeat measurements, is on the order of 1 dB [17]. The variability in received power due to antenna placement within the environment is on the order of 10 dB, much higher than the expected uncertainty. As a result, we do not report measurement uncertainties on our graphs.

Figures 9.5 and 9.6 show measured received-power data at frequencies of 50, 162, and 448 MHz acquired while the transmitters were carried by foot through the Hazel-Atlas and Greathouse tunnels, respectively. The signals were sampled at approximately 48 kHz and the power averaged over 1-second intervals. The left and right halves of the graph show measurements made walking into and out of the tunnels, respectively, and thus mirror each other. The vertical dashed lines on the graph correspond to the entrance (#1, #5), turn (#2, #4), and turn-around point (#3), as shown in Fig. 9.4(a) and (b).

In Fig. 9.5, the small increases in received power shown as bumps between points 2–3 and 3–4 in the Hazel-Atlas tunnel illustrate an alternative signal path through one of the air vents located in the small chambers off the main tunnel. The size of the air vent relative to the wavelength determines how significant this additional path is. The small increases in received signal power in the Greathouse tunnel, shown between points 2–3 and 3–4 in Fig. 9.6, are caused by an additional signal path encountered at the junction of the main tunnel and the stairwell 100 m into the tunnel (denoted by an "S" in Fig. 9.4(b)). This additional signal path was encountered after the hairpin, as the transmitter returned to the junction labeled #2/#4 in Fig. 9.4(b). The horizontal lines in the graphs indicate that the received signal levels are below the noise floor of the receiver; that is, less than approximately −130 dBm.

We see from both Figs. 9.5 and 9.6 that the lower frequencies drop off more rapidly as the transmitter moves deeper in the tunnel within the first 100 m of the test (between points 1–2 and 4–5). This rapid attenuation is due to the lossy waveguide effect described in references [5, 9–11]. The signals for the 448 MHz carrier frequency (Figs. 9.5(c) and 9.6(c)) exhibit less attenuation, and this is where the models of [9] may apply. Signals may travel even further at higher frequencies, as discussed in [9–11]. This frequency dependence may play a significant role in

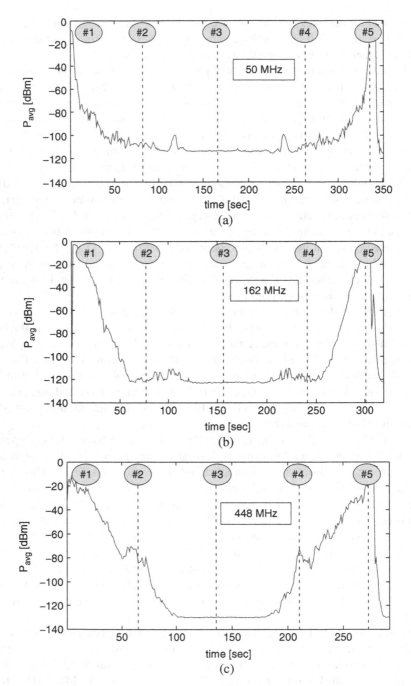

Fig. 9.5 Received-power data in the Hazel-Atlas Mine for three carrier frequencies: (**a**) 50 MHz, (**b**) 162 MHz, (**c**) 448 MHz. In each case the #2 and #4 vertical *dashed lines* correspond to the turn at 100 m: once on the way into the tunnel and once on the way out. The #3 *dashed line* represents the end point. For the Hazel-Atlas mine tunnel, the end was at 200 m, shown in Fig. 9.4(a)

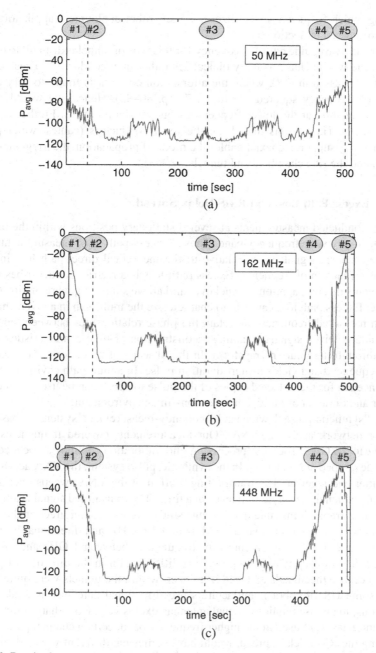

Fig. 9.6 Received-power data in the Greathouse Mine tunnel for three carrier frequencies: (a) 50 MHz, (b) 162 MHz, (c) 448 MHz. In each case the #2 and #4 vertical *dashed lines* correspond to the turn at 100 m: once on the way into the tunnel and once on the way out. The #3 *dashed line* represents the end point. For the Greathouse mine tunnel, the end point was approximately 350 m into the tunnel, as shown in Fig. 9.4(b)

deciding which frequencies to utilize in US&R robot deployment applications, as will be discussed in Section 9.4.

The exact waveguide cut-off frequency for this type of tunnel is difficult to define, because the walls behave as lossy dielectrics rather than conductors. These conditions are discussed in [18], where the attenuation constant is found to vary as the inverse of frequency squared (Section 2.7, pp. 80–83). Hence, we would expect higher attenuation at the lower frequencies but no sharp cut-off. Further complications in the Hazel-Atlas tunnel are the axial conductors (cables, water pipes, rails) that may support a coaxial-cable-like mode of propagation, the irregular cross-section, and the side chambers and tunnels.

9.3.2.2 Excess Path Loss and RMS Delay Spread

We also conducted measurements at several stationary positions within the tunnels covering a very wide frequency band. These "excess-path-loss" measurements provide the received signal power relative to the theoretical direct-path loss in free-space as a function of frequency. Excess path loss is a metric that describes signal impairments in a propagation channel over and above simple signal reduction due to distance. Excess path loss can help to characterize the multipath in a given channel: At each measured frequency, we retain the phase relationships between the transmitted and received signals, enabling reconstruction of time characteristics of the signal through the Fourier Transform. In the absence of reflections, the measured wide frequency band yields a short-duration pulse. In a multipath environment, the period needed for the reflected copies of the pulse to decay can be used to study the number and duration of multipath reflections in an environment.

Our "synthetic-pulse," wideband-frequency-measurement system is based on a vector network analyzer (VNA). Our measurements covered frequencies from 25 MHz to 18 GHz. The post-processing and calibration routines associated with it were developed at NIST [19]. In the synthetic-pulse system, the VNA acts as both transmitter and receiver. The transmitting section of the VNA sweeps over a wide range of frequencies a single frequency at a time. The transmitted signal is amplified and fed to a transmitting antenna. For this study, we used omnidirectional discone antennas for frequencies between 25 MHz and 1.6 GHz, and directional horn-type transmitting and receiving antennas for frequencies between 1 GHz and 18 GHz. We used directional antennas to provide additional gain in the direction of propagation, because the signals received from deep within the tunnels were quite weak. While some US&R robot manufacturers use directional antennas in weak-signal conditions, many use omnidirectional antennas exclusively. Note that if omnidirectional antennas were used in the higher frequency band, certain channel parameters, including the RMS delay spread, would have somewhat different values than those measured here.

The received signal was picked up over the air in the tunnel by the receiving antenna and was relayed back to the VNA via a fiber-optic cable. The fiber-optic cable maintains the phase relationships between the transmitted and received signals, enabling post-processing reconstruction of time-domain effects associated with the received signal such as the power-delay profile. The broad range of frequencies and time-domain representation provide insight into the reflective multipath nature of the tunnel that cannot be captured by use of single-frequency measurements. The receiving antenna must remain fixed during each measurement, so these tests are carried out at discrete locations, unlike the single-frequency tests.

We measured excess path loss every 20 m starting approximately 10 m from the transmitting antenna, as shown in Fig. 9.7(a) and (b). The VNA was located at the Hazel-Atlas portal (Fig. 9.7(a)) and in the Greathouse chamber (Fig. 9.7(b)). The transmitting antenna was located at the portal for the Hazel-Atlas tunnel and at the junction of the chamber and the tunnel for the Greathouse tunnel.

Figure 9.8(a)–(c) show measured excess path loss over a frequency band from 25 MHz and 1.6 GHz for increasing distances into the Hazel-Atlas tunnel. These graphs are all at distances less than 100 m; that is, before the right-angle turn. The top curve in each graph represents the received power level, referenced to the

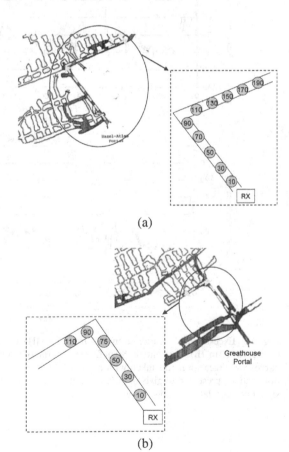

(a)

(b)

Fig. 9.7 Data-collection locations for the synthetic-pulse measurements.
(**a**) Hazel-Atlas mine tunnel.
(**b**) Greathouse mine tunnel

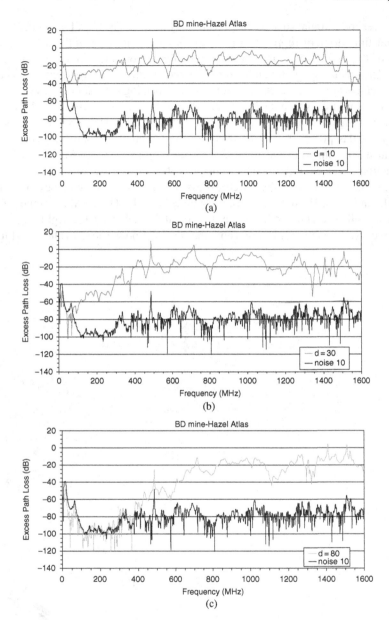

Fig. 9.8 Excess path loss measurements from 25 MHz to 1.6 GHz carried out at different distances: (**a**) 10 m, (**b**) 30 m, and (**c**) 80 m from the portal of the Hazel-Atlas mine tunnel. The "noise" curves were measurements taken with no transmitted signal. These were not made at every location, and are presented to give an indication of the dynamic range of each measurement (received signal power relative to noise power)

calculated free-space path loss at that location. The bottom curve represents the noise floor of the measurement system, to provide an indication of the dynamic range of each measurement.

The graphs of Fig. 9.8 show data starting from 0 Hz; however, the valid (calibrated) measurement range is stated for each graph. A rough estimate for the uncertainty in this measurement based on the VNA manufacturer's specifications is on the order of 0.2 dB. The variability in received power due to antenna placement within the environment is on the order of 10 dB, higher than the expected uncertainty. Thus, we do not report measurement uncertainties on our graphs.

Figure 9.8 shows that in a line-of-sight condition, the spectrum of the received signal displays significant frequency dependence. At frequencies between 25 MHz and 1.6 GHz, the lossy waveguide effect is shown by the rapidly decreasing signal on the left-hand side of the graph. We see that a carrier frequency higher than approximately 700 MHz would suffer less loss compared to lower frequencies in this particular tunnel. The same type of low-frequency attenuation was seen in the Greathouse tunnel as well.

Figure 9.9(a) and (b) show the excess path loss for frequencies from 1 GHz to 18 GHz in the Greathouse tunnel. Again, this is the path loss or gain that would exceed the free-space path loss at each location. For this measurement, the transmitter was located within the tunnel itself and, unlike for the Hazel-Atlas tunnel shown in Fig. 9.8, the reflections from the tunnel actually increase the power at the location of the receiver for some frequencies, shown by the excess path loss greater than 0 dB.

Figure 9.9(a) shows well defined nulls and peaks, corresponding to a direct path plus one or more strong reflections, when a line-of-sight path exists. This is characteristic of a "Rician" fading profile. Figure 9.9(b) shows that once the receiving antenna turns the corner, the signal takes on a more random variation with frequency, because transmission consists of only reflected signals. This is characteristic of a "Rayleigh" fading profile. We see that the average received signal level is relatively constant with frequency, but the peaks and nulls are significant.

Finally, we present the RMS delay spread for the two mine tunnels in Table 9.1 for frequencies from 25 MHz to 1.6 GHz and 1 to 18 GHz. An error analysis for these data is in process; consequently, we report no uncertainties in the RMS delay spread. We see that the shortest delay spreads are found by use of the directional antennas, as expected since reflected signals arriving from directions located behind the receive antenna are not received. A comparison of the effect on the RMS delay spread of using omnidirectional versus directional receive antennas in high multipath environments can be found in [20]. The delay spread in the line-of-sight case is nearly the same as for the non-line-of-sight case because of the strong multipath in the line-of-sight condition. In many environments, the line-of-sight delay spread is shorter because of a strong direct-path signal. The complete set of excess-path-loss data is given in [5].

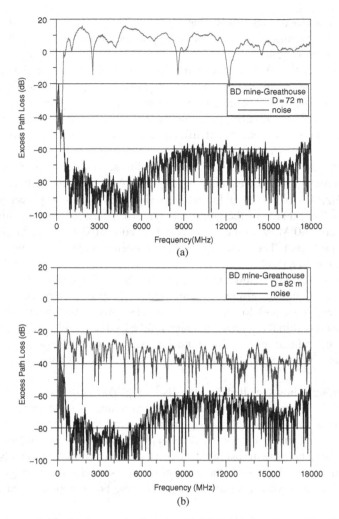

Fig. 9.9 Excess path loss for frequencies from 1 to 18 GHz in the Greathouse tunnel (**a**)72 m into the tunnel in a line-of-sight condition, and (**d**) 82 m into the tunnel in a non-line-of-sight condition

9.3.2.3 Tests of Robot Communications

We also carried out tests on a commercially available robot in the Hazel-Atlas tunnel. Control and video were as-built for the commercial product. We used the omnidirectional antennas that came with the system for all tests in order to assess the default capabilities of this robot. The robot we used is controlled with a spread-spectrum, frequency-hopping protocol, which was configured to transmit in the unlicensed 2.4 GHz industrial, scientific, and medical (ISM) band. The control channel utilizes a modulation bandwidth of approximately 20 MHz. The output power of the bidirectional control link is nominally 500 mW.

Table 9.1 RMS delay spread for the Hazel-Atlas and Greathouse mine tunnels

Distance (m)	Hazel Atlas tunnel		Greathouse tunnel	
	Low frequencies (ns)	High frequencies (ns)	Low frequencies (ns)	High frequencies (ns)
0	31.0	14.4	–	–
10	25.3	17.6	22.7	3.2
20	18.5	7.6	14.3	5.0
30	15.9	15.0	15.2	3.8
40	17.0	11.5	17.6	4.0
50	15.5	13.1	21	19.3
60	19.7	20.6	18.1	7.3
70	17.2	11.1	23.1	11.6
80	15.2	10.0	14.2	3.8
90	15.2	8.4	–	–
100	15.7	9.6	10.0	3.7
110	x	7.5	19.8	4.1

Left columns: Frequencies from 25 MHz to 1.6 GHz measured with omnidirectional antennas. Right columns: Frequencies from 1 to 18 GHz measured with directional antennas. The gray-shaded areas represent a non-line-of-sight propagation condition. The "x" at 110 m in the Hazel-Atlas tunnel indicates that the received signal was too weak to calculate the RMS delay spread

The robot transmits video by use of one of ten channels between 1.7 and 1.835 GHz. The robot we tested transmitted at 1.78 GHz by use of an analog modulation format that was non-bursted and non-frequency-agile. The video channel utilized approximately 6 MHz of modulation bandwidth. The output power was nominally 2 W.

The robot controller was located at the entrance to the tunnel, shown in Fig. 9.10. We positioned the robot inside the tunnel after the first bend in a non-line-of-site condition. The robot was moved through the tunnel on a cart, shown in Fig. 9.2(b), so that we could check the control link even after video was lost. Every 10 m, the video quality and control link were checked. Video was rated qualitatively by the

(a)   **(b)**

Fig. 9.10 (a) Robot operator positioned at the entrance to the Hazel-Atlas mine tunnel. (b) The robot was operated in a non-line-of-sight condition more than 100 m inside the tunnel

robot operator, and control was checked by the ability of the operator to move the robot arm, and verified by a researcher in the tunnel. No attempt was made to provide more granularity in these tests; that is, we assumed that moving the arm up was equivalent to moving it down or rotating it.

Table 9.2 shows the results of our tests. We were able to communicate with the robot in a non-line-of-sight condition deep within the tunnel. This is consistent with the results of Figs. 9.8 and 9.9, which indicate that signals in the low gigahertz range should propagate farther than those at lower frequencies.

Table 9.2 Results of robot wireless communication link tests carried out inside the Hazel-Atlas tunnel at Black Diamond Mines Regional Park

Distance in tunnel (m)	Video quality (1.7 GHz)	Control of arm (2.4 GHz)
100	good	yes
110	good	yes
120	poor (intermittent)	yes
130	poor (intermittent)	yes
140	very poor	yes
150	none	yes
160	none	delay experienced
170	none	intermittent control
180	none	delay experienced
190	none	delay experienced
200	none	delay experienced
205	none	none

Table 9.2 also shows that control of the robot was possible much deeper into the tunnel than where we were able to receive video, even though the output power of the video channel was higher (2 W for video vs. 0.5 W for control). A much higher data rate is necessary to maintain high-quality video transmission, as opposed to the relatively small amount of data needed to control the robot. Transmitting this large amount of data requires a higher received signal strength than for the control channel; therefore, failure of the video before the control is not unexpected. The delay experienced in controlling the robot when it was deep in the mine indicates packet loss and resend for error correction under weak-signal conditions.

9.4 Modeled Results

9.4.1 Single-Frequency Path Gain Models

To study the extent of signal attenuation and waveguiding in these tunnels, we implemented an analytical model that simulates signal propagation in tunnel environments having various physical parameters [9, 10, 21]. Briefly, the model assumes a single dominant mode in a lossy rectangular waveguide with the attenuation α in dB/m expressed for vertical polarization as

$$\alpha = \alpha_{\text{TUNNEL}} + \alpha_{\text{ROUGHNESS}} + \alpha_{\text{TILT}}, \tag{9.1}$$

where

$$\alpha_{\text{TUNNEL}} = 4.343\lambda^2 \left(\frac{1}{a^3 \sqrt{\varepsilon_R - 1}} + \frac{\varepsilon_R}{b^3 \sqrt{\varepsilon_R - 1}} \right), \tag{9.2a}$$

$$\alpha_{\text{ROUGHNESS}} = 4.343\pi^2 h^2 \lambda \left(\frac{1}{a^4} + \frac{1}{b^4} \right), \tag{9.2b}$$

$$\alpha_{\text{TILT}} = 4.343 \frac{\pi^2 \theta^2}{\lambda}, \tag{9.2c}$$

λ is the wavelength, a is the width of the tunnel, b is the height of the tunnel, and h is the roughness, all in meters. Other parameters include ε_R, the dielectric constant of the rock walls, and θ is the angle of the tunnel-floor tilt in degrees.

We set the parameters of the model to approximate the Hazel-Atlas tunnel, given below in Table 9.3. This model works well only for frequencies well above the cutoff frequency; that is, for wavelengths significantly less than the dimensions of the tunnel [9, 10]. Hence, in Fig. 9.11 we compare measured and modeled results for only 448 MHz.

Table 9.3 Parameters used in the tunnel model

Parameter	Value
Width	2 m
Height	2 m
Wall roughness	0.3 m
ε_r	6
Tilt	1°

In Fig. 9.11, the increase in measured signal strength at a distance of around 80 m, is caused by signal propagation through an air vent as well as through the tunnel, as was seen in Fig. 9.5. The agreement between the measured and modeled data led us to conclude that waveguiding plays a significant role in radio propagation in these tunnels.

The model also lets us explore which frequencies may be optimal for robot or other wireless communications in the tunnel. Figure 9.12 compares a number of commonly used emergency responder frequencies as a function of distance within the tunnel.

As discussed in [9, 10], the frequency-dependent behavior of the tunnel leads to a "sweet spot" in frequency. Below the sweet spot, signals do not propagate well, due to the effect of waveguide-below-cutoff attenuation and wall loss. Above the sweet spot, free-space path loss (which increases with frequency) and α_{TILT} dominate, and signals do not propagate well. Again, models such as these may enable a choice

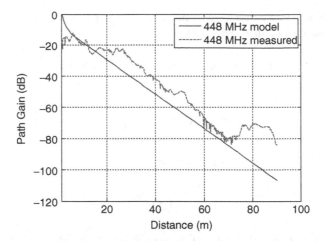

Fig. 9.11 Comparison of measured and modeled data for the Hazel-Atlas tunnel. The carrier frequency is 448 MHz. The modeled data simulate waveguide propagation for a waveguide whose physical parameters approximate those of the tunnels

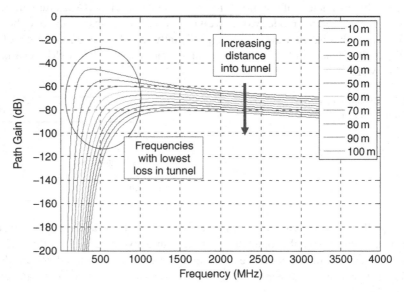

Fig. 9.12 Path gain versus frequency for various distances in a tunnel having physical characteristics similar to those of the Hazel-Atlas tunnel. Frequencies of approximately 400 MHz–1 GHz propagate further than either lower or higher carrier frequencies

of appropriate frequency for US&R robot communications in tunnel environments. Note that these results are valid only for a tunnel with these dimensions, wall materials, and surface roughness. The curves would need to be recalculated for other types of tunnels.

We also used the model to investigate the video performance of the robot, described in Section 9.3.2. The frequency-hopping control channel would need to be

Fig. 9.13 Path gain curves for a tunnel with a right-angle turn at 100 m (*solid*) and the flat earth (*dotted*) environment. The curve labeled "ELM = 0 dB" indicates where the excess link margin calculation predicts loss of signal. As shown, this occurs approximately 150 m into the tunnel

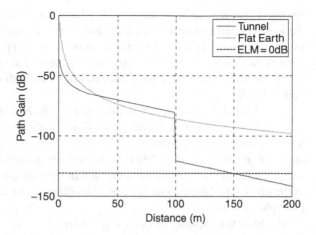

modeled by use of other methods, because it consists of several narrowband channels that frequency hop within a wide modulation bandwidth. In Fig. 9.13, we plot the estimated path gain at a carrier frequency of 1.78 GHz for the tunnel environment with a right-angle turn 100 m from the receiver. We used the parameters in Table 9.3 for the model. A path gain of −40 dB was used as an approximation for the turn in the tunnel at 100 m, based on work done by Lee and Bertoni in [22]. We plot the flat-earth path gain for comparison. The flat-earth model [23] is commonly used to represent line-of-sight propagation in a low-multipath, outdoor environment. In this model, signals propagate between the transmit and receive antennas along two paths only: a direct path and a single reflection off of the ground. This is in contrast to the high-multipath, waveguiding channel encountered in the tunnel.

Figure 9.13 also shows the theoretically computed excess link margin (ELM). The ELM is the difference between the received signal strength and the minimum receiver sensitivity. The receiver sensitivity is determined by the thermal noise of the receiver and the receiver's front-end amplifier noise (5 dB, as a rule of thumb). The thermal noise is given by $N = kTB$, where k is the Boltzmann constant, T is the temperature in kelvins, and B is the bandwidth of the receiver. In order for a wireless link to be maintained, the ELM usually must be greater than zero dB.

The ELM plotted in Fig. 9.13 agrees well with the measured results from Table 9.2, which show that the video completely drops out between approximately 140 and 150 m. Given the fluctuation in signal strength due to multipath in this tunnel environment, once the link margin drops below 10 dB at approximately 120 m, the video quality degrades and the picture becomes intermittent.

9.4.2 Channel Capacity Model

In general, received RF power and modulation bandwidth effectively place an upper bound on the capacity of a communications link. That is, a link exists between the capacity, bandwidth, and signal-to-noise ratio in any propagation environment. The

Shannon channel capacity theorem [16] helps to explain how these factors affect
the useful distance over which a robot can return a wideband signal such as a video
image or control signal. For example, in order to compare robot X that uses four
cameras with robot Y that uses two cameras, the user should understand that if both
robots use the same transmission bandwidth, robot Y should transmit video further
than robot X.

The channel capacity estimate provided by the Shannon theorem will be crude
for a tunnel environment because the Shannon theorem is based on the assumption of a Gaussian noise (no multipath) environment, while the distribution of the
received signal in the tunnel beyond the corner is closer to Rayleigh (high multipath). To account for additional reduction in channel capacity due to the multipath
in the tunnel, the Shannon model can be modified using techniques described in, for
example, [23, 24].

The Shannon capacity theorem is given by

$$C = B\log_2 (1 + S/N),$$ (9.3)

where C is the channel capacity in b/s, B is the channel bandwidth in hertz, S is the
received signal power in watts, and N is the measured noise power in watts. The
capacity represented by this equation is an upper limit. In reality, this capacity is
difficult to attain with real hardware, and actual capacity of an uncoded signal can be
closer to 50 % of the Shannon limit in a Gaussian noise environment. As mentioned
above, since the tunnel environment is a high-multipath Rayleigh environment, the
capacity may be reduced much lower than this, easily to 25% or less of the Shannon
limit when the received signal is weak.

Because our robot used an analog video signal, Shannon's limit cannot be applied
directly to estimate the channel capacity. However, the robot's control link was
based on an IEEE 802.11b standard for digital transmission. The modulation bandwidth for an 802.11b transmission is 20 MHz. Figure 9.14 shows the Shannon channel capacity from Equation (9.3) for a system having a 20 MHz modulation bandwidth, a 2.44 GHz carrier frequency (the actual carrier frequency was somewhere
between 2.412 and 2.462 GHz), 500 mW output power, and omnidirectional antennas (0 dBi gain). As before, we assume that the corner introduced 40 dB of attenuation. Figure 9.14(a) shows the Shannon capacity estimate along a 200 m path in
the tunnel, with a right-angle turn at 100 m, and Fig. 9.14(b) is a close-up of the last
100 m only.

In an 802.11b system, the transmitted data rate reduces dynamically as the channel degrades, with a lower limit of 1 Mb/s when the received signal is weak or a
great deal of interference exists. We see that at 160 m, where Table 9.2 shows that
our robot started experiencing intermittent control, the data rate is on the order of
1 Mb/s if approximately 15% of the Shannon capacity is transmitted. We used for-
mulas from [24] to find the additional reduction in capacity due to Rayleigh fading
associated with the non-line-of-sight condition. This corresponds to a carrier-to-
noise ratio of around −5 dB, which is close to the excess link margin of −5.8 dB
at 160 m computed for this case. Note that the range the robot can travel can be

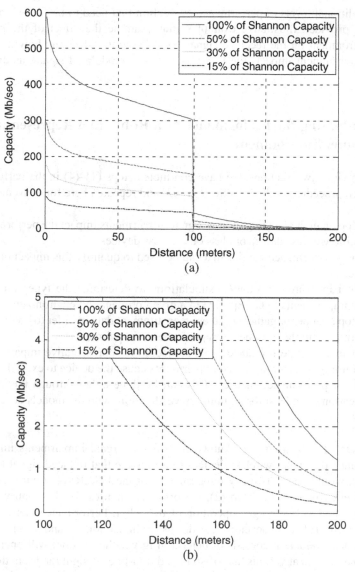

Fig. 9.14 Channel capacity predicted by the Shannon theorem for a carrier frequency of 2.4 GHz and a modulation bandwidth of 20 MHz. At 160 m, where we experienced intermittent video, 15% of the Shannon limit is ~1 Mb/s, which is the minimum data rate specified for an 802.11b signal before it fails

extended by using coding, signal processing techniques that include error correction. However, there is a limit to how far such signal processing can extend the range.

The above discussion presents a framework that may help the end user understand how to establish a bound for predictive planning. In practice there may be many additional levels of performance evaluation that need to be carried out. Here we have provided illustrative examples that examine the effect of the propagation environment on the received-signal level (Section 9.4.1) and how the propagation environment impacts the transmission of a modulated signal, as discussed here.

9.5 Evaluating the Performance of a Robot in a Representative Tunnel Environment

In the previous two sections, we have conducted steps (1)–(4) in the performance evaluation procedure outlined in Section 9.2, and repeated here for convenience:

1. Develop an understanding of how signal impairments impact the performance of a specific wireless device or class of wireless devices.
2. Develop performance metrics that can be used to quantify this impact on performance.
3. Conduct measurements and/or simulations to determine the type and level of signal impairments to be expected in a given propagation environment.
4. Develop a model or gather sufficient measurement data in order to predict device performance in the presence of representative impairments.
5. Evaluate device performance when subjected to representative impairments by determining whether the signal impairments cause the device to exceed specified values of performance metrics. This can be done either through measurement verification or, at least for preliminary verification, with the models developed in step 4.

For step (1), we used prior knowledge of how signal impairments impact the performance of typical wireless devices to anticipate that reduced signal level and multipath would be the two key impairments for the US&R robot wireless link in a tunnel environment. For step (2), performance metrics for the control channel were identified as "go/no-go" operation of the robot. Performance metrics for the video link are still being developed, as discussed in Section 9.2 and [6, 7]. Thus, our performance evaluation consists of determining whether a robot will operate with certain parameters at various line-of-sight and non-line-of-sight ranges in the tunnel environment.

Step (3) was illustrated in Section 9.3, where we described measurements that enabled determination of the type and level of signal impairments in the tunnel environment. We saw that the received signal level was impacted by both standard free-space path loss signal attenuation and also by a lossy waveguide effect that significantly reduced received signal levels at the lower frequencies. Multipath was clearly seen in the form of peaks and nulls in the received signal across frequency. When a line-of-sight condition existed, structured deep nulls and peaks of the received signal across frequency could be seen as the direct-path signal and one or more strong reflected signals added destructively or constructively. In a non-line-of-sight condition, the peaks and nulls took on a more random appearance. At the higher frequencies, the received signal level at times increased above the value that would be received in a free-space condition. This effect was again due to the waveguiding properties of the tunnel.

Step (4) was carried out in Section 9.4, where we used parameters of the specific tunnel environment in which we conducted our tests to predict the performance of the robot, both in terms of received signal power and in terms of channel capacity. The model results were verified by measurements of robot performance in the tunnel where the signal impairments were measured. The agreement between the model predictions and the robot measurements gave us confidence that the model could be used to predict robot performance in other tunnels; for example, those that are more representative of a typical emergency response scenario. This leads us to Step (5), where we try to determine whether the signal impairments in a representative environment would cause the robot to fail.

To predict and evaluate robot performance in a more representative tunnel, we used the model of Section 9.4.1 for a smooth-walled tunnel having dimensions of 6 × 4.5 m, similar to those of a subway tunnel. The simulation, first presented in [21], was based on a scenario in which a subway train proceeds through a 1,500 m (approximately 5,000 ft) under-river tunnel, passing through a 200 m straight portion, through a large radius curve for the next 200 m, and then along a straight section. The subway train undergoes a major explosion when it is one-third of the way from the destination station in the direction of travel. The subway train stops at this location due to the explosion and a robot is deployed to search for victims (Fig. 9.15).

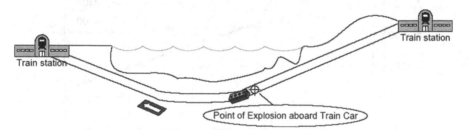

Fig. 9.15 Representative tunnel environment consisting of a 1500 m smooth-walled underground tunnel with a large-radius curve. An environment such as this could be specified for the development of test methods to evaluate the performance of robots for US&R applications (drawing not to scale)

We demonstrate a method to predict and evaluate the received-signal level for a robot deployed in this environment using two models, one for the straight sections of the tunnel and one for the curvature in the tunnel. In Section 9.4, we verified the use of the model for the straight section. This model was also verified for use in a large roadway tunnel in [11]. Thus, we have a high degree of confidence that this model will allow us to evaluate the use of the robot in a subway tunnel. The model for the tunnel curvature was first presented in [21]. Because the curvature is around a large-radius bend, the 90° turn from our measurements cannot be used. The model of [21], used here, is based on a physical representation of the tunnel, but to verify its performance, measurements or additional simulations would need to be conducted. As a result, the example presented in this section illustrates the method for predicting and evaluating robot performance, but additional work needs to be done before these results are used in practice.

Figure 9.16 shows the predicted path gain for each of the responder frequencies of interest using the path-loss model discussed in Section 9.4.1 combined with the model for the tunnel curvature from [21]. We plot only the first 1,000 m for clarity, where 0 m corresponds to the departure station. The propagation characteristics in each of the three tunnel sections (line-of-sight, curved, and non-line-of-sight) introduce different types and levels of signal impairments into the received signal. In reality, additional loss may be anticipated in some tunnels due to dampness of the walls and additional roughness from the track-bed and conduits, which will tend to absorb energy and increase path loss.

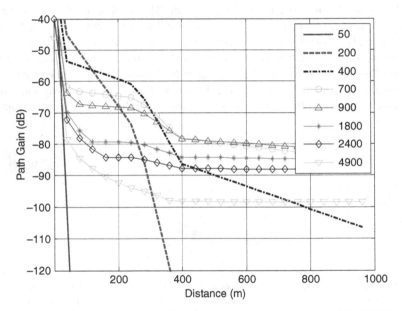

Fig. 9.16 Path gain calculated for the 6 m × 4.5 m subway tunnel shown in Fig. 9.15 for eight different frequencies (in MHz). Three distinct propagation regions can be identified: Line-of-sight (less than 200 m), large-radius curve (200–400 m), and a straight, non-line-of-sight section (greater than 400 m) [15]

For short distances into the tunnel, where a line-of-sight condition exists, the lowest loss is seen between the 200 and 900 MHz frequencies. These results agree with the generalized trend that was seen in [11]; in that the frequencies in the middle range tend to provide the lowest path loss for shorter distances into tunnels of this size.

In the curved section of the tunnel, the loss tends to increase as the frequency decreases. From Fig. 9.16 we see that the 400 MHz signal decreases significantly in the large-radius bend. The rate of loss would increase in bends having a smaller radius of curvature.

For distances farther into the tunnel, beyond the curve, the model shows little difference in average signal loss between the frequencies of 700 MHz and 2.4 GHz. However, note that once the robot is in a non-line-of-sight condition, based on our observations in Fig. 9.9(b) and [5], we would expect the rapid variation in signal amplitude due to multipath is greater at higher frequencies. Dudley et al. [11] concluded that in both straight and curved tunnels there is little benefit in using increasing frequencies beyond a point where the attenuation (or ELM) flattens as frequencies increase. This will depend to some degree on the dimensions of the tunnel, but for the purpose of subway-sized tunnels, there is little to be gained in operating above 1.5 GHz.

This scenario illustrates a representative tunnel environment for US&R robot deployment that could be used for evaluating the performance of the wireless link for robots used in stand-off, tunnel-based applications where the operator is located in a non-line-of-sight condition from the incident. The scenario contains a number of key environmental elements that are specific to tunnels, including a line-of-sight portion where waveguiding effects occur, a curved section where frequency-dependent loss occurs, and a non-line-of-sight section where significant multipath occurs. The received signal level predicted by this model is only one component of a comprehensive performance evaluation; however, it serves to illustrate the performance evaluation method effectively.

A standardized test method, such as the non-line-of-sight test method described in Section 9.2 that captures the key signal impairments presented by this scenario, could be developed. The performance of robots for use in US&R applications could then be evaluated under these conditions. This would be a natural evolution for the ASTM standard described above and would complete Step (5) in the performance evaluation procedure.

9.6 Conclusion

We have presented a framework for evaluating the performance of the wireless link used in urban search and rescue robots, using a subterranean tunnel as an example of a representative responder environment. The evaluation method is based on extraction of the type and level of key signal impairments in a tunnel environment through measurement of the propagation characteristics of the tunnel. A model is then developed so that robot performance can be predicted in a representative class of (tunnel) environments. Using the model, representative values of key signal impairments can be replicated in a test environment to evaluate the expected performance of robots in a class of propagation environments; that is, in other tunnels. Real-time performance evaluation can also enable a robot to compensate for degradation of channel characteristics by, for example, automatically deploying repeaters or changing the digital modulation format to one that is optimized for a given environment.

Results showed effects of waveguide-below-cutoff propagation and wall attenuation in the tunnels we measured, which agree with previously published results. We saw frequency-dependent peaks and nulls in the channel due to strong multipath reflections and attenuation in the tunnel. In non-line-of-sight conditions, we saw classic small-scale fading, manifested in noise-like multipath effects.

We implemented models of radio propagation and channel capacity within the tunnel environment and discussed how the models could be verified by measuring the performance of a robot within the tunnel and comparing the measured and modeled results. Note that more comprehensive performance evaluation procedures may also be carried out, where detailed models of the so-called physical layer would be constructed, including the effects of modulation, coding, equalization, power control, rate adaptation, etc. The goal of the work presented here was to illustrate some simple methods to predict and/or evaluate the expected over-the-air performance of a robot in a tunnel for other, more representative tunnel environments. An example of this was presented for a subway tunnel containing a large-radius curve. Such an example could provide the basis for a standardized test method to evaluate the line-of-sight and non-line-of-sight range performance of robots in tunnel environments for US&R applications.

Acknowledgments We gratefully acknowledge the contributions of the following: Chriss Grosvenor of the Electromagnetics Division of NIST and Dr. Robert Johnk of the Institute for Telecommunications Science (formerly of NIST) for assistance with the measured data; Dr. Bert Coursey, Director of the DHS Office of Standards and Elena Messina and Adam Jacoff of NIST's Manufacturing Engineering Laboratory for funding and supporting this work; Dr. Alex Bordetsky of the Naval Postgraduate School for facilitating the measurements during recent interagency marine interdiction operation system tests; Bill Dunlop, Steve MacLaren, and Dave Benzel of Lawrence Livermore National Laboratory for logistical and technical support; Roger Epperson, Park Supervisor and Gary Righettini, Mine Manager of the East Bay Regional Park District and the Black Diamond Mines Regional Preserve, for allowing us to conduct measurements; Frederick M. Remley for details on the NTSC video standard.

References

1. E.Messina, "Performance Standards for Urban Search and Rescue Robots," *ASTM Standardization News,* August 2006. http://www.astm.org/cgi-bin/SoftCart.exe/SNEWS/AUGUST_2006/messina_aug06.html.
2. National Institute of Standards and Technology, *Statement of Requirements for Urban Search and Rescue Robot Performance Standards-Preliminary Report.* http://www.isd.mel.nist.gov/US&R_Robot_Standards/Requirements%20Report%20(prelim).pdf.
3. K.A. Remley, G. Koepke, E. Messina, A. Jacoff, and G. Hough, "Standards development for wireless communications for urban search and rescue robots," *Proceedings of the International Symposium on Advanced Radio Technol.,* pp. 66–70, Boulder, CO, Feb. 2007.
4. C.L. Holloway, W.F. Young, G. Koepke, K.A. Remley, D. Camell, and Y. Becquet, "Attenuation of Radio Wave Signals Coupled Into Twelve Large Building Structures," *National Institute of Standards and Technology Note 1545,* Apr. 2008.
5. K.A. Remley, G. Koepke, C.L. Holloway, C. Grosvenor, D. Camell, J. Ladbury, D. Novotny, W.F. Young, G. Hough, M.D. McKinley, Y. Becquet, and J. Korsnes, "Measurements

to support broadband modulated-signal radio transmissions for the public-safety sector," *National Institute of Standards and Technology Note 1546*, Apr. 2008.

6. M.H. Pinson, S. Wolf, and R.B. Stafford, "Video performance requirements for tactical video applications," *Proceedings of IEEE Conference on Homeland Security Applications*, 2007, pp. 85–90.

7. C. Ford and A. Webster, "Introduction to Objective Multimedia Quality Assessment Models," *Proceedings of the International Symposium on Advanced Radio Technology*, Boulder, CO, March 7–9, 2006, pp. 8–15.

8. http://www.3gpp.org/ftp/Specs/latest/Rel-7/25_series/. See 3GPP TS 25.101, Annex B.

9. A.G. Emslie, R.L. Lagace, and P.F. Strong, "Theory of the propagation of UHF radio waves in coal mine tunnels," *IEEE Transactions on Antennas and Propagation*, vol. 23, no. 2, Mar. 1975, pp. 192–205.

10. M. Rak and P. Pechac, "UHF propagation in caves and subterranean galleries," *IEEE Transactions on Antennas and Propagation*, vol. 55, no. 4, April 2007, pp. 1134–1138.

11. D.G. Dudley, M. Lienard, S.F. Mahmoud, and P. Degauque, "Wireless propagation in tunnels," *IEEE Antennas and Propagation Magazine*, vol. 49, no. 2, April 2007, pp. 11–26.

12. C. Nerguizian, C.L. Despins, S. Alles, and M. Djadel, "Radio-channel characterization of an underground mine at 2.4 GHz," *IEEE Transactions on Wireless Communication*, vol. 4, no. 5, Sept. 2005, pp. 2441–2453.

13. M. Ndoh, G.Y. Delisle, and R. Le, "An approach to propagation prediction in a complex mine environment," *Proceedings of the 2003 International Conference on Applied Electromagnetics and Communication (ICECom)*, Oct. 2003, pp. 237–240.

14. M. Ndoh, and G.Y. Delisle, "Underground mines wireless propagation modeling," *Proceedings of the 2004 IEEE Vehicular and Technology Conference*, Sept., 2004, pp. 3584–3588.

15. D. Didascalou, J. Maurer, and W. Wiesbeck, "Subway tunnel guided electromagnetic wave propagation at mobile communications frequencies," *IEEE Transactions on Antennas and Propagation*, vol. 49, no. 11, Nov. 2001, pp. 1590–1596.

16. C.E. Shannon, "Communication in the presence of noise," *Proceedings of IRE*, vol. 37, no. 1, Jan. 1949, pp. 10–21.

17. M. Rütschlin, K.A. Remley, R.T. Johnk, D.F. Williams, G. Koepke, C. Holloway, A. MacFarlane, and M. Worrell, "Measurement of weak signals using a communications receiver system," *Proceedings of the International Symposium on Advanced Radio Technology*, Boulder, CO, March 2005, pp. 199–204.

18. P. Delogne, *Leaky Feeders and Subsurface Radio Communications*, IEEE Press, New Jersey, 1982.

19. B. Davis, C. Grosvenor, R.T. Johnk, D. Novotny, J. Baker-Jarvis, and M. Janezic, "Complex permittivity of planar building materials measured with an ultra-wideband free-field antenna measurement system," *NIST Journal of Research*, vol. 112, no. 1, Jan.–Feb., 2007, pp. 67–73.

20. K.A. Remley, G. Koepke, C. Grosvenor, R.T. Johnk, J. Ladbury, D. Camell, and J. Coder, "NIST tests of the wireless environment in automobile manufacturing facilities," *National Institute of Standards and Technology Note 1550*, Oct. 2008.

21. G. Hough, "Wireless robotic communications in urban environments: issues for the fire service," Thesis, Naval Postgraduate School, March 2008. http://theses.nps.navy.mil/08Mar_Hough.pdf.

22. J. Lee and H. Bertoni, "Coupling at cross, T, and L junctions in tunnels and urban street canyons," *IEEE Transactions on Antennas and Propagation*, vol. 51, no. 5, May 2003, pp. 926–935.

23. A. Goldsmith, *Wireless Communications*, Cambridge University Press, Cambridge, MA, 2005.

24. W.C.Y. Lee, "Estimate of channel capacity in Rayleigh fading environment," *IEEE Transactions on Vehicular Technology*, vol. 39, no. 3, Aug. 1990, pp. 187–189.

Chapter 10
Quantitative Assessment
of Robot-Generated Maps

C. Scrapper, R. Madhavan, R. Lakaemper, A. Censi, A. Godil, A. Wagan,
and A. Jacoff

Abstract Mobile robotic mapping is now considered to be a sufficiently mature
field with demonstrated successes in various domains. While much progress has
been made in the development of computationally efficient and consistent mapping
schemes, it is still murky, at best, on how these maps can be evaluated. We are motivated by the absence of an accepted standard for quantitatively measuring the performance of robotic mapping systems against user-defined requirements. It is our belief
that the development of standardized methods for quantitatively evaluating existing
robotic technologies will improve the utility of mobile robots in already established
application areas, such as vacuum cleaning, robot surveillance, and bomb disposal.
This approach will also enable the proliferation and acceptance of such technologies
in emerging markets. This chapter summarizes our preliminary efforts by bringing
together the research community towards addressing this important problem which
has ramifications not only from researchers' perspective but also from consumers',
robot manufacturers', and developers' viewpoints.

10.1 Introduction

Mobile robots permitting collaborative operations of man and machine present a
new frontier of research with almost limitless possibilities by serving as an indispensable aid in hazardous, unstructured environments and in performing repetitive
or mundane tasks that require a high level of accuracy. In the future, robots will play
an increasingly vital role in assisting humans in a variety of domains ranging from
innocuous daily chores around the household to potentially harmful situations. The
use of robots, either tele-operated or autonomous, can not only save lives in dangerous work environments but also can improve productivity (e.g. factory floors) and in

C. Scrapper (✉)
The MITRE Corporation, McLean, VA 22102, USA
e-mail: cscrapper@mitre.org

R. Madhavan et al. (eds.), *Performance Evaluation and Benchmarking
of Intelligent Systems*, DOI 10.1007/978-1-4419-0492-8_10,
© Springer Science+Business Media, LLC 2009

some cases provide solutions which are not achievable by humans alone (e.g urban search and rescue in extremely confined spaces). It is not hard to see that the ability to build a map of the working environment is a desirable capability for robots operating in many domains of interest. For example, in a disaster scenario concerning extrication of victims, a robot-generated map will serve as in invaluable tool that will assist responders in assessing the overall situation by identifying victims and hazardous conditions in the environment.

Not surprisingly, the development of efficient robotic mapping systems have received their due attention from roboticists. A myriad of approaches have been proposed and implemented, some with greater success than others. The capabilities and limitations of these approaches vary significantly depending not only on the operational domain and on-board sensor limitations, but also on the requirements of the end user: Will a 2D map suffice as an approximation of a 3D environment? Is a metric map really needed, is it enough to have a topological representation for the intended tasks, or do we need a hybrid metric-topological map [38]? Therefore, it is essential for both the developers and the consumers of robotic systems to understand the performance characteristics of employed mapping methodologies that will allow them to make an informed decision.

To the authors' knowledge, there is no accepted standard for quantitatively measuring the performance of robotic mapping systems against user-defined requirements; and furthermore, there is no consensus on what objective evaluation procedures need to be followed to deduce the performance of these systems. For instance, the current methodologies for assessing the quality of maps is generally based on qualitative analysis (i.e. visual inspection). This approach does not allow for better understanding of what errors specific systems are prone to and what systems can meet the performance criteria required for a specific task. In addition, it has become common practice in the literature to compare newly developed mapping algorithms with prior methods by presenting images of generated maps. This procedure turns out to be suboptimal, particularly when applied to large-scale maps.

The lack of reproducible and repeatable test methods have precluded researchers working towards a common goal from exchanging and communicating results, intercomparing robot performance, and leveraging previous work that could otherwise avoid duplication and expedite technology transfer. This lack of cohesion in the community hinders the progress in many domains, such as manufacturing, service, search, rescue, and security. Providing the robotic mapping community access to standardized tools, reference data sets, and an open-source library of solutions will help them to evaluate the cost and benefits associated with available technologies. The development of standardized methods for quantitatively evaluating existing robotic technologies will not only improve the utility of mobile robots in already established application areas, such as vacuum cleaning, robot surveillance, and bomb disposal, but will also enable the proliferation and acceptance of such technologies in other emerging markets.

Some members of the robotic mapping community have recognized the deficiency in the quantitative evaluation of robotic mapping and are attempting to address it through several different programs. For example, the Robotics Data Set

Repository (Radish) provides a collection of standard robotics data sets [33]. The OpenSLAM repository contains collections of source codes of various Simultaneous Localization and Mapping (SLAM) algorithms [32]. While this is a step in the right direction, they do not address objective evaluation methodologies and the replication of algorithms is not straightforward. Emerging standard test methods for emergency response robots, developed by the National Institute of Standards and Technology (NIST) and the Department of Homeland Security (DHS), provide the research community with an efficient way to test their algorithms. The suite of test methods developed can be proliferated widely to minimize the costs associated with maintaining functional robots and traveling to one of the permanent test sites for validation and practice [2].

There are approaches that measure map quality beyond visual inspect. In [9], map quality is assessed using conditional random fields. The assessment is proposed as an introspective inspection of workspace representations towards analyzing the reliability/plausibility of the representation. A single 3D laser map is segmented into planar patches based on neighboring relations into 'plausible' and 'suspicious' using a context-sensitive classification framework. The proposed framework can be thought of as a qualitative assessment based on quantitative metrics. Though it is not clear how to extend this method to assess and compare quality of two maps of the same area, it does provide a innovative approach that could be incorporate as an automated tool in an evaluation framework.

Many researchers have suggested using vision rather than laser range finders for mapping purposes [44, 1, 45]. In addition to being a passive sensor, cameras are attractive due to their low consumption of power, relatively low cost, and ability to provide large bandwidth of information. In [1], the authors developed a testbed infrastructure as a vision SLAM benchmark using synchronized inertial measurement unit (IMU), global positioning system (GPS) and stereo images in an outdoor setting. The proposed benchmark may provide a mechanism for comparing two maps generated by using the data collected via the proposed infrastructure and geographically referenced aerial images as ground truth. While feasible in some cases, the assumption of known ground truth is, in general, overly restrictive.

The RoboCup Rescue competitions [34] have proved to be a good forum to evaluate task-based performance of robots. An image similarity metric and a cross entropy metric are outlined in [41] to measure the quality of occupancy grid maps. The metric gives an indication of distortion of the map with respect to a ground truth map in the presence of noise and pose errors. This metric is embedded in the Jacobs Map Analysis Toolkit [21] and has been tested for comparing maps in the RoboCup context.

While contributions by individual researchers are important steps in the right direction to overcome technological barriers to robotic mapping, a concerted effort among all interested parties is crucial. The primary focus of our effort is to bring together researchers, consumers, and vendors to define objective methodologies for quantitatively evaluating robot-generated maps [27, 37].

Standard test methods are vital in establishing a confident connection between developers and consumers regarding the expectations and performance objectives of robotic technologies. They consist of well-defined testing apparatuses, procedures, and objective evaluation methodologies that isolate particular aspects of a system in known testing conditions [3]. This provides developers with a basis for understanding the objective performance of a system and allows consumers to confidently select systems that will meet their requirements.

In order to ensure the integrity of the test methods, it is essential to use a developmental cycle, shown in Fig. 10.1, that continuously reassesses the validity of the test methods. This process starts with a comprehensive analysis of the application domain to identify requirements with associated metrics and the range of performance, starting from a baseline threshold to the objective "best-case" performance. This analysis provides the basis for developing test methods, procedures, and testing scenarios that are intentionally abstract so as to be repeatable across a statistically significant set of trials and reproducible by other interest parties.

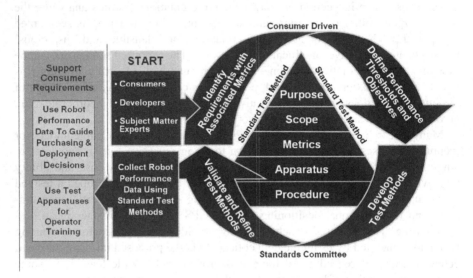

Fig. 10.1 The standard test method developmental cycle, modeled by NIST based on a diagram by Bert Coursey from the Department of the Homeland Security, relies on a requirement-driven approach to iteratively assess the validity and integrity of the standard. The inclusion of all interested parties at the onset of the process is essential to the development of comprehensive standards that provides the basis for evaluating the performance of robotic technologies

The remainder of this chapter is dedicated to defining test methods for robotic mapping and identifying techniques that can be used to evaluate the performance and assess the integrity of these systems using the cycle described above. Section 10.2 provides a brief description of the emerging test apparatuses used to challenge robotic mapping in specific ways. Section 10.3 develops a theoretical approach, using the Cramér-Rao Bound (CRB), to assess the thresholds and objective performance of pose and mapping estimates. Sections 10.4 and 10.5 outline two experimental techniques used to quantitatively assess the quality of robot-generated maps. Section 10.6 provides conclusions and continuing research.

10.2 Developing Test Scenarios for Robotic Mapping

The performance of any given robotic mapping system is largely dependent on its ability to reliably accomplish two fundamental tasks. First is the ability of the system to make accurate measurements of its surrounding environment. Second is the ability of the system to reliably determine valid correspondences between observations, for example, associating an object in one observation with its counterpart in another. The type and the conditions of the environment strongly influence the ability of the system to accomplish either task. Furthermore, subtle differences in relatively similar environments may have very different effects on the overall performance of the system.

As noted earlier in Section 10.1, the evaluation of robot-generated maps is often based on qualitative approaches that do not take into account how specific environmental conditions impact the performance of the system. While this type of analysis provides some indication of the overall performance, it does not allow researchers to understand what errors a specific system is prone to, how these errors impact the overall performance of that system, and how performance of that system compares with competing approaches. When developing standard test methods for evaluating robot-generated maps, it is important to develop repeatable and reproducible testing scenarios that isolate potential failure conditions in a controlled environment.

The remainder of this section summarizes a suite of test apparatuses designed to classify the performance of robotic mapping systems over a range of application requirements. With each test apparatus focused on challenging the system with varying levels of environmental complexities, this suite is intended to provide a comprehensive evaluation that will serve as the baseline for comparison and will help developers refine the capabilities of their systems (or address the limitations). Section 10.2.1 provides a brief description of the process used to develop the test apparatuses used in this suite. Sections 10.2.1.1, 10.2.1.2, and 10.2.1.3, introduce the resulting prototype apparatuses currently used to characterize mapping systems.

10.2.1 Performance Singularity Identification and Testing

Identifying *performance singularities*, or the point where the mapping system fails to be well-behaved, is essential for understanding the impact of the environment on the overall performance of the mapping and what environmental conditions are problematic for robotic mapping systems in general. *Performance singularity identification and testing* [35, 36] defines a two-pronged approach by which one can systematically evaluate the impact of environmental conditions (that may contribute to the occurrence of performance singularities) and analyze the impact of these singularities on the overall performance of the system.

The first step in this approach is to evaluate the performance at the system level to identify divergences in the performance of the mapping system. Using ground truth information about the location of the robot and the surrounding environment, the *performance evaluation* step facilitates the decomposition of the errors arising in the pose estimate. This enables the discovery of irregularities and helps identify environmental situations where a performance singularity has occurred.

The second step in this approach is to analyze performance of a mapping system at the algorithmic level to gain insight into the cause and repercussions of the performance singularity previously identified. *Performance analysis* takes advantage of the ground truth to measure the error in the pose estimate at each discrete observations. This produces a convergence profile elucidating the convergence characteristics, such as the stability and accuracy of the pose estimate.

10.2.1.1 The Maze: Scenarios with Distinct Features

The Maze testing apparatus, seen in Fig. 10.2, provides a scenario that limits environmental complexities to provide the best-case scenario, where mapping systems should perform optimally. This apparatus, whose overall dimension is 15 m by 10 m, uses a closed set of distinct mapping features and vertical walls that produces unique

Fig. 10.2 The Maze, seen in the *top right* image, is a testing apparatus that limits complexities in the environment in order to evaluate the objective performance of particular mapping system. The *top left* shows a configuration of the maze whose dimensions are 15 m by 10 m, with *black lines* representing the maze layout and the *gray arrows* showing the configuration of 15° pitch and roll ramp flooring, shown again in the *bottom left* image. The maze utilizes various materials and shapes to produce additional mapping features, such as concave and convex surfaces, shown in the *bottom right* image

observations. The non-flat flooring, comprised of 15° ramps, makes the vehicle pitch and roll as it traverses the maze. This reduces the uncertainty in the mapping system when associating features and increases the likelihood of the in determining valid correspondences. Additionally, perpendicular surfaces appear in almost every scan. This increases the ability of the system to make accurate measurements of the surrounding environment and accentuates the displacement of features between two consecutive scans. Limiting the environmental complexities allows developers to tune their systems and establishes a baseline for comparison. The modularity of the apparatus enables the randomization of the maze configurations for repetitive testing. Common building materials makes it a low-cost and easy to replicate.

10.2.1.2 The Tube Maze: Scenarios with Occluded Features

The Tube Maze testing apparatus, shown in Fig. 10.3, challenges the mapping system's ability to determine valid correspondences. This maze, whose overall size is 6 m by 10 m, also contains a closed set of distinct features that are partially observable depending on the location and orientation of the system at a given time. In this testing apparatus, nearby features may periodically occlude more distant features as the robot moves through the environment. This produces a situation where consecutive observations may not contain the same set of features, increasing the likelihood of correspondence errors. However, the unobstructed features close to the robot enable the system to make accurate measurements of the immediate vicinity and should help the system avoid catastrophic failures. This montage of partially observable features is indicative of situations commonly found in unstructured environments; and therefore, provides a critical testing scenario when developing stable and robust mapping systems. Similarly, the modularity of the apparatus and common building materials provide a simple way to validate successful mapping.

Fig. 10.3 The Tube Maze apparatus designed to test the ability of the system to reliably determine valid correspondences. This apparatus, whose overall size is 6 m by 10 m, consists of continuous 15° pitch and roll ramp flooring and reconfigurable tubing (preferably 10 cm in diameter) arranged to restrict the robot to a predetermined path. The density and height of the tubes are adjusted to test the mapping system's ability to handle occlusions and to consider three dimensional objects

10.2.1.3 The Tunnel: Scenarios with Minimal Features

The Featureless Tunnel apparatus implements the degenerative case for mapping algorithms. As shown in Fig. 10.4, this apparatus presents a symmetric and featureless environment that is designed to inhibit the ability of the system to make accurate measurements of its environments and to complicate the systems ability to determine valid correspondences. The lack of distinct features increases the potential for catastrophic errors by preventing the convergence of the pose estimate in the mapping system. While this situation does not occur commonly (except in culverts, sewers, and tunnels), this testing apparatus is essential to understanding how the system fails.

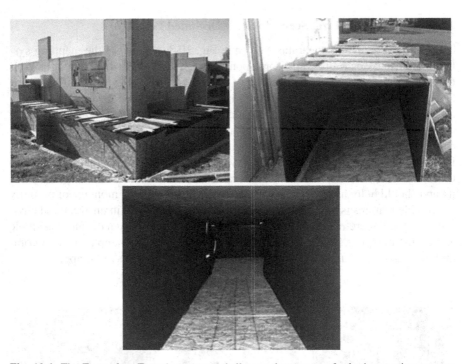

Fig. 10.4 The Featureless Tunnel apparatus challenges the aspects of robotic mapping systems that require correspondences to estimate the robot's pose. The apparatus is 15 m with a single turn, 15° roll ramp flooring, and black felt covered interior walls

10.3 Assessing Objective Performance Using Theoretical Analysis

The test methods, as defined in Section 10.2, provide the basis for evaluating a range of application requirements with associated metrics, thresholds of performance, and best-case performance objectives. Theoretical analysis of mapping systems can provide insight into the range of performance in application domains and how these systems can be improved.

We use the following notation. Let $x_k \in$ SE(2) be the robot pose at time k; let $\delta_k \in$ SE(2) be the incremental motion of the robot, such that $x_{k+1} = x_k \oplus \delta_k$, where $\oplus$ is the pose composition operator (group operation on SE(2)). Let w represent the "world" or "map". Let y_k be the sensor readings. The sensor model can be specified either in a functional form such as $y_k = f(x_k,w) + \varepsilon_k$, where ε_k is a noise term, or alternatively using the distribution $p(y_k|x_k)$, w). With this notation, we formalize *localization* as the problem of estimating x_k given the observations $y_{1:k}$ and a *known* map w. We call *pose-tracking* the problem of recovering the robot displacements δ_k from the sensor readings, without knowing the map. Pose tracking using range scans is commonly called *scan matching*. We call *mapping* the problem of recovering w given known poses. Finally, SLAM is the problem of recovering both poses and map at the same time. This classification is useful in this particular analysis; but, in practice, distinctions are blurred between these problems. For example, some algorithms such as the Iterative Closest Point (ICP) are applicable to both localization and pose tracking, and a complete solution may decompose the SLAM problem in smaller sub-problems involving pose-tracking, mapping, and (re)localization.

Industrial deployment of localization/SLAM algorithms implies matching the algorithm properties to the operational requirements.*Computational properties,* such as speed and memory consumption, are easy to define and measure directly. *Statistical properties* make sense for most algorithms that work in a probabilistic formulation of the problem. Such properties are the accuracy (estimate covariance), the presence of a bias, and the consistency (whether the algorithm has a good estimate of its actual accuracy). These properties are easy to define mathematically, but might be hard to measure in practice because a ground truth is needed. Equally important are the *robustness properties*, which refer to the ability for the algorithm to work even if the assumptions on which it relies are slightly violated. For example, a localization algorithm should not fail completely if the provided map is only slightly different than the actual environment. Likewise, it should not fail if the sensor has a covariance slightly larger than the assumed one. The robustness properties are hard to define analytically, because by definition they refer to unknown violations of the assumptions. Finally, the output of some algorithms, such as environment maps, might be used by both machines and humans, but this *"user-friendliness"* cannot be defined mathematically.

All these desirable properties are sometimes contrasting. For example, speed versus accuracy is an obvious trade-off in many algorithms. Other typical trade-offs include robustness versus accuracy (for example, this appears in choosing the percentage of measurements to discard as outliers) and generality versus accuracy (an algorithm which makes more assumptions about the environment can be more precise than one that works in more situations).

There are essentially two ways to verify these properties: either using benchmarks, or using theoretical analysis. These two are complementary activities: a theoretical analysis can be done only on some kind of idealized model of the system and the algorithm; the benchmarks can verify whether the assumptions made for the actual implementation in the actual environment. Conversely, benchmarks are incomplete without analysis, because they do not explain *why* the algorithm behaves in a certain way, and *whether* and *how* the algorithm can be improved.

10.3.1 The Case for Statistical Bounds

In practice, it might be infeasible to prove analytically that a particular localization/SLAM algorithm has one of the above mentioned properties. In estimation theory, there are a number of ready-made results for the canonical estimators, such as the maximum-likelihood estimator, regarding their accuracy and consistency. However, these kinds of results are not easily transferred to the actual algorithms, because of the ad-hoc approximations that are necessary in the implementations.

For example, in the Bayesian framework, the solution to the filtering problem is given by a recursive formula which has a closed form. Most SLAM papers start with this uncomputable formula, and start simplifying it using various assumptions, until an approximation which can be computed efficiently is obtained. The problem is that, once the symbol "$\simeq$" is used, one loses any guarantee about the properties of the resulting algorithm. Another canonical example of theoretically sound, but hard to analyze, estimators are particle filters. The asymptotic behavior of particle filters provides strong results as the number of particles goes to infinity, but nothing is guaranteed for a finite computation [10]. The answer to the question of how many particles does one actually need is usually very fuzzy [11]. These problems are specific to "dense" algorithms. In the "discrete" case, for Extended-Kalman Filter-based methods, it is easier to do theoretical analysis because the complexity of actual sensors has been abstracted away into bearing/range observations of landmarks: hence we know that the EKF is inconsistent, where the inconsistency originates, and what to do about it [12, 19, 17].

Because an explicit white-box analysis of a localization/SLAM algorithm might be infeasible, a possible first step is to investigate the problem itself, for example by considering the statistical bounds to the problem.

The theory of statistical bounds is well developed and offers many tools [40]. The Cramér-Rao Bound (CRB) is a classic one which is easy to derive and to use. In the nonlinear case with additive Gaussian noise ($y = \mathbf{f}(x) + \varepsilon \oplus$), one defines the Fisher Information Matrix (FIM) as $\mathcal{I}[x] = \frac{\partial \mathbf{f}^{\mathrm{T}}}{\partial x} \sum^{-1} \frac{\partial \mathbf{f}}{\partial x}$, with Σ being the covariance of the noise. Then the CRB establishes that, for any unbiased estimator, $\mathrm{cov}[\hat{x}] \geq (\mathcal{I}[x])^{-1}$; if the estimator is biased, $\mathrm{cov}[\hat{x}] \geq [I + \frac{d}{dx}b_{\hat{x}}(x)](\mathcal{I}[x])^{-1}[I + \frac{d}{dx}b_{\hat{x}}(x)]^{\mathrm{T}}$, where $b_{\hat{x}}$ is the bias of the estimator. In general, the CRB is not tight, except in special cases, such as when $\mathbf{f}$ is linear; the CRB is approximately tight at high signal-to-noise ratios.

The CRB is tool with many uses. It provides a lower bound for the accuracy that is a baseline for comparing the actual experimental results. It allows to verify the realism of accuracy claims, and the proper execution of experiments. After it has been proved to be tight, it can also be used to predict the actual covariance.

Nevertheless, it is worth pointing out some intrinsic limitations of this kind of analysis. This theory applies only when localization is modelled probabilistically, and it only models the effect of stochasticity in the readings, which is only one of the many sources of error in the algorithms (others are, e.g., convergence to local minima). Moreover, this theory only gives negative results; establishing positive results of guaranteed accuracy must still be done on a case-by-case basis.

10.3.2 The CRB for Range-Finder Localization

We define localization as estimating the pose given a perfect map and a range scan. This case has been considered in the paper [7], from which we recall the main results. Assume the pose of the robot is $x = (t, \theta)$, and the output of the range finder is $\{\tilde{\rho}_i\}_{i=1}^n$ where $\{\tilde{\rho}_i\}$ is the i-th ray along direction φ_i. The FIM is a 3×3 symmetric semi-definite positive matrix which can be computed as follows:

$$
\mathcal{I}[x] = \sum_{i=1}^{n} \frac{1}{\sigma_i^2 \cos^2 \beta_i} \begin{bmatrix} v(\alpha_i)v(\alpha_i)^{\mathrm{T}} & r_i \sin(\beta_i)v(\alpha_i) \\ * & r_i^2 \sin^2(\beta_i) \end{bmatrix} \tag{10.1}
$$

In this expression, $v(\alpha_i)$ is the versor corresponding to the surface normal direction α_i, $\beta_i = \alpha_i - (\theta + \varphi_i)$, is the incidence angle, and r_i is the distance to the obstacle. The FIM depends both on the environment, and the particular pose of the robot in the environment: there are parts of the environment where localization is easier than in others.

By computing the CRB as $(\mathcal{I}[x])^{-1}$, one obtains the achievable accuracy in a particular environment and pose. Experiments show that the CRB is approximately strict; in localization, the high signal-to-noise condition corresponds to having the sensor standard deviation σ negligible with respect to the size of the environment, which is usually the case. Given the FIM for "one shot" localization, the covariance over a trajectory can be evaluated by propagating the CRB through the system dynamics, according to a standard procedure [40].

The FIM can be used also in a more qualitative way to study the observability of the problem: the FIM drops rank in under-constrained situations (corridor or circular environment).

Thus it is possible to do a fairly complete characterization of localization. The reason is that, being a finite dimensional problem, all is needed is a straightforward use of the basic tools of statistics. However, this cannot be so easily extended to pose tracking.

10.3.3 The CRB for One-Shot Pose Tracking

We now consider "one-shot" pose tracking, in which we estimate the robot displacement given two sensor readings. Let $x_1 \in \mathrm{SE}(2)$ be the first pose, $\delta \in \mathrm{SE}(2)$ be the robot motion, and therefore $x_2 = x_1 \oplus \delta \in \mathrm{SE}(2)$ be the second pose. The sensor model reads $y = \mathbf{f}(x, w) + \varepsilon$, with w now an unknown map. Because the map is unknown, pose-tracking is qualitatively different from localization, and conceptually closer to full SLAM. Moreover, the problem is ill-posed if a prior for the map is not specified. Because of the unknown map, using the CRB is inconvenient, as one should: (1) choose a particular (differentiable) parametrization of w; (2) define the prior for w; (3) use a variant of the CRB such as the Bayesian Cramér-Rao bound [40].

However, the paper [8] shows that there is a "trick" one can use to obtain accuracy bounds without considering the prior distribution of the map. The result is that a lower bound for the FIM of δ is:

$$\mathcal{I}[\delta] \leq ((\mathcal{I}[x](x_1))^{-1} + (\mathcal{I}[x](x_2))^{-1})^{-1} \tag{10.2}$$

In this expression, $\mathcal{I}[x](x_1)$ and $\mathcal{I}[x](x_2)$ is the FIM for localization of x, evaluated at the poses $x = x_1$ and $x = x_2$ respectively. This bound is significant because it depends neither on the representation, nor on the prior used for the map. Therefore, it allows reducing the analysis of pose-tracking, an infinite-dimensional problem that involves both w and x, to the analysis of localization, a finite dimensional problem that involves only x.

The bound in (10.2) can be very optimistic, but it can also be shown that this is also the "best possible" bound, in the sense that there is always a certain prior for the map such that Equation 10.2 holds with equality. Moreover, it can be shown that (10.5) holds with equality in the limit $\delta \to 0$. This means that, for small steps, the prior for the map is unimportant; the data itself is the model.

10.3.4 The CRB for Pose Tracking Over a Trajectory

Let us consider now the problem of evaluating the accuracy of pose tracking over a trajectory. It might be counter-intuitive, but it is not necessarily true that the error of pose tracking grows linearly with the number of steps. The reason is that scans are matched pairwise: scan y_n is used for estimating both $\delta_{n-1} = \ominus x_{n-1} \oplus x_n$ and $\delta_n = \ominus x_n \oplus x_{n+1}$. The errors on δ_n and δ_{n-1} are now correlated, and because of this correlation, the covariance of the cumulative estimate $\delta_1 \oplus \cdots \oplus \delta_n$ is not just the sum of the uncertainties anymore. This effect must be taken into account when propagating the uncertainty [30], and it is likely to be an important effect to consider when deriving accuracy bounds [29].

As it turns out, for relative sensors this is actually a positive effect: errors tend to cancel out. A one-dimensional toy example can show this point. Suppose the robot is moving on a line; $x \in \mathbb{R}$, and there is a single wall at point $w \in \mathbb{R}$. The range-finder then measures a single reading $y = w - x$ which is the distance to the wall. Assume now that there are n steps going from pose x_1 to x_{n+1}, with $n + 1$ range-finder readings defined as: $y_i = (w - x_i) + \varepsilon_i$. Assuming we are doing pose tracking, we would first estimate the incremental displacement, by computing

$$\delta_i = y_{i+1} - y_i \tag{10.3}$$

Then, we would combine the incremental estimates as to obtain the cumulative estimate

$$\delta_{1:n} = \delta_1 + \delta_2 + \cdots + \delta_n \qquad (10.4)$$

What is the covariance of $\delta_{1:n}$? A wrong answer would be the following. From (10.3), it is clear that cov $[\delta_i] = 2\sigma^2$ if σ^2 is the covariance of the single reading. The naive way would be to sum covariances and obtain $\text{cov}[\delta_{1:n}] = 2n\sigma^2$. But this is not correct because the δ_i are correlated. In fact, Equation (10.4) can be rewritten as: $\delta_{1:n} = \delta_1 + \delta_2 + \cdots + \delta_n = (y_2 - y_1) + (y_3 - y_2) + \cdots + (y_{n+1} - y_n) = y_{n+1} - y_n$. Because the ys cancel out, the errors compensate exactly, and one then can conclude that cov $[\delta_{1:n}] = 2\sigma^2$, which is independent of the number of steps. This is an extreme example which corresponds to a trivial system. For nonlinear systems, the errors will not cancel perfectly. Still, this example shows that, in general, the bound over a trajectory is

$$\text{cov}(\delta_1 \oplus \cdots \oplus \delta_n) \geq (\mathcal{I}[x](x_1))^{-1} + (\mathcal{I}[x](x_{n+1}))^{-1} \qquad (10.5)$$

This is, however, very optimistic for actual scan matching; but it is the best one can do without considering the map.

10.3.5 The CRB for Mapping and SLAM

We now turn to the problem of mapping (estimating the map given the readings of known poses) and SLAM (where also the poses are unknown). Establishing accuracy bounds on map estimation is more laborious.

The dominant phenomenon is that asking what is achievable of a mapping mapping system, in terms of accuracy and metric consistency, is an ill-posed question if the prior (marginal probability) of the map distributions is not specified. Consider the following toy example. Suppose that the world is allowed to have only two shapes: triangle ($\triangle$) and circle ($\bigcirc$), and that they are equally likely. Formally, we set the world set $W = \{\triangle, \bigcirc\}$ and the prior $p(w = \bigcirc) = p(w = \triangle) = 0.5$. A decent sensor can distinguish exactly between the circle and the triangle. Therefore, the error of mapping is zero (or conversely, the accuracy is infinite): once one decides which one of the two shapes is correct, the reconstruction is perfect. The same holds even in the case where W is a much larger set, but $\triangle, \bigcirc$ are the only objects having non-zero prior. Therefore, the achievable accuracy depends arbitrarily on the prior.

Another example is the following: an "unstructured" environment has many more degrees of freedom than a "structured" one, which, for the most part, could be described even by a finite-dimensional representation. Therefore, it can be seen intuitively that, if the same sensor is used in both kinds of environments, the achievable accuracy in reconstructing the map is higher in the structured environment. However, to quantify this intuition, one needs to: (1) formally define the set S of structured environments; (2) formally define the set $\mathcal{U}$ of unstructured environments; (3) embed both in a larger set $\mathcal{W}$, for example the set of all closed curves; (4) restate the structured/unstructured hypotheses by defining appropriate structured/unstructured priors on $\mathcal{W}$; (5) apply one of the Bayesian bounds to derive that, yes, indeed, mapping is easier in a structured environment.

This formal reasoning about shapes has not been used in SLAM research yet, while it is used in other fields such as computer vision and stochastic geometry. An "intrinsic" theory of shape can be used to discuss the properties of shapes and shape distributions independently of their representation; see, e.g., the classical work [25] for points distributions and for curves [28, 22]. Such an approach is necessary for establishing meaningful bounds on the accuracy of mapping and SLAM.

10.4 Evaluating Local Metric Consistency of Robot-Generated Maps Using Force Field Simulation and Virtual Scans

In this section, we discuss how the integration of low-level spatial cognition processes (LLSC) and mid–level spatial cognition processes (MLSC) can help to improve the performance of robot mapping, and how a LLSC/MLSC system can be used for map evaluation.

In robot cognition, MLSC processes infer the presence of mid–level features from low-level data based on regional properties of the data. In our example case, we detect the presence of simple mid-level objects, i.e. line segments and rectangles. The MLSC processes model world knowledge, or assumptions about the environment. The example assumes the presence of (collapsed) walls and other man made structures. If possible, wall-like elements or elements resembling rectangular structures are detected, our system generates the most likely ideal model as a hypothesis, called 'Virtual Scan'. Virtual Scans are generated from the ideal, expected model in the same data format as the raw sensor data, hence Virtual Scans are added to the original scan data indistinguishably for the low-level alignment process; the alignment is therefore performed on an augmented data set.

In robot cognition, LLSC processes usually describe feature extraction processes based on local properties like spatial proximity. An example is metric inference on data points (laser scanner reflection points). In our system, laser scans (virtual or real) are aligned to a global map using mainly features of local proximity using the LLSC core process of 'Force Field Simulation' (FFS). FFS was recently introduced to robotics [24].

In FFS, each data point can be assigned a weight, or value of certainty. It also does not make a hard, but a soft decision about the data correspondences as a basis for the alignment. This is achieved by computation of a correspondence probability to multiple neighboring points, based on weight, distance and direction of underlying linear structures. Mainly these features makes FFS a natural choice over its main competitor, Iterative Closest Point (ICP) [6, 31], for the combination with Virtual Scans (however, the general idea of Virtual Scans is applicable to both approaches). The weight parameter can be utilized to indicate the strength of hypotheses, represented by the weight of virtual data.

FFS is an iterative alignment algorithm. The two levels (LLSC: data alignment by FFS, MLSC: data augmentation) are connected by a feedback structure as shown in Fig. 10.5, which is repeated in each iteration:

- The FFS-low-level-instances pre-process the data. They find correspondences based on low-level features. The low-level processing builds a preliminary version of the global map, which assists the mid-level feature detection
- The mid-level cognition module analyzes the preliminary global map, detects possible mid-level objects and models ideal hypothetical sources. These can be seen as suggestions, fed back into the low-level system by Virtual Scans. The low-level system in turn adjusts its processing for re-evaluation by the mid-level systems.

In such a system, MLSC processes steer LLSC processes introducing higher knowledge to enable spatial inferences the LLSC system is not able to draw by itself. However, the MLSC system also needs assistance of the LLSC for two reasons: MLSC systems concentrate on higher information which needs LLSC pre-processed data (e.g. a set of collinear points is passed to the MLSC as a single line segment). But also LLSC processes have to support the suggestions stated by the MLSC. Since MLSC introduces higher knowledge, it is dangerous to focus on spatial mid-level inferences too early. Feedback with the LLSC system enables more careful evaluation of plausibility.

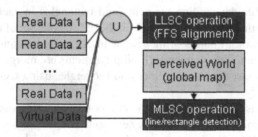

Fig. 10.5 This figure illustrates the feedback loop between the low-level spatial cognition module (LLSC) and the mid-level spatial cognition module (MLSC). The LLSC module works on the union of real scans and the Virtual Scan. The MLSC module in turn re-creates a new Virtual Scan based on the result of the LLSC module

The potential of MLSC has been largely unexplored in robotics, since recent research mainly addressed LLSC systems. Although this work on sophisticated statistical and geometrical models like extended Kalman Filters (EKF),e.g. [18], Particle Filters [14] and ICP [6, 31] utilized in mapping approaches show impressive results, their limits are clearly visible. However, having these well-engineered low-level systems at hand, it is natural to connect them to MLSC processes to mutually assist each other. In [5], the importance of 'Mental Imagery' in (Spatial) Cognition is emphasized and basic requirements of modeling are stated. Mental Images invent or recreate experiences resembling actually perceived events or objects. This is closely related to Virtual Scans.

10.4.1 Scan Alignment using Force Field Simulation

We assume the scans to be roughly pre-aligned. FFS alignment, described in detail in [24], is able to iteratively refine such an alignment based on the scan data only. In FFS, each single scan is seen as a non-deformable entity, a "rigid body". In each iteration, a translation and rotation is computed for each single scan simultaneously. This process minimizes a target function, the "point potential", which is defined on the set of all data points (real and Virtual Scans: FFS can *not* distinguish).

FFS solves the alignment problem as an optimization problem utilizing a gradient descent approach motivated by simulation of dynamics of rigid bodies (the scans) in gravitational fields, but "replaces laws of physics with constraints derived from human perception" [24]. The gravitational field is based on a correspondence function between all pairs of data points, the 'force' function. FFS minimizes the overlaying potential function induced by the force and converges towards a local minimum of the potential, representing a locally optimal transformation of scans. The force function is designed in a manner that a low potential corresponds to a visually good appearance of the global map. As scans are moved according to the laws of motion of rigid bodies in a force field, single scans are not deformed.

10.4.2 Augmenting Data Using Virtual Scans

The analysis module detects line segments and rectangles in each iteration of the FFS alignment. Both detection steps work on the entire point set of the current global map, i.e. the union of all points of the real scans. A preprocessing step transforms the point-based data to line segments. Similar segments are merged. This simplified data set allows for fast detection of lines and rectangles using techniques based on [16, 23] respectively.

A Virtual Scan is a set of virtual laser scan points, superimposed over the entire area of the global map. The detected line segments and rectangles are 'plotted' into the Virtual Scans, i.e. they are represented by point sets as if they would be detected by a laser scanner.

An important feature of a Virtual Scan is, that each point in the Virtual Scan is assigned a weight, being the strength of hypothesis of the virtual structure it represents. Figure 10.6 shows an example. This data set consists of 60 single laser scans. The scans resemble the situation of an indoors disaster scenery, scanned by multiple robots. We used an initial rough guess of robot poses as global map for two different runs of FFS, once with Virtual Scans, once without. The experiment was performed to demonstrate the increase in alignment performance using Virtual Scans. The increase in performance was evaluated by visual inspection, since for this data set no ground truth data is available. Comparing the final global maps of both runs, the utilization of Virtual Scans leads to distinct improvement in overall appearance and mapping details, see Fig. 10.7. Overall, the map is more 'straight' (compare e.g. the top wall), since the detection of globally present linear structures (top and left wall in Fig. 10.7) adjusts all participating single segments to be collinear. These

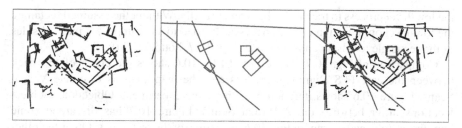

Fig. 10.6 The progression of Virtual Scans in the early stages of FFS. The image on the *left* shows the initial approximation of the global map using pose estimates and range image information received from the robot. The Virtual Scans, shown in the *center* image, represents lines and rectangles detected in the global map. The image on the *right* visualizes the superimposition of the global map and the Virtual Scans, which will be used the next iteration of FFS

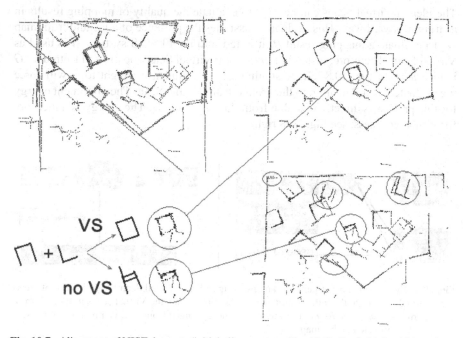

Fig. 10.7 Alignment of NIST data sets (initial alignment see Fig. 10.6). *Top left*: after 10 iterations with detected line and rectangular objects plotted into the Virtual Scan (solid lines). *Top right*: Final result using Virtual Scans, after 100 iterations. MLSC objects are not shown for clarity of display. Compare to *Bottom Right*: final result of alignment without Virtual Scans. Encircled areas show examples of improvement using Virtual Scans. *Bottom left*: The center rectangle could only be aligned correctly using MLSC information

corrections advance into the entire structure. More objectively, the improvements can be seen in certain details, the most distinct encircled in Fig. 10.7, bottom right. In particular, the rectangle in the center of the global map is an excellent example of a situation where correct alignment is not achievable with low-level knowledge only. Only the suggested rectangle from the Virtual Scan (see Fig. 10.7, top left) can force the low-level process to transform the scan correctly. Without the assumed rectangle the low-level optimization process necessarily tried to superimpose 2 parallel sides of the rectangle to falsely appear as one (Fig. 10.7, bottom right).

Comparison of Fig. 10.7, top left, and Fig. 10.6 shows the effect of feedback between the core FFS alignment process and the map analysis to create Virtual Scans. Figure 10.6 shows iteration 5 of the same experiment. Objects and object locations differ between the 2 Virtual Scans. Figure 10.7 has discarded some hypotheses (objects) present in Fig. 10.6, e.g. some of the rectangles. Other hypotheses are modified, e.g. the top wall is adjusted.

10.4.3 Map Evaluation Using Virtual Scans

The idea of Virtual Scans can be used to evaluate the quality of mapping results in a straightforward way. This evaluation assumes the presence of a ground truth map G. To evaluate a mapping result R, it is fed into the FFS/VS system. G is used as Virtual Scan. Therefore, instead of *creating* a Virtual Scan, the ground truth data G is *inserted*, see Fig. 10.8 left. Assigning a high confidence weight to G will force the evaluated data R to align to the ground truth Virtual Scan. The alignment energy for this process is directly readable from the FFS module. The energy is a measure for visual closeness, see Fig. 10.8 right.

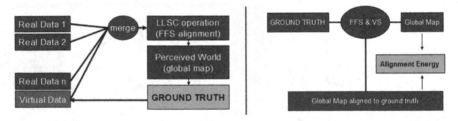

Fig. 10.8 Evaluation of robot-generated maps using FFS and Virtual Scans. *Left*: Instead of creating a Virtual Scan from the initial estimation of the global map, the Virtual scans will result from ground truth global maps. *Right*: The alignment energy gained from FFS is a measure for visual fitness of the evaluated global map

This evaluation procedure is adjustable to local ground truth data, since the adjustment energy in regions of interest can be emphasized. The energy computed in FFS is a symmetric measure, i.e. aligning R to G leads to the same measure as aligning G to R. This can be used for 'inverse evaluation' (evaluating the ground truth G with R) in the following manner: G can be manually split into local regions of interest (room, hallway, etc.). These regions are represented as single scans and used as input for the FFS system, while the map R is used as Virtual Scan. Such a setting has huge advantages, since, on one hand, it is more independent of the actual data representation of R. On the other hand, and more important, the manual split of G defines regions which can be assigned independent evaluation scores.

Note that in the Virtual Scan evaluation approach, ground truth data representation is not limited to *physical objects*, but can consist of *geometric properties* (e.g. evaluate how well map represents lines or rectangles). In this case, the original Virtual Scan approach is utilized, instead of insertion. The properties are then defined by means of MLSC-analysis modules. This is especially interesting if no ground truth data is available.

10.5 Evaluating Global Metric Consistency of Robot-Generated Maps

Evaluating the global metric consistency of maps is probably the most logical and most intuitive method for assessing the quality of robot–generated maps. Metric maps provide a representation where the spatial relationship between any two objects in the map is proportional to the spatial relationship of the corresponding objects in the actual environment. Therefore, assessing the quality of the global metric maps is based on the spatial consistency of features, such as walls and hallways, between the map produced by the robot and the actual operational environment. Using geographically registered images as the underlying representation for these maps, provides a basis for assessing spatial consistency between the robot-generated map and the ground truth map of the environment as shown in Fig. 10.9.

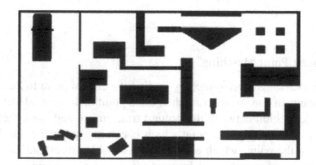

Fig. 10.9 Metric maps maintain the spatial relationship between features in the environment using geo-registered images

This section introduces preliminary work to develop novel methods to assess the global metric consistency of robot-generated maps. In order to identify and measure the consistency between the structural features in the environment, this approach uses three different algorithms to detect corresponding features found in the maps: Harris Corner Detector (described in Section 10.5.1), Hough-Transform (Section 10.5.2), and Scale Invariant Feature Transform (discussed in Section 10.5.3). Each of these algorithms will provide a measure of map accuracy that will be discussed in Section 10.5.4.

10.5.1 Harris-Based Algorithm

The first algorithm developed to assessing the global consistency of robot–generated maps is based on the Harris Corner Detector [15, 39]. Corner capturing algorithms have proven to be effective means for identifying points of interest that may be used to mark corresponding features between two maps. Additionally, corner detector algorithms are inherently invariant to rotation, scale, illumination variation and image noise. This is a desirable metric which will enable us to deal with minor noise, rotation and scale problems in the map. However, this approach relies on two fundamental assumptions. First, global metric maps are at the minimum represented as a binary images that shows traversable and non-traversable space as different colors. Second, many of the operational environments contain some structural components resulting in angles of the incidents between two features.

Let us assume we are given two map images to compare, namely, X and Y. In order to compare these maps, the Harris Corner Detector is used to identify points of interest in each of the map images. After calculating the interest point using the Harris corner detector, we use the closest point matching process to generate the vector maps which are used for calculating the quality metric (described in Section 10.5.4). To generate the vector map, we first find the corners which are closest to the point under consideration. We then use that point in the robot–generated map and find its closest point in the ground truth map and eliminate those points from both maps. This increases the number of true points that are matched to arrive at the map quality.

The closest point matching process and the vector map generation are described below:

10.5.1.1 Closest Point Matching

Closest point matching is performed by finding the closest point to the corresponding interest points in one image to another. Each point in the ground truth is mapped in a one to one fashion between the ground truth image and the target image. To keep points from matching to a point which is extremely far, the matching is performed only for the points which exist below a specified threshold. So it generates a displacement map for each point from one image to another image. The obvious benefit is the localized identification of the object interest points.

The closest point match can be described by:

$$Match = Dis(FV(P(x,y))) - FV(P_\theta(x,y))) \tag{10.6}$$

Equation 10.6 describes that the match is the point which is equivalent to the point in one map to the corresponding region in another map under a specified threshold, where FV is the feature vector of $P(x,y)$ and Dis is the distance between two corresponding feature vectors. Only in the case of the SIFT features the comparing criteria is based on the calculated descriptors.

10.5.1.2 Vectorial Space

The displacement or vector map calculated in the last step provides much more information regarding the kind of distortion which appeared in the image. This way the vector map is a localized distortion map in the image. This can be done in both directions to identify the missing features which were not captured and extra features which do not really exist. Figure 10.10 shows the displacement of closest points in ground truth as compared to the vectorial space of the test image.

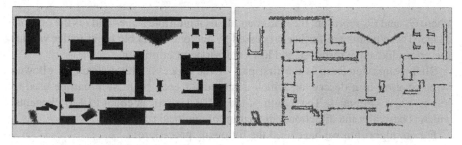

Fig. 10.10 The vector map calculated from the ground truth map, on *left*, and the robot-generated map, on *right*, shows the correspondence between the points of interested identified in both maps

10.5.2 Hough-Based Algorithm

To account for the structural detail we have used Hough transform [20, 4] to transfer the map from Euclidean space to Hough space. This has the benefit of identifying lines in the image. These lines are compared according to the position of lines as points in the Hough space. Hough space is created by exchanging the Euclidean coordinates with the parameterized values from the parametric form of the equation of the line.

$$r_\theta = x \times \cos\theta + y \times \sin\theta$$

This helps in identifying lines easily as in the case of Hough space the points with large values will be highly likely to represent the lines. This same process can be repeated to generate the space for circles and other geometrical objects detection. A variation of the Hough transform which is known as the generalized Hough transform, can be used to detect different types of arbitrary shapes in images. This can be used to detect lines, squares (e.g. rooms), circle (e.g. roundabouts) in the map which will be a more generalized way to calculate the map quality. After detection of these features the matching features can be located in the ground truth map and compared for map quality.

10.5.3 Scale Invariant Feature Transform

Scale Invariant Feature Transform (SIFT) was introduced by the David Lowe in [26]. Since then, SIFT based localized features have gained prominence among researchers due to their invariability to rotation scale and even dynamic changes. To assess the map quality we have proposed an algorithm based on SIFT. SIFT feature are calculated from extrema detection by finding the extrema points from difference in Gaussian images as shown in Equation 10.7:

$$DOG(I) = (G_m \times I) - (G_n \times I) \qquad (10.7)$$

where G_m and G_n represent the Gaussian filters at multiple scales and I is the original image. These points are further processed to find out the stable point under various conditions like edge response and low contrast point elimination.

SIFT points detection is the first part of the process. This is usually followed by calculation of a descriptor, which is stored for each point so it can be used to compare across images. The length of the SIFT detector is equal to 128 elements, which is basically the directional histogram of the local region.

For our algorithm we have used the following procedure:

1. First the entropy [13] of the image is calculated so that important regions with high entropy are identified. As our maps are binary images it is necessary to convert them into multiple scales with more information so that useful features are calculated.
2. This image is passed on to the SIFT for feature detection and descriptor calculation, see Fig. 10.11 for an example of SIFT features.

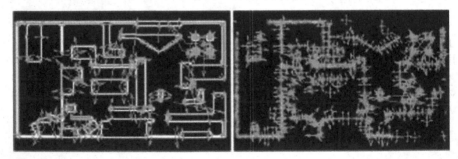

Fig. 10.11 Using Scale Invariant Feature Transform (SIFT) extract points of interest in the ground truth map and the robot-generated maps provides features to evaluating the metric quality of maps. The figure on the *left* shows the points of interest extracted from the underlying ground truth map using SIFT, where the image on the *right* shows the points of interested extracted from the robot-generated map

10.5.4 Quality Measure

The map quality measure is calculated using the ratio between the set of features. The map quality can be defined mathematically as

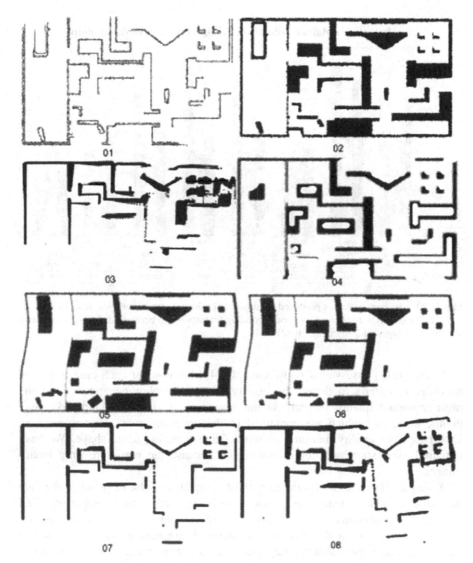

Fig. 10.12 The set of test images evaluated using the the methodologies described in this chapter. The results of the evaluation are shown in Fig. 10.13. In an attempt to account for boundary conditions, distorted ground truth maps (images 01, 02, and images 04–06) were used in conjunction actual maps produced by robots (images 03 and images 07–08)

$$q = \frac{RMF}{GTF} \tag{10.8}$$

where *RMF* is the number of valid feature points found in the robot generated map and *GTF* is the number of feature points in the ground truth map.

The set of test images are shown in Fig. 10.12 and the corresponding map quality values are depicted in Fig. 10.13.

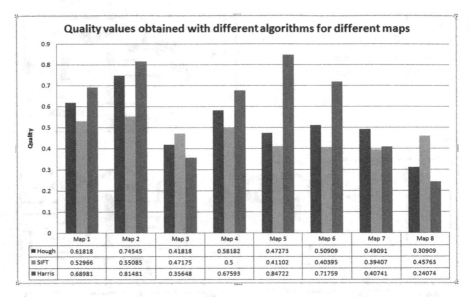

Fig. 10.13 The quality values produced by different evaluation methodologies described in this section. The quality value is a ratio of corresponding features found between the ground truth map and the test images (shown in Fig. 10.12)

Image quality assessment is difficult [42, 43] because for each case there can be different criteria to define the quality. For these robot generated maps the most important quality measure is the amount of features or landmarks (e.g. points, lines, etc) which are contained in the generated map. That is why we have based our quality measure on those features having same shape. We have not used the texture and color information because the maps are only binary images.

A very subtle issue is with the finding of the quality of the maps when they are the subset of a larger map. The ground truth is assumed to be the super-set of all the maps so it contains all the features and information. So to assess the quality of the map which is smaller than the ground truth, we have to identify the subset from ground truth for which the map was generated. This remains an issue with this algorithm although for maps which are equivalent to the ground truth the algorithm gives fairly accurate results.

Another remaining issue is the utilization of the threshold. Utilization of threshold can be a problem because we will not be able to match features if the maps are not aligned as in the case of Harris and Hough transform but this is not the case for SIFT based detector because it can detect matches even if they are far away, independent of scale, rotation and dislocation. Although for the Harris and Hough alignment of the map remains an important point. Alignment can be achieved by a start-up marker that identifies a stable point between the robot generated map and the ground truth. A map can be considered more accurate if it consistently shows good performance in all three measures.

10.5.5 Limitations

This system is only suitable for offline measurement for the quality of the maps. The measure of quality is very difficult to define because the requirements on which the map quality is based can be changed according to the need.

This algorithm measures the quality only on the basis of the information content of the image. These maps only contain bi-level images without any additional information. Map distortions and noise are not considered because the information is intact even with the added noise.

Some of the limitations which are observed are due to the type of maps used for processing. If the map has noise, such as a jagged line or map with distortions, most likely the Harris corner detector will find lots of corners which could give erroneous results. Also Hough transform will fail for the case when point cloud data is separated quite far apart. Similarly, for the SIFT case, if there is too much noise in the maps, this will introduce additional features which can cause problems during comparison of the features, because closely related features will give similar results.

10.6 Conclusion

Motivated by the absence of a comprehensive and consistent methodology for evaluating robot-generated maps in a quantitative sense, our efforts have focused on bringing together the research community to collectively address this problem. This Chapter discussed our recent efforts in addressing this problem based on the aforementioned recursive developmental cycle that encompasses standard test methods and objective evaluation methodologies.

Based on our previous work on performance singularity identification and testing, three test scenarios were developed by accounting for environmental conditions that robots typically encounter in unstructured domains. These scenarios considered the cases where there were distinct features readily available for reliable correspondence determination, occluded features that provided a considerable degree of difficulty, and a pathological case where establishing correspondences was extremely difficult. By varying the degree of difficulty, one can evaluate and analyze the robustness of mapping approaches.

Theoretical measures were developed using the Cramer-Rao Bound to arrive at a lower bound accuracy to compare experimental results. It was shown how the CRB bounds can be useful for assessing localization, pose tracking, and mapping estimates. Force Field Simulation was proposed as a methodology for assessing consistency of maps. By using the concept of Virtual Scans within the iterative FFS algorithm and experimental data, it was demonstrated how to evaluate the consistency of map quality at a local (metric) level. Three measures were then proposed to quantitatively assess global (metric) map quality using features extracted by three different detectors namely, the Harris Corner Detector, the Hough Transform and the Scale Invariant Feature Transform with respect to a ground truth map.

Our continuing research efforts will focus on how we can better refine the test methods and improve the map evaluation process. By working closely with the research community to develop standardized tools via standard test methods and reference data sets, researchers and consumers alike will be able to better evaluate the costs and benefits associated with robotic mapping technologies based on the end-user constraints.

Reference

1. Abdallah, S., Asmar, D,. Zelek, J.: Towards benchmarks for vision SLAM algorithms. In: Proceedings of the IEEE International Conference on Robotics and Automation, pp. 3834–3839 (2006)
2. ASTM Committee E54.08.01 on Homeland Security Operations; Operational Equippment; Robots. http://www.astm.org/COMMIT/SUBCOMMIT/E5408.htm
3. ASTM International: In: Form and style for ASTM Standards. (2007)
4. Ballard, D.H.: Generalizing the Hough Transform to Detect Arbitrary Shapes, Los Altos, pp. 714–725 (1987)
5. Bertel, S., Barkowsky, T., Engel, D., Freska, C.: Computational modeling of reasoning with mental images: basic requirements. In: D. Fum, F. del Missier, A. Stocco (Eds.), Proceedings of the 7th International Conference on Cognitive Modeling ICCM06 (2006)
6. Besl, P., McKay, N.: A method for registration of 3.d shapes. IEEE PAMI, 14(2) (1992)
7. Censi, A.: On achievable accuracy for range-finder localization. In: Proceedings of the IEEE International Conference on Robotics and Automation (ICRA), pp. 4170–4175. Rome, Italy (2007). DOI 10.1109/ROBOT.2007.364120. URL http://purl.org/censi/2006/accuracy. See also expanded version at http://purl.org/censi/2007/accuracy
8. Censi, A.: On achievable accuracy for pose-tracking. In: Proceedings of the IEEE International Conference on Robotics & Automation (ICRA) (2009).URL http://purl.org/censi/2006/icpcov
9. Chandran-Ramesh, M., Newman, P.: Assessing Map Quality using Conditional Random Fields.k In: Proceedings of the Field and Service Robotics (2007)
10. Crisan, D., Doucet, A.: A survey of convergence results on particle filtering for practitioners. IEEE Transactions on Signal Processing 50(3), 736–746 (2002).URL citeseer.ist.psu.edu/crisan02survey.html
11. Fox, D.: Adapting the sample size in particle filters through KLD-sampling. International Journal of Robotics Research 22(12) (2003)
12. Frese, U.: A discussion of simultaneous localization and mapping. Autonomous Robots 20(1), 25–42 (2006)
13. Gonzalez, R., Woods, R., Eddins, S.: Chapter 11, pp. 714–725 (2003)
14. Grisetti, G., Stachniss, C., Burgard, W.: Improving grid-based slam with rao-blackwellized particle filters by adaptive proposals and selective resampling. ICRA (2005)
15. Harris, C., Stephens, M.: A combined corner and edge detection. In: Proceedings of The Fourth Alvey Vision Conference, pp. 147–151 (1988).
16. Hough, P.V.C.: Methods and means for recognizing complex patterns. US patent 3,069,654 (1962)
17. Huang, G., Mourikis, A., Roumeliotis, S.: Analysis and improvement of the consistency of Extended Kalman Filter based SLAM. In: Proceedings of the IEEE International Conference on Robotics & Automation (ICRA), pp. 473–479 (2008). DOI 10.1109/ROBOT.2008.4543252
18. Huang, S., Dissanayake, G.: Convergence analysis for extended kalman filter based slam. IEEE International Conference on Robotics and Automation (2006)

19. Huang, S., Dissanayake, G.: Convergence and consistency analysis for Extended Kalman Filter based SLAM. IEEE Transactions on Robotics **23**(5), 1036–1049 (2007). DOI 10.1109/TRO.2007.903811
20. Illingworth, J., Kittler, J.: A Survey of the Hough Transform. Computer Vision, Graphics, and Image Processing **44**(1), 87–116 (1988)
21. Jacobs Map Analysis Toolkit. http://usarsim.sourceforge.net/
22. Joshi, S., Kaziska, D., Srivastava, A., Mio, W.: Riemannian structures on shape spaces: A framework for statistical inferences. Statistics and Analysis of Shapes pp. 313–333 (2006)
23. Lagunovsky, D.S.A.: Fast line and rectangle detection by clustering and grouping. Proc. of CAIP'97, Kiel, Germany (1997)
24. Lakaemper, R., Adluru, N., Latecki, L., Madhavan, R.: Multi robot mapping using force field simulation. Journal of Field Robotics, Special Issue on Quantitative Performance Evaluation of Robotic and Intelligent Systems (2007)
25. Le, H., Kendall, D.G.: The Riemannian structure of Euclidean shape spaces: A novel environment for statistics. Annals of Statistics **21**(3), 1225–1271 (1993)
26. Lowe, D.G.: Distinctive Image Features from Scale-Invariant Keypoints. International Journal of Computer Vision **60**(2), 91–110 (2004)
27. Madhavan, R., Scrapper, C., Kleiner, A.e.: Quantitative Performance Evaluation of Navigation Solutions for Mobile Robots. Workshop Proceedings, Robotics: Science and Systems (RSS) Conference, Zurich, Switzerland; http://kaspar.informatik.uni-freiburg.de/~rss/(2008)
28. Michor, P.W., Mumford, D.: Riemannian geometries on spaces of plane curves. Journal of the European Mathematical Society **8**, 1–48 (2003)
29. Mourikis, A.: Personal communication (2008)
30. Mourikis, A., Roumeliotis, S.: On the treatment of relative-pose measurements for mobile robot localization. In: Proceedings of the IEEE International Conference on Robotics & Automation (ICRA). Orlando, FL (2006)
31. Nuechter, A., Lingemann, K., Hertzberg, J., Surmann, H., Pervoelz, K., Hennig, M., Tiruchinapalli, K., Worst, R., Christaller, T.: Mapping of rescue environments with kurt3d. Proceedings of the International Workshop on Safety, Security and Rescue Robotics (SSRR '05),Kobe, Japan (2005)
32. OpenSLAM. http://www.openslam.org/
33. Radish: The Robotics Data Set Repository. http://radish.sourceforge.net/
34. RoboCup Rescue. http://http://www.robocuprescue.org
35. Scrapper, C., Madhavan, R., Balakirsky, S.: Stable Navigation Solutions for Robots in Complex Environments. In: IEEE International Workshop on Safety, Security, and Rescue Robotics (SSRR2007) (2007)
36. Scrapper, C., Madhavan, R., Balakirsky, S.: Performance analysis for stable mobile robot navigation solutions. pp. 696206-1-696206-12. SPIE (2008)
37. Scrapper, C., Madhavan, R., Balakirsky, S.O.: Special Session on 'Quantitative Assessment of Robot-generated Maps'. Proceedings of the Performance Metrics for Intelligent Systems (PerMIS) Workshop, R. Madhavan and E. Messina (eds.); http://www.isd.mel.nist.gov/PerMIS_2008/ (2008)
38. Thrun, S.: Learning Metric-Topological Maps for Indoor Mobile Robot Navigation. Artificial Intelligence pp. 21–71 (1998)
39. Trajkovic, M., Hedley, M.: Fast corner detection". Image and Vision Computing **16**, 75–87 (1998)
40. Trees, H.L.V., Bell, K.L.: Bayesian Bounds for Parameter Estimation and Nonlinear Filtering/Tracking. Wiley-IEEE Press (2007)
41. Varsadan, I., Birk, A. and Pfingsthorn, M.: Determining Map Quality through an Image Similarity Metric. In: Proceedings of the RoboCup Symposium (2008)
42. Wang, Z., Bovik, A.: A Universal Image Quality Index. In: IEEE Signal Processing Letters, vol. 9, pp. 81–84 (2002)

43. Wang, Z., Bovik, A.C.: Why is Image Quality Assessment so Difficult. In: in Proc. IEEE Int. Conf. Acoust., Speech, and Signal Processing, pp. 3313–3316 (2002)
44. Weingarten, J., Siegwart, R.: EKF-based 3D SLAM for Structured Environment Reconstruction. In: Proceedings of the IEEE/RSJ International Conference on Intelligent Robots and Systems, pp. 3834–3839 (2005)
45. Zhou, W., Miro, J., Dissanayake, G.: Information Efficient 3D Visual SLAM in Unstructured Domains. In: Proceedings of the International Conference on Intelligent Sensors, Sensor Networks and Information, pp. 323–328 (2007)

Chapter 11
Mobile Robotic Surveying Performance for Planetary Surface Site Characterization

Edward Tunstel, John M. Dolan, Terrence Fong, and Debra Schreckenghost

Abstract Robotic systems will perform mobile surveys for scientific and engineering purposes as part of future missions on lunar and planetary surfaces. With site characterization as a task objective various system configurations and surveying techniques are possible. This chapter describes several examples of mobile surveying approaches using local and remote sensing configurations. A geometric measure of area coverage performance is applied to each and relative performance in surveying a common area is characterized by expected performance trends. Performance metrics that solely express geometric aspects of the robotic task are limited in utility as decision aids to human mission operators. As such, the importance of enriching such metrics by incorporating additional attributes germane to surveying on planetary surfaces is highlighted. Examples of enriched metrics employed by recent NASA research work on human-supervised robotic surveying are provided.

11.1 Introduction

On Earth, the exploration and settlement of uncharted territories requires prior prospecting and surveying for useful resources. The same is true for exploration and eventual human settlement of outposts on planet surfaces. Space agencies rely on precursor robotic missions to acquire the data and information necessary to understand planetary surface regions and the feasibility of sending human explorers on future missions. Surveying refers to the systematic method or technique of making measurements essential for accurately determining the geo-spatial location of commodities of interest in a designated area. Prospecting refers to the methodical and qualitative physical search or exploration for the commodity. Intelligent robots will need to be equipped with effective techniques for performing these tasks in

E. Tunstel (✉)
Space Department, Johns Hopkins Univ. Applied Physics Laboratory, Laurel, MD 20723, USA
e-mail: edward.tunstel@jhuapl.edu

R. Madhavan et al. (eds.), *Performance Evaluation and Benchmarking
of Intelligent Systems*, DOI 10.1007/978-1-4419-0492-8_11,
© Springer Science+Business Media, LLC 2009

preparation for the return of astronauts to the Moon and exploration of planetary surfaces beyond.

Task-oriented algorithms that support systematic mobile surveys using science instruments are needed for planetary surface characterization on science missions. They are also needed for in-situ resource prospecting and mapping on robotic missions that serve as precursors to human exploration missions. Typical objectives of site surveys include sensor coverage of designated areas. Area coverage problems for mobile robotic survey systems commonly employ sensing devices requiring close proximity to or contact with the measured phenomenon. Examples of such "local sensing" devices include ground penetrating radar, metal detectors for humanitarian de-mining, fluorescence imagers for organic molecule detection, and various spectrometer types. Mobile robotic vehicles, or rovers, that carry survey systems comprised of local sensing devices must physically cover most, if not all, of the terrain in the designated survey area.

Remote sensor-based area coverage contrasts with these more common area coverage problems for mobile robotic surveys. Remote sensing instruments can acquire measurements at significant distances away from the measured phenomenon (e.g., based on radar or optical devices such as lasers). Measurements along a line-of-sight to detect airborne phenomena such as near-surface gas emissions, for example, account for coverage of terrain below that line-of-sight. This permits a two-dimensional search over terrain using discrete linear measurements from a distance (similar to scanning laser rangefinders).

Remote sensor-based methods are not applicable to all surveying tasks. For surveys in which they are not a better solution, they are often excellent complements to local sensor-based methods. That is, remote sensor-based surveys can serve as an efficient means to cover wide areas with the purpose of localizing smaller areas at which local sensor-based surveys of higher resolution are appropriate.

Mobility algorithms for surveying provide a means to systematically acquire measurements covering an area to be surveyed by transporting sensors and instruments to multiple locations and vantage points. Algorithms employing parallel-line transects or parallel swaths are commonly used to address robotic area coverage problems by single robots [1, 2] and multiple robots [3] when using local sensing devices. Full and partial coverage planners have also been proposed for rovers that survey terrain using local sensing devices [4]. Random walk and chaotic coverage trajectories are also possible [5, 6] but less popular for systematic coverage tasks. Remote sensor-based survey approaches for rovers have recently been proposed for single- and two-rover systems performing measurements through the near-surface atmosphere [7, 8].

This chapter examines several examples of both mobile survey types and applies a geometric measure of their area coverage performance. It further advocates the importance of additional attributes germane to surveying tasks for planetary surface exploration and presents representative examples from NASA-funded robotics research. The additional attributes are intended to enrich the effectiveness and relevance of basic geometric measures or support formulation of new metrics for intelligent/autonomous robotic survey systems in planetary surface domains.

11.2 Local Sensor-Based Surveying

Due to required proximity to measured phenomena and relatively small footprints of sensitivity, local sensing devices typically necessitate dense coverage of a designated *survey region* by the host mobile platform. As such, local sensor-based mobile surveys seek to acquire measurements that cumulatively cover all or most of the survey region. Associated survey sensors or instruments are typically mounted on a rover body or deployed on a rover-attached boom or manipulator arm. Rover mobility serves to transport the footprint of the survey instrument(s) over terrain along trajectories that fill the survey region. Figure 11.1 depicts this scenario.

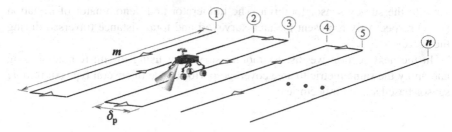

Fig. 11.1 Local sensor-based surveying along n parallel transects of length m

Among alternative survey trajectories for the mobile platform, parallel transects, spirals, and random walks have been proposed. Parallel transects are most commonly applied for coverage tasks although a spiral coverage approach was proposed for mobile mining of regolith on the lunar surface as part of a so-called spiral mining system [9]. Recent field tests, focused on planetary surface site characterization, used a rover to perform surveys using a ground penetrating radar (GPR) along parallel transects covering a 700×700 m survey region [10]. This instrument was used to map the subsurface structure at the site. A total traverse distance of 20.5 km was sufficient to cover the survey region using densely spaced parallel transects (with no overlap of the GPR sensor footprint on adjacent transects) [10]. North-south and east-west transects were planned and executed. Parallel transect trajectories were also suggested for systematic resource prospecting of wide areas on the lunar surface [11].

Basic geometric measures are often used to measure area coverage performance of such local sensor-based survey algorithms. Examples include measures of distance traveled, rover drive- and run-time [10], and percent of total area covered [2]. Variants of the latter have been proposed based on distribution of measurement samples within the cells of a tessellated grid representation of the survey region [4]. Here, we apply a basic geometric measure comprising a combination of such attributes. It is referred to as the *quality of performance, QoP*, defined as a ratio of area covered to distance traveled [12]. Applying this metric to the recent field test result mentioned above would yield a QoP of 24 based on the survey region area and total traverse distance (note that the north-south and east-west transects performed in that field test covered the survey region twice, effectively). In theory, an optimal value for this metric might be associated with the minimum distance traversed while acquiring non-overlapping survey sensor measurements; although in practice, it may depend on a variety of sensor, system, and task characteristics. In general, the QoP for a local sensor-based survey along parallel transects (Fig. 11.1) is computed as

$$Q_p = \frac{m(n-1)\delta_p}{[mn + (n-1)\delta_p]} \qquad (11.1)$$

where m is the transect length, n is the number of transects traversed, and δ_p is the separation distance between adjacent transects and is assumed here to be comparable to the survey sensor footprint. The numerator and denominator of Equation (11.1) respectively represent the area surveyed and total distance traversed during the survey.

In the next section we discuss mobile surveying using in-situ remote sensing and apply the same metric to area coverage performance of several types of remote sensor-based survey trajectories.

11.3 Remote Sensor-Based Surveying

Mobile remote sensor-based surveys can be performed by measurement systems whose components are separated by a distance across terrain. Such systems are comprised of an active/passive instrument component on one end and a passive/active component on the other end. One end could be stationary while the other is mobile (fixed-mobile) or both ends could be mobile (mobile–mobile). Both are considered here beginning with a fixed-mobile configuration, which is suitable for single-site surveys (unless the fixed component is also transportable to other sites).

11.3.1 Single-Site Remote Sensing Surveys

Consider a fixed-mobile configuration comprised of an active rover-mounted instrument, a passive receiver or retroreflector at a fixed location a distance away, a rover pan-tilt unit to point the instrument at the passive component for measurements, and the rover itself to move the instrument spatially over terrain. The passive component would remain stationary at a position central to a designated survey region. This configuration is similar to those of Total Station systems commonly used by civil engineers for land surveys and comprised of a theodolite on one end and stationary 360° retroreflector on the other. Like a civil engineer, a rover using such a survey system can acquire measurements from any radial direction when the fixed component is within line-of-sight and measurement range.

For mobile surveys, measurements are coordinated with rover mobility to survey terrain via a series of measurements across a distance d, which varies with rover position relative to the fixed component (Fig. 11.2). Such mobile robotic systems are under development for planetary surface surveying to achieve optical measurements at maximum distances of hundreds of meters [13]. The long-range measurement capability coupled with rover mobility enables wide-area surveys.

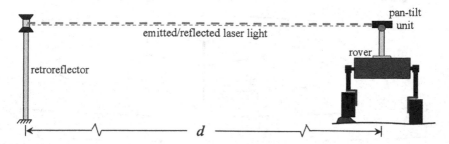

Fig. 11.2 Example of a distributed fixed-mobile measurement configuration

Concentric circular or spiral trajectories are compatible with distributed fixed-mobile configurations for remote sensor-based surveying. A designated survey region with a fixed instrument component at its center can be covered by traversing concentric circular trajectories as follows. The location of the fixed component is known and considered to be the origin of an inertial coordinate system in which the survey region and task is defined. Rover pose during surveys is estimated relative to this coordinate system. Beginning at a designated radial distance from the fixed component, the rover moves in arc-increments stopping periodically on the trajectory to acquire measurements. We refer to these measurement locations as *m-nodes*. Measurements along a line-of-sight between the rover-mounted active instrument and the fixed component account for 2-D coverage of terrain below the line-of-sight. Such measurement techniques are used on Earth with laser-based spectrometers to probe for and detect gas emissions during environmental site surveys [14], and they are being developed for the same fundamental use on Mars [13]. An accumulation of such linear measurements from discrete radial locations and distances achieves survey region coverage.

The following four parameters are used to configure a concentric circular trajectory covering a given survey region (Fig. 11.3): innermost circle radius, ρ_1; radial distance, δ_c, between circumferences of consecutive circles; arc length, s, between consecutive m-nodes on a circle; and positive integer, n, designating the nth or outermost circle including the survey region. The algorithm assumes that the rover is already within the survey region and that the fixed component is within line-of-sight from the rover [7]. If it is not, then no measurement is made. The survey completes when the nth circular trajectory is followed.

A fixed-mobile configuration performs a spiral survey in a similar manner, differing only in that the rover drives in arcs along a trajectory of continuously increasing radius and needs no specific maneuvers to transition between successive spiral branches at larger radii [7]. Figure 11.4 illustrates the spiral survey trajectory, which is parametrically similar to a circular survey.

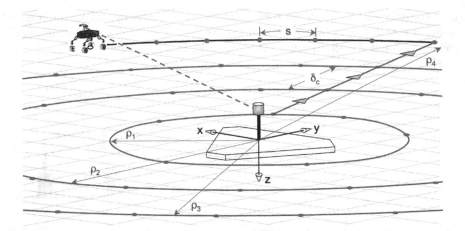

Fig. 11.3 Concentric circular remote sensing survey and parameters

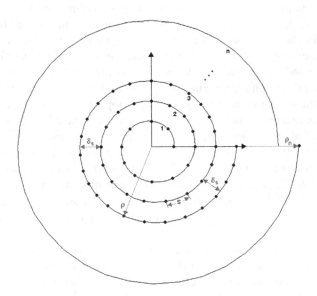

Fig. 11.4 Overhead view of spiral remote sensing survey and parameters

Both surveys are configured in a flexible manner to achieve desired degrees of measurement and area coverage resolution using the four parameters (ρ, δ, s, n). The surveys are primarily constrained by rover kinematic limitations, instrument effective minimum and maximum ranges, and terrain topography in the survey region whether executed radially inward or outward.

11.3.2 Single-Site Remote Survey Performance

Related research on distributed surveying [12] introduced the quality of performance metric defined earlier. We also apply this metric here as a basis for comparing expected performance of the concentric circular and spiral trajectories for distributed surveying.

The area of a survey region covered by either a concentric circular or linear spiral trajectory is equal to or roughly the same as $A = \pi \rho_n^2$, where ρ_n is the radius of the outermost circle or spiral branch. Areas within the survey region that are occupied by the stationary instrument component (at the origin of the survey coordinate system) and non-traversable obstacles are neglected. The total traverse distance D_c required for a concentric circular survey is the sum of distances traveled on each circumference and the radial separation distances, δ_c, between them:

$$D_c = 2\pi \left(\sum_{i=1}^{n} \rho_i \right) + (n-1)\delta_c \qquad (11.2)$$

yielding the following QoP,

$$Q_c = \frac{\pi \rho_n^2}{2\pi \left(\sum_{i=1}^{n} \rho_i \right) + (n-1)\delta_c}. \qquad (11.3)$$

For each linear spiral branch traversed (every $\theta = 2\pi$ radians), the spiral radius ρ increases by δ_s (Fig. 11.4), i.e., $\rho = (\delta_s/2\pi)\theta$. It can be shown [15] that the total traverse distance D_s required for a linear spiral trajectory is then expressed as

$$D_s = \frac{\delta_s}{4\pi}\theta_n^2 \qquad (11.4)$$

where θ_n is the maximum spiral angle reached. The resulting QoP is then

$$Q_s = \frac{4\pi^2 \rho_n^2}{\delta_s \theta_n^2}. \qquad (11.5)$$

Based on the QoP metric the two fixed-mobile configurations for remote sensor-based surveying can be compared. With roughly the same survey region area, their QoPs are distinguished by distance traveled. If the spiral begins and ends as shown in Fig. 11.4, then $\theta_n = 2n\pi$, and $D_s = \delta_s \pi n^2$. For closest equivalence between the two trajectories, let the first circle radius be equal to the initial spiral radius, ρ_1, and let $\rho_1 = \delta_c = \delta_s$. Under these conditions, $\rho_2 = \rho_1 + \delta_c = 2\delta_c$, $\rho_3 = 3\delta_c$, $\rho_4 = 4\delta_c$, and so on. The summation term in Equation (11.2) then becomes a function of an arithmetic series and simplifies as follows.

$$\sum_{i=1}^{n} \rho_i = \delta_c + 2\delta_c + 2\delta_c + \ldots + n\delta_c \qquad (11.6)$$

$$= \delta_c(1 + 2 + 3 + \ldots + n)$$

$$= \delta_c \left[\frac{1}{2} n(n + 1) \right]$$

Using this result in Equation (11.2), we have

$$D_c = 2\pi \delta_c \left[\frac{1}{2} n(n + 1) \right] + (n - 1)\delta_c \qquad (11.7)$$

$$= \delta_c \left[\pi n^2 + (\pi + 1)n - 1 \right].$$

Therefore, $D_c > D_s$, independent of an equivalent separation distance. A rover executing a concentric circular survey of n circles would need to traverse over $(4n-1)\delta_c$ meters more to cover the same area as it could with a spiral trajectory of n branches. As an example, to traverse a survey trajectory of $n = 3$ concentric circles separated by $\delta_c = 10$ m, a rover would drive a linear distance of 397 m; to survey a roughly equivalent area using a spiral trajectory of $n = 3$ branches separated by $\delta_s = 10$ m it would drive a linear distance of 283 m, or 29% less.

11.3.3 Multiple-Site Remote Sensing Surveys

An example of a mobile-mobile, or tandem, survey system configuration is illustrated in Fig. 11.5. Both rovers could carry the active and passive components of the distributed survey instrument (e.g., for redundancy), or each rover could carry the companion component to the other's payload. Such survey configurations are suitable for multiple-site surveys due to the mobility of both platforms. The same dual mobility enables this tandem configuration to perform a number of approaches

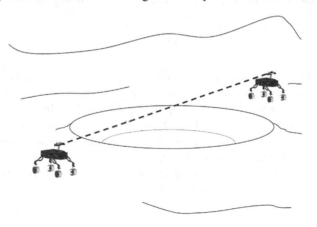

Fig. 11.5 Tandem distributed remote sensing configuration

to remote sensor-based mobile surveying including, as special cases, the approaches described above for single-site surveys.

11.3.4 Multi-Site Remote Survey Performance

A tandem-robot system was proposed in [8, 12] in which one robot carried an active instrument and the other carried the instrument's passive receiver. The robots would cooperatively perform remote sensor-based surveys using either parallel-swath or circular patterns where the width of a swath is the separation distance between the robots. The QoP metric was applied to compare several variants of these survey patterns including those illustrated below in Fig. 11.6. These survey approaches are referred to here as tandem-parallel and tandem-circular, and the QoPs for each are:

$$Q_{tp} = \frac{mn\delta_{tp}}{[2mn + 2(n-1)\delta_{tp}]} \tag{11.8}$$

$$Q_{tc} = \frac{\pi(n\delta_{tc}/2)^2}{\frac{\delta_{tc}}{2}\left(\pi n^2 + 2n - 4\right)} \tag{11.9}$$

where m is the length of a surveyed parallel swath (see Fig. 11.6), δ_{tp} and δ_{tc} are the robot separation/measurement distances during a survey, and n (in both equations) is the total number of swaths surveyed [12]. The denominators of Equations (11.8)

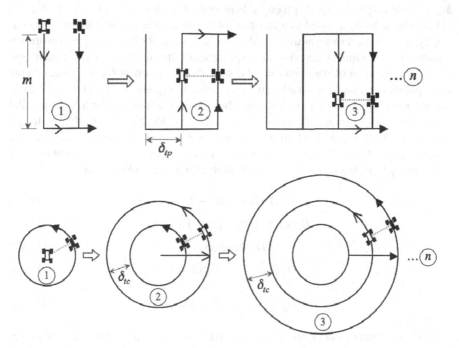

Fig. 11.6 Tandem parallel and circular remote sensing survey trajectories

and (11.9) express the total distances, D_{tp} and D_{tc}, traversed by both robots for the respective tandem survey types.

11.4 Characteristic Performance of Mobile Surveys

The use of a common performance metric to comparatively rank a set of options provides a valuable basis or choosing the best option for a given task. However, direct one-to-one comparisons of the mobile surveying approaches discussed above are not straightforward given their respective differences in survey system configuration and survey trajectories. In fact, they can be prescribed for considerably different types of survey tasks in practice. Nonetheless, somewhat common grounds for comparison would provide a general sense for the relative performance of each approach. As a useful compromise a characteristic comparison is made here, still using a common metric, but based on assumptions that serve to equalize differences across the set of options. We use the QoP as the common metric here but any of a variety of metrics could be used instead.

Consider a mobile survey of a designated region with common area, A, and performed by each of the local and remote sensor-based methods discussed above. Recall that the metric for each survey type is the total area covered divided by the total distance traversed, D, during the survey (QoP $= A/D$). Since A is considered to be a common survey region size, differences in the QoP are functions of differences in D only. Further assume that each survey is designed for a comparable area coverage resolution; this would call for equal values of the parameters δ_p, δ_c, δ_s, δ_{tp}, and δ_{tc}. For the respective QoPs given by Equations (11.1), (11.3), (11.5), (11.8), and (11.9), the distance traveled is a function of the area coverage resolution. For simplicity, we assume a value of unity as the common value for the respective resolution parameters. With this assumption the expression of distance traversed for each QoP becomes a function of parameter n only, with the exception of Q_p and Q_{tp}, i.e., the local parallel and tandem-parallel survey metrics in Equations (11.1) and (11.8). If we assume a square survey region, then the length of a transect for a local parallel survey becomes $m = (n-1)\delta_p = (n-1)$, since δ_p is set to unity. Similarly, the length of a swath for a remote tandem-parallel survey becomes $m = n\delta_{tp} = n$, since δ_{tp} is set to unity. With all traverse distances as functions solely of the n-parameter for each survey type, the following characteristic expressions, $D(n)$, apply

$$D_p(n) = n^2 - 1 \tag{11.10}$$

$$D_c(n) = \pi n^2 + (\pi + 1)n - 1 \tag{11.11}$$

$$D_s(n) = \pi n^2 \tag{11.12}$$

$$D_{tp}(n) = 2n^2 + 2(n - 1) \tag{11.13}$$

$$D_{tc}(n) = \frac{1}{2}\pi n^2 + n - 2. \tag{11.14}$$

Note that Equations (11.10), (11.11), (11.12), (11.13), and (11.14) are only characteristic of the respective distances traversed due to the different meanings for n

for each survey type, i.e., the number of parallel transects, concentric circles, spiral branches, and parallel swaths. For this discussion we will generally refer to n as the number of survey *passes*. Figure 11.7 shows the characteristic trends of traverse distances required by each mobile survey configuration to cover the same area with comparable coverage resolution. The characteristic trends are shown for up to 10 survey passes only. They effectively reveal the sensitivity of the QoP metric to D, based on assumptions made above.

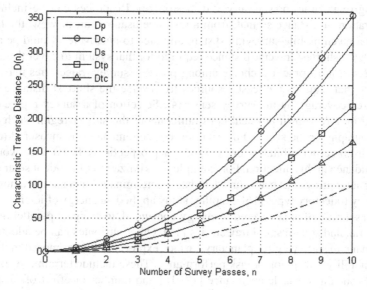

Fig. 11.7 Characteristic traverse distance trends for mobile survey configurations as a function of the number of survey passes

As the number of survey passes increases, the characteristic distances for the circular and spiral remote sensor-based surveys increase at the fastest and roughly the same rate. Circular remote-sensing surveys require the longest characteristic traverse distances. The QoPs for these fixed-mobile surveys would be expected to decrease most rapidly with high numbers of survey passes, while higher QoPs would be expected for spiral surveys with low numbers of passes. For the tandem configurations, parallel circular surveys would be expected to perform better than parallel swath surveys, which would require characteristically longer traverse distances. The QoPs for tandem approaches would be expected to be higher than those of fixed-mobile approaches for the same survey region and coverage resolution. This reflects the advantage of using more than one rover for distributed remote sensing survey tasks [8]. The popular parallel transect survey approach for single rovers with local sensing devices has the slowest trend of increasing distance as the number of survey passes increases. Thus, its QoP would be expected to be high and least impacted by n, relative to the other approaches, for surveying the same area at a comparable coverage resolution.

11.5 Enriching Metrics for Surveys on Planetary Surfaces

Thus far we have considered evaluation of coverage performance for surveys based on simple geometric measures of area and distance. Other important measures should be considered to enrich performance evaluation in the domain of planetary robotic systems and otherwise resource-constrained field robot systems. In fact, the selection of a candidate survey trajectory type for a given survey region can be based on metrics that provide some measure of relative resource usage during trajectory following (e.g., time, power, survey data storage). Resources can be included in consideration of distance so that, from a resource usage point of view, the highest QoP for a given mobile surveying system and area to be covered would be associated with the shortest traverse distance required for that system to cover the area. In that case, the most practical choice among possible survey trajectories may be the trajectory that requires the least resources to traverse the shortest distance resulting in area coverage by the survey sensor(s). Selection of a survey type can also be based on any known physical constraints about the survey region such as terrain topography or the size and distribution of phenomena to be measured (e.g., as derived from earlier orbital reconnaissance or prospecting efforts). Past work provides guidance in this direction. For example, maximization of incident solar energy on rover solar arrays has been considered as a determinant for selection among several survey trajectory types [16]. Another study applied an energy efficiency metric, defined as a ratio of area covered to energy consumed, to evaluate parallel line, circular spiral, and square spiral trajectories [15]. Other attributes can be adopted that are germane to surveying for planetary surface exploration but proposed in the context of mobility/navigation or task performance. These include terrain traversability measures such as obstacle abundance [17, 18] and number or effort of rovers and human operators involved [11].

While identifying domain specific attributes that would enrich the relevance of existing geometric measures is helpful, the manner in which they would be incorporated into a metric formulation is also worth considering. Metric formulations that are computationally complicated or multi-dimensional can be difficult to apply, hard to interpret, or both. In the domain of planetary site characterization, a number of system, mission, and/or environmental constraints will affect the performance of mobile surveying tasks. Performance metrics based solely on geometric aspects of the task do not capture other important performance impacts of the task, and therefore are particularly limited as decision aids. Human mission operators or task supervisors will be better equipped to select appropriate survey methods when using metrics that account for a broader range of performance impacts that include system resources, terrain information or task constraints in addition to geometric measures like the QoP.

As one example of how a resource attribute can change and influence the effectiveness of a basic metric consider the following. In [15], energy consumed by robot wheel motors was considered based on an empirically derived model of DC motors. Differences in energy efficiency were attributed in part to the required amount of turns the robot must execute to follow the search pattern. Depending on how a rover mobility system is kinematically constrained to execute a traverse, many turns-in-place may be necessary throughout a survey. Such maneuvers are not captured by a metric like the QoP and unless energy consumption (at least) is considered, the overall performance of a survey could be obscured. The study considered energy consumed during accelerations in addition to during turns, which led to a conclusion that circular spiral surveys were most efficient for larger survey areas while parallel line scans were most efficient for small areas [15]. This conclusion is based on the fact that the robot continuously moves without stopping and turning when executing spiral trajectories, thus consuming less energy over longer distances required for spiral surveys. Finally, consider that while distance traversed and energy consumed are correlated in most cases, if terrain traversability is ignored then a mobile surveying metric will not capture the distance or energy impacts of surveying a rough and rocky terrain cluttered with obstacles. Due to such considerations, we advocate for enrichment of metrics for mobile surveying tasks to improve their utility as decision aids for actual mission operations.

11.5.1 Consolidated Metric for Human-Supervised Robotic Prospecting

NASA-funded research on human-supervised autonomy and execution of robotic tasks in the context of lunar surface operations has focused on approaches for both scientific and engineering surveying. A Robot Supervision Architecture (RSA) for human tele-supervision of a fleet of mobile robots was presented in [11]. The RSA was focused on the scientific task of autonomously searching an area for in-situ resources/minerals to demonstrate human-robot interactions during tele-supervised prospecting and to validate prospecting performance. Figure 11.8 depicts a typical deployment scenario: one or more human astronauts supervise multiple robots from

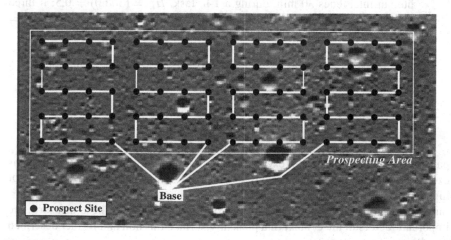

Fig. 11.8 Robot paths within prospecting area

a climate-controlled habitat ("Base"). In the depicted example, four robots are dispatched from the Base and prospect at fixed intervals using a parallel transect path as described in Section 11.2. Surface sensing and other prospecting science would be performed at discrete prospecting sites.

Many factors potentially relate to the performance of a human-supervised multirobot system of this sort. Instead of using multiple individual and independent metrics, such as those described in [19], we propose a consolidated performance metric based on the following notions: (1) greater area, accuracy, terrain difficulty, and safety of prospecting coverage mean increased performance; (2) greater effort and time required mean decreased performance. The consolidated performance metric is thus expressed as follows [11]:

$$P = \frac{ACTS}{(R/w + H_E)t} \tag{11.15}$$

where:

> P: performance in units of (area accurately and safely prospected)/(effort-time)
> A: area covered
> C: prospecting accuracy; $C = 1$ corresponds to the highest possible accuracy and $C = 0$ corresponds to the lowest possible accuracy
> T: terrain difficulty factor ($T \geq 1$) with $T = 1$ corresponding to the easiest terrain (a flat surface without obstacles)
> S: safety factor ($S \geq 1$) with $S = 1$ corresponding to the least safe task performance, i.e., one requiring 100% extra-vehicular activity (EVA) for the astronauts
> R: number of robotic rovers (integer)
> H_E: human effort defined as the degree of human task intervention in full-time equivalents ($0 \leq H_E \leq H$; H is the integer number of humans); e.g., if one human intervenes 30 min. during a 1-h. task, $H_E = (30/60) = 0.5$; if three humans intervene 15, 30, and 45 min. respectively during a 1-h. task, $H_E = (15/60) + (30/60) + (45/60) = 1.5$
> w: factor allowing commensurability of human and rover time by giving the relative value of the former to the latter; e.g., $w = 4$ sets human time to be four times as valuable as rover time
> $R/w + H_E$: combined human-rover effort
> $t = $ time required to cover A.

Note that for the purposes of the consolidated metric, P, the type of prospecting path used is of little relevance (except to the extent that it impacts t); different paths executed in comparable time periods would yield similar performance values. Of higher relevance, however, are the individual measures of prospecting accuracy, terrain difficulty, and safety included in the consolidated metric. These measures are important factors in the context of a planetary surface mission and further described as follows:

- *Prospecting accuracy*: This quantifies how accurately the system can identify regions of geological interest in the covered area. If the system covers a large area with little human effort, but fails to identify any of the existing regions of interest, its overall performance should be zero. This is reflected in the factor $C = (1 - E)$ in the numerator of the Equation (11.15). This factor ranges between 0 and 1, where 0 corresponds to all directly observed regions being misestimated with the highest possible error (hence $E = 1$), and 1 corresponds to all directly observed regions being correctly identified (hence $E = 0$). E is a normalized error metric over regions that were directly measured by the rovers.
- *Terrain difficulty*: This is reflected primarily or exclusively in a reduced average speed of the rovers. One could, for example, take as inputs the average obstacle size and density, the average terrain slope, and the average terrain traction (and maybe other inputs) and calculate a "terrain difficulty factor" T whereby the average rover speed is decreased with respect to a flat surface without obstacles, for which $T = 1$. All other factors being equal, a higher terrain difficulty factor T gives a higher performance, as reflected by its presence in the numerator of Equation (11.15).
- *Safety*: A reasonable measure of safety is related to the percentage of EVA time the astronauts spend during the task. Let us define $V = (1 - \%$ EVA time), so that $V = 1$ corresponds to no EVA time, and $V = 0$ corresponds to all time being EVA time. Safety is difficult to make commensurable with performance as defined above, but a possibility is to define $S = (1 + (\beta - 1)V)$, in which case $V = 0$ corresponds to $S = 1$, or baseline minimally safe performance, and $V = 1$ corresponds to $S = \beta$, or the safest possible performance. The subjective factor β must be provided by human judgment and represents how much more valuable safe performance is than unsafe performance. If $\beta = 4$, for example, then full safety with all other factors being equal results in a performance 4 times better than that achieved with minimal (maximum-EVA) safety.

A given performance target, P_{target}, calculated using Equation (11.15) can be met in a variety of ways by a human-supervised multi-robot system. Such flexibility increases the utility of the metric as a decision aid to mission operators. As an illustration, consider a 100×100 m prospecting area, A, and assume maximum prospecting accuracy with minimum safety and terrain difficulty so that $C = T = S = 1$. As a subjective choice for this hypothetical prospecting task, let the number of humans controlling be $H = 2$, and let human time be considered 5 times as valuable as rover time, i.e., $w = 5$. The performance target for this task becomes $P_{target} = 10$ km^2/(effort-hour) $= A / [(R/5 + H_E)t]$, which can then be achieved under the following scenarios for various amounts of area coverage time and human and robot effort:

1. $H_E = 1, R = 0, t = 1$ h: One human does the task alone in an hour.
2. $H_E = 2, R = 0, t = \frac{1}{2}$ h: Two humans do the task with no rovers in $\frac{1}{2}$ h.
3. $H_E = 0, R = 5, t = 1$ h: Five rovers do the task without human help in an hour.

4. $H_E = 0$, $R = 10$, $t = \frac{1}{2}$ h: Ten rovers do the task without human help in a half-hour.
5. $H_E = 1/5$, $R = 4$, $t = 1$ h: Four rovers do the task with 12 min (1/5*1 h) of human supervision in an hour.
6. $H_E = 6/5$, $R = 4$, $t = \frac{1}{2}$ h: Four rovers do the task with 36 min (6/5*$\frac{1}{2}$ h) of human supervision in $\frac{1}{2}$ h or 30 min (both humans involved).
7. $H_E = 6/5$, $R = 9$, $t = 1/3$ h: Nine rovers do the task with 24 min (6/5*1/3 h) of human supervision in 1/3 h or 20 min (both humans involved).

Scenarios 1 and 2 together and 3 and 4 together show that doubling human effort or rover effort halves the time required for equivalent performance. Scenarios 1 and 3 together and 2 and 4 together show that one human doing the task is equivalent in performance to five rovers doing the task. Scenarios 3 and 5 show that, for the same time required, the removal of one rover's effort requires the provision of only 1/5 of a human's effort. Scenarios 5 and 6 show that the addition of one human's effort with no increase in rovers halves the time, whereas scenarios 5 and 7 show that the addition of one human's effort along with five rovers' effort reduces the time by a factor of three.

11.5.2 Metrics for Real-Time Assessment of Robot Performance

Beyond assessing performance in a purely analytic manner (i.e., for *a priori* planning or post-mortem analysis), we believe that it is also important to develop metrics that can be used *during* survey operations. In particular, we anticipate that future lunar surface surveys will be conducted with robots that are remotely supervised initially by Earth-based operators and later by operators situated in lunar habitats as discussed above. Ground control teams (including scientists and engineers) or operators at telesupervisor workstations in lunar habitats will need to monitor robot performance in order to dynamically adjust robot plans.

Some of our work [19, 20] has addressed defining and computing performance metrics in real-time. Our approach is to monitor data streams from robots, compute performance metrics in-line, and provide Web-based displays of these metrics. Our displays provide summary views of current performance as well as timeline plots, which are useful for spotting trends and key events. We have identified three categories of metrics that are particularly germane to real-time monitoring:

- *Mission*: metrics that describe the robot's contribution to the mission. For example, "Work Efficiency Index" [21] describes the ratio of productive time on task to overhead time. When WEI exceeds 1.0, the robot is spending more time accomplishing mission objectives, i.e., more productive.

- *Task*: metrics that describe the rover's performance with respect to the task (e.g., survey) plan. For example, "Percentage Distance Complete" summarizes the percentage of the planned distance that has been traveled by the rover.
- *Robot*: metrics that describe the robot's health and status. For remotely operated, or supervised, robots, two useful metrics inspired by reliability techniques are "Mean Time to Intervene" (MTTI) and "Mean Time Between Interventions" (MTBI) [22].

We evaluated our approach for real-time performance assessment during a field test at Moses Lake Sand Dunes (in Moses Lake, WA USA) in June 2008 [23]. In this test, we used two NASA Ames K10 planetary rovers (Fig. 11.9) to perform scouting and survey operations. The K10's were equipped with 3D scanning lidar, panoramic and high-resolution terrain imagers, and ground-penetrating radar. We simulated remote lunar science operations with a ground control team located at the NASA Johnson Space Center (Houston, TX USA).

Fig. 11.9 NASA Ames K10 planetary rovers at Moses Lake Sand Dunes

Figure 11.10 shows two real-time performance displays from the Moses Lake test. The "dashboard" provided ground control with summary metrics including robot task performance (drive time, % distance complete, etc.), instrument performance (run time, completed measurements, etc.) and communication quality (e.g., data gaps). The timeline display provided operators with a time-based plot of the history of performance values. A detailed description these displays and how metrics were computed, interpreted and used is presented in [20].

11.5.2.1 Considerations for Real-Time Robot Performance Assessment

Understanding the intent of rover activities provides an important basis for grounding the computation and interpretation of performance metrics. In particular, we assert that metrics, whether used for real-time monitoring or post-mortem analysis,

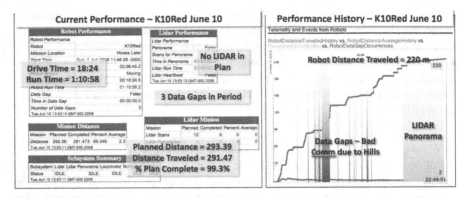

Fig. 11.10 Real-time robot performance displays. Left, "dashboard" summary; right: timeline plot

can only be interpreted in context, i.e., with respect to expectations of performance for a particular task, activity plan, robot mode of operation, etc.

For example, the computation of metrics during autonomous operations should be distinguishable from metrics during direct teleoperation because expected performance can be quite different in these different control modes. In addition, the computation of metrics during anomalous operations should be separable from nominal operations. This is because a robot can be considered to perform "well" if it recovers quickly from a contingency, while at the same time performing poorly on nominal mission objectives.

For operations that involve human supervision, we also contend that performance assessment should consider only time periods when the rover is intended to be performing tasks [20]. Thus, we define a concept called a "duty period" that corresponds to a contiguous interval of operations, such as from rover startup until a planned ground-control break period. This helps reduce bias caused by idle time in the computation of metrics.

11.6 Summary and Conclusions

Science and engineering surveys will need to be conducted by mobile robotic systems to characterize sites on planetary surface missions. Various system configurations and surveying techniques are possible and performance metrics provide a basis for evaluating the options. Geometric metrics are a useful starting point for assessing expected survey performance in terms of coverage. However, other measures, such as system resources, terrain information, or mission-related task constraints should be considered in order to make a complete evaluation. While this chapter has focused on survey by wheeled rovers it should be noted that the discussion applies equally well to walking or flying robots.

This chapter illustrates how the QoP metric can provide a basis for performance comparison of different types of mobile survey trajectories, and additional metrics that can supplement such coverage metrics are discussed. When computing these metrics for a rover in the field or on a mission, however, it is important to consider how the environment, the equipment, and the mission affect the interpretation of computed values. For example, terrain traversability can impact both the distance traveled and the energy expended by the rover.

Through our work, we have found that expectations about nominal performance can be strongly influenced by environmental conditions. In addition, nominal performance can be degraded over the course of multiple missions due to component wear or systemic problems. In general, the limits imposed by component use and subsystem problems define the expectations about robot degraded mode performance. Robot operations can be further constrained by flight rules, i.e. mission-imposed constraints on operation. Mission constraints define safe rover behaviors. For robotic survey, this would include constraints such as the minimum separation between rovers, or the maximum slope traversable by a rover.

When comparing the values of metrics computed for different rovers or the same rover over multiple duty periods, it is important to establish a basis of comparison that identifies which of these performance dimensions predominate and how these dimensions combine to establish performance expectations. Considerations include changes to or differences between the rover equipment, features of the environment that significantly affect rover operations, and changes in safety constraints.

Acknowledgments We thank Hans Utz, Tod Milam, David Lees, and Matt Deans for assisting with the development and implementation of real-time robot performance monitoring.

This work was partially supported by the NASA Astrobiology Science and Technology Instrument Development program, the NASA Exploration Systems Technology Development Program under the "Human-Robotic Systems" project and the NASA Small Business Innovative Research (SBIR) program.

References

1. Garcia E, Gonzalez de Santos P (2004) Mobile-robot navigation with complete coverage of unstructured environments. Rob Auton Syst 46:195–204.
2. Wong SC, Middleton L, MacDonald BA (2002) Performance metrics for robot coverage tasks. In: Proceedings of the Australasian Conference on Robot and Automation, Auckland, New Zealand, pp. 7–12.
3. Mei Y, Lu Y-H, Hu YC, Lee CSG (2006) Deployment of mobile robots with energy and timing constraints. IEEE Trans Robot 22(3):507–522.
4. Fong T, Bualat M, Edwards L, Fluckiger L et al. (2006) Human-robot site survey and sampling for space exploration. In: Proceedings of the AIAA Space Conference, San Jose, CA.
5. Carpin S, Pillonetto G (2003) Robot motion planning using adaptive random walks. In: Proceedings of IEEE International Conference on Robotics and Automation, Taipei, Taiwan, pp. 3809–3814.
6. Nakamura Y, Sekiguchi A (2001) The chaotic mobile robot. IEEE Trans Robot Autom 17(6):898–904.
7. Tunstel E, Anderson G, Wilson E (2007) Autonomous mobile surveying for science rovers using *in situ* distributed remote sensing. In: Proceedings of IEEE International Conference on Systems, Man, and Cybernetics, Montreal, pp. 2348–2353.

8. Anderson GT, Hashemi RR, Wilson E, Clark M (2000) Application of cooperative robots to search for water on Mars using distributed spectroscopy. In: Proceedings of International Symposium on Robotics and Applications WAC, Maui, HI.

9. Schmitt HH, Kulcinski GL, Sviatoslavsky IN, Carrier III WD (1992) Spiral mining for lunar volatiles. In: Proceedings of the 3rd International Conference on Engineering, Construction and Operations in Space III, Denver, CO, pp. 1162–1170.

10. Fong T, Allen M, Bouyssounouse X, Bualat MG et al. (2008) Robotic site survey at Haughton Crater. In: Proceedings of the 9th International Symposium on Artificial Intelligence, Robotics and Automation in Space, Los Angeles, CA.

11. Elfes A, Dolan J, Podnar G, Mau S, Bergerman M (2006) Safe and efficient robotic space exploration with tele-supervised autonomous robots. In: Proceedings of the AAAI Spring Symposium, Stanford, CA, pp. 104–113.

12. Hashemi RR, Jin L, Anderson GT, Wilson E, Clark M (2001) A comparison of search patterns for cooperative robots operating in remote environments. In: Proceedings of the International Symposium on Information Technology: Computing and Coding, Las Vegas, NV, pp. 668–672.

13. Wilson EW, Tunstel EW, Anderson GT (2007) BioGAS spectrometer for biogenic gas detection and location on the surface of Mars. In: Proceedings of the AIAA Infotech@Aerosp Conference and Exhibition, Rohnert Park, CA.

14. Hashmonay RA, Yost, MG (1999) Innovative approach for estimating fugitive gaseous fluxes using computed tomography and remote optical sensing techniques. J Air Waste Manag Assoc 49(8):966–972.

15. Mei Y, Lu Y-H, Hu YC, Lee CSG (2004) Energy-efficient motion planning for mobile robots. In: Proceedings of the IEEE International Conference on Robotics and Automation, New Orleans, LA, pp. 4344–4349.

16. Shillcutt KJ (2000) Solar based navigation for robotic explorers. Doctoral Thesis, Robotics Institute, Carnegie Mellon University, Pittsburgh, PA, CMU-RI-TR-00-25.

17. Tunstel E (2007) Operational performance metrics for Mars Exploration Rovers. J Field Robot 24(8–9):651–670.

18. Sukhatme GS, Bekey GA (1996) Multicriteria evaluation of a planetary rover. In: Proceedings of the IEEE International Conference on Robotics and Automation, Minneapolis, MN.

19. Schreckenghost D, Fong T, Milam T (2008) Human supervision of robotic site surveys. In: Proceedings of the 6th Conference on Human/Robotic Technology and Vision for Space Exploration, STAIF, Albuquerque, NM.

20. Schreckenghost D, Fong T, Milam T, Pacis E, Utz H (2009) Real-time assessment of robot performance during remote exploration operations. In: Proceedings of the IEEE Aerospace Conference, Big Sky, MT.

21. Gernhardt M (2005) Work efficiency indices. Presentation at NASA Johnson Space Center, Houston, TX.

22. Arnold J (2006) Towards a framework for architecting heterogeneous teams of humans and robots for space exploration. MS Thesis, Department Aeronaut & Astronaut, Massachusetts Institute of Technology, Cambridge, MA.

23. Fong T, Bualat M, Deans M et al. (2008) Field testing of utility robots for lunar surface operations. In: Proceedings of the AIAA Space Conference, San Diego, CA, AIAA-2008-7886.

Chapter 12
Performance Evaluation and Metrics for Perception in Intelligent Manufacturing

Roger Eastman, Tsai Hong, Jane Shi, Tobias Hanning, Bala Muralikrishnan, S. Susan Young, and Tommy Chang

Abstract An unsolved but important problem in intelligent manufacturing is dynamic pose estimation under complex environmental conditions—tracking an object's pose and position as it moves in an environment with uncontrolled lighting and background. This is a central task in robotic perception, and a robust, highly accurate solution would be of use in a number of manufacturing applications. To be commercially feasible, a solution must also be benchmarked against performance standards so manufacturers fully understand its nature and capabilities. The PerMIS 2008 Special Session on "Performance Metrics for Perception in Intelligent Manufacturing," held August 20, 2008, brought together academic, industrial and governmental researchers interested in calibrating and benchmarking vision and metrology systems. The special session had a series of speakers who each addressed a component of the general problem of benchmarking complex perception tasks, including dynamic pose estimation. The components included assembly line motion analysis, camera calibration, laser tracker calibration, super-resolution range data enhancement and evaluation, and evaluation of 6DOF pose estimation for visual servoing. This Chapter combines and summarizes the results of the special session, giving a framework for benchmarking perception systems and relating the individual components to the general framework.

12.1 Introduction

Real-time three-dimensional vision has been rapidly advancing over the past 20 years, leading to a number of successful laboratory demonstrations, including real-time visual servoing [1, 2, 3], autonomous vehicle navigation [4], and real-time people and vehicle tracking [5]. However, the advances have frequently not yet made the transition to commercial products, due in part to a lack of objective methods for

R. Eastman (✉)
Loyola University Maryland, Maryland, MD, USA
e-mail: reastman@loyola.edu

R. Madhavan et al. (eds.), *Performance Evaluation and Benchmarking of Intelligent Systems*, DOI 10.1007/978-1-4419-0492-8_12,

empirical performance evaluation. To ensure a new algorithm or sensor system is reliable and accurate enough for commercial application in a safety or performance critical environment, the new system must be tested again rigorous standards and benchmarks.

In several areas of computer vision, there are well-established benchmarks for static problems where the solution need not run in real time, as well as challenge problems for real-time processing. The Middlebury Stereo Dataset [6] and the NIST Face Recognition Grand Challenge [7] have succeeded well in advancing their respective research algorithms by providing well-defined challenge tasks with ground truth and evaluation metrics. The success of these efforts led to workshop series such as BenCOS (Benchmarking Automated Calibration, Orientation and Surface Reconstruction from Images) and CLEAR (Classification of Events, Activities and Relationships) in video tracking. The DARPA Grand Challenge series for autonomous vehicles demonstrated that clear and well-motivated benchmark tasks could lead a research community to assemble disparate, incomplete solutions into successful full solutions in a few years.

Despite these successes, there are a number of computer vision tasks for which well-defined benchmarks do not yet exist, or do not exist to the level of commercially required precision and robustness. Of interest in this chapter is dynamic pose estimation under complex environmental conditions—tracking an object's pose and position as it moves in an environment with uncontrolled lighting and background. This is a central task in robotic perception, and a robust, highly accurate solution would be of use in a number of applications. Creating a standard benchmark for this task would help advance the field, but is difficult because the number of variables is very large and the development of ground truth data is complex. The variables in such a task include the size and shape of the object, the speed and nature of the object motion, the complexity and motion of background objects, the lighting conditions, among other elements.

The PerMIS 2008 Special Session on "Performance Metrics for Perception in Intelligent Manufacturing," held August 20, 2008, brought together academic, industrial and governmental researchers interested in calibrating and benchmarking vision and metrology systems. The papers each addressed an individual problem of interest in its own right, but taken together the papers also fit into a framework for the benchmarking of complex perception tasks like dynamic pose estimation.

A general framework for the evaluation of perception algorithms includes three steps:

(1) Model the conditions and requirements of a task to best understand how to test sensor systems for that task.
(2) Design and calibrate a standard measurement system to gather system data and ground truth under controlled conditions.
(3) Establish metrics for evaluating the performance of the system under test with respect to the ground truth.

The PerMIS special session papers, collected here in summary form, fit well into this framework.

The first paper, in Section 12.2, looks at the conditions for a pose estimation task on the automotive manufacturing line. In the paper "Preliminary Analysis of Conveyor Dynamic Motion," Dr. Jane Shi [8] analyzes the motion of a standard assembly line conveyance mechanism. This work helps understand the operating conditions and requirements for a sensor measuring object motion.

The second paper, in Section 12.3, looks at the calibration of a camera and a range sensor. In the paper, "Calibration of a System of a GrayValue Camera and an MDSI Range Camera," Dr. Tobias Hanning et al. [9] present a method for cross-sensor calibration. Cross-sensor calibration is critical for setting up the controlled conditions for data collection from a perception system under test and the standard reference system.

The third paper, in Section 12.4, looks at the calibration of a laser tracker. In the paper "Performance evaluation of laser trackers," Dr. Bala Muralikrishnan et al. [10] present current standards for evaluating the static performance of a laser tracker that can be used to establish ground truth. Standards and test procedures for dimensional metrology are well-established and highly accurate for static measurements, with coordinate measuring machines and laser trackers giving position measurements in microns. Computer vision work on benchmarks can well benefit from this technology and related expertise.

The fourth paper, in Section 12.5, looks at the evaluation of an algorithm for enhancement of static range data. In the paper "Performance of Super-Resolution Enhancement for LADAR Camera Data," Hu et al. [11] present techniques for the evaluation of a super-resolution algorithm that produces enhanced range data. In this work, the focus is on the quality of raw data from a range sensor and the sharpening of that data for greater resolution.

The fifth and final paper, in Section 12.6, looks at the evaluation of a system for dynamic robotic servoing. In the paper "Dynamic 6DOF Metrology for evaluating a visual servoing system," Chang et al. [12] bring together various issues in dynamic sensor system evaluation: requirements modeling, calibration, performance metrics, data collection and data analysis.

This work specifically addresses performance metrics. In this study 6DOF position and orientation data is collected simultaneously from the sensor under test and a ground truth reference system. Metrics for comparison can then include average or maximum differences in the two data streams, or percentage of time over a threshold difference.

12.2 Preliminary Analysis of Conveyor Dynamic Motion

In automotive general assembly (GA), vehicle bodies are carried on a continuous moving assembly line through hundreds of workstations where parts are assembled. In order to eliminate these automation stop stations and enable robots to interact with continuously moving conveyors, it is essential that the conveyors' dynamic motion behaviors at the assembly plant floor are characterized and quantified. This section presents the method to collect the conveyor motion data, illustrates collected raw data, presents the motion frequency analysis results, and summarizes basic statistical analysis results for both linear acceleration and computed linear speed.

12.2.1 Conveyor Motion Data Collection Method

Line tracking of a moving vehicle body for assembly automation is mainly concerned with the vehicle body moving speed in the 3D space with respect to a fixed reference frame. An ideal method and its associated instrument to collect the vehicle body motion data needs to be simple to set up, easy to initialize for data recording, and non-intrusive to operators on the production line. In addition, minimum secondary processing of recorded raw data is desired. After investigating several available motion sensors, a 6° of freedom (DOF) accelerometer was chosen based on its embedded data logging capabilities of all 6 axes data: three linear accelerations and three angular velocities. Figure 12.1 illustrates the use of the accelerometer to collect conveyor motion data at an assembly plant.

The accelerometer was placed on a flat surface of a conveyor with its X direction aligned with the main conveyor travel direction as shown. The accelerometer's Z direction is defined as up. The accelerometer can record about 30 min of 10 bit data of linear accelerations and angular velocities at a 500 Hz sampling rate to its embedded flash memory. An external toggle switch simply turns on or off the data logging. Table 12.1 below summarizes the measurement range and the resolution.

Fig. 12.1 Conveyor motion data collected using a 6DOF accelerometer

Table 12.1 Accelerometer measurement range and resolution

Parameter		Measurement range	Resolution	
Linear Acceleration	X acceleration	2.23 g	0.005 g	4.902 cm/s^2
	Y acceleration	2.28 g	0.005 g	4.902 cm/s^2
	Z acceleration	2.30 g	0.005 g	4.902 cm/s^2
Angular Velocity	X rotation	78.5 deg/s	0.0766 deg/s	
	Y rotation	77.3 deg/s	0.075488 deg/s	
	Z rotation	79.6 deg/s	0.077734 deg/s	

12.2.2 Raw Motion Data Collected

Raw conveyor motion data were collected for a total of six different types of conveyors from four different assembly plants. For each conveyor type, the accelerometer was riding through a workstation or a section of multiple workstations four or five times so that the randomness of conveyor motion could be studied. During the active data recording period, any external disturbances to the natural conveyor movement, such as an operator stepping on the conveyor carriers or heavy parts loaded onto the vehicle body, were noted so that any exceptional signal in the raw data was detected. Figure 12.2 shows three acceleration data (X—blue, Y—red, Z—green) collected from a platform type of conveyor with stable and smooth motion (left) and a chain type of conveyor with rough and jerky motion (right).

Please note the scale of the right graph is significantly larger than the scale of the left graph. The jerky motion mainly occurred in the X direction resulting from a significant acceleration of short duration.

During the data collection period, the vehicle body could stop at a random time and for a random period of time, and then start again. Since these stops and re-starts occur randomly, their motion characteristics need to be understood in order for a

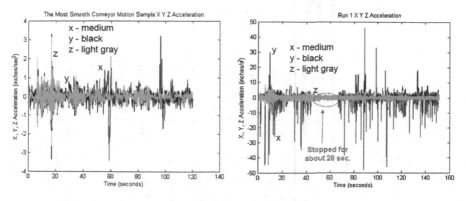

Fig. 12.2 Sample linear accelerations of a stable conveyor (*left*) and a jerky conveyor (*right*)

Table 12.2 Statistics of three linear accelerations

Acceleration	Mean (X) cm/s²	Mean (Y) cm/s²	Mean (Z) cm/s²	StdDev(X) cm/s²	StdDev(Y) cm/s²	StdDev(Z) cm/s²	Max(X) cm/s²	Max(Y) cm/s²	Max(Z) cm/s²	Min(X) cm/s²	Min(Y) cm/s²	Min(Z) cm/s²
Conveyor 1 Sample	0.0	0.0	0.0	1.0	0.8	0.25	3.6	7.9	1.0	−3.6	−13.5	−2.3
Conveyor 2 Sample	0.0	0.0	0.0	1.8	1.5	1.8	9.7	7.1	9.9	−14.5	−7.9	−7.6
Conveyor 3 Sample	0.0	0.0	0.0	14.2	4.3	6.4	127.3	42.4	59.2	−79.0	−32.5	−51.6
Conveyor 4 Sample	0.0	0.0	0.0	10.4	2.5	2.0	101.1	13.0	9.1	−85.9	−13.2	−12.2
Conveyor 5 Sample	0.0	0.0	0.0	9.9	3.0	1.5	98.0	57.7	9.9	−106.7	−19.6	−12.7
Conveyor 6 Sample	0.0	0.0	0.0	6.9	2.0	3.6	32.3	11.9	38.1	−30.0	−22.6	−22.6

future robot to follow the vehicle during stops and starts. These random starts and stops were captured for future analysis.

12.2.3 Statistical Analysis of Linear Accelerations

Basic statistical analysis of three linear accelerations—the mean, the standard deviation, maximum and minimum value—were performed and the results are summarized in Table 12.2 for one set of data samples on each conveyor type.

The conveyor motion stability can be evaluated simply by the value of standard deviation. The smaller the standard deviation value, the more stable the conveyor motion. Conveyor 1 is significantly stable compared with Conveyor 3 as shown by their standard deviation as well as maximum and minimum values. In the future, more sophisticated statistical analysis can be performed to further analyze the dynamic motion data to detect any quantifiable relationships among data sets.

12.2.4 Fast Fourier Transformation Analysis

In order to detect any frequency patterns of the acceleration data, Fast Fourier Transformation (FFT) analysis was performed. Figure 12.3 left is a graph of FFT results for a smooth conveyor and Fig. 12.3 right is a graph of FFT results for a jerky conveyor. The horizontal X axis is the frequency in Hz and the vertical Y axis is the power spectrum distribution as a function of frequency.

Table 12.3 provides a summary of the frequencies for the three highest peaks of FFT analysis results for the six types of conveyors.

When the highest dominant peak occurs at zero frequency, it means that there is zero net acceleration in Conveyor 1's case, and the conveyor motion speed is constant. When multiple and equally strong peaks occur at different frequencies, it means that the conveyor acceleration varies at multiple frequencies in Conveyor 5's

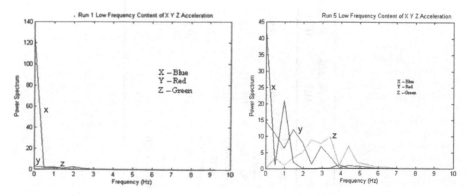

Fig. 12.3 FFT results of a smooth conveyor (*left*) and a jerky conveyor (*right*)

Table 12.3 FFT Frequency of three linear accelerations

FFT analysis	X frequency (Hz)	X frequency (Hz)	X frequency (Hz)	Y frequency (Hz)	Y frequency (Hz)	Y frequency (Hz)	Z frequency (Hz)	Z frequency (Hz)	Z frequency (Hz)
Conveyor 1 Sample	0.0	0.5	1.0	0.0	0.5	1.0	1.0	0.0	–
Conveyor 2 Sample	1.5	0.0	0.5	0.5	1.5	0.0	1.5	0.0	0.5
Conveyor 3 Sample	2.0	1.5	1.0	0.5	2.0	1.5	2.4	3.4	2.9
Conveyor 4 Sample	1.0	0.5	2.9	1.5	1.0	2.0	3.9	3.4	1.5
Conveyor 5 Sample	1.5	1.0	0.5	1.5	0.0	2.0	3.9	3.4	2.4
Conveyor 6 Sample	0.5	1.0	1.5	0.0	0.5	1.0	1.0	0.5	0.0

case. Based on the FFT analysis results in Table 12.3 , it can be concluded that jerky motion along the conveyor main travel direction is of random nature because the frequencies are not at fixed values and their magnitudes are varying among different samples of the same conveyor type.

12.2.5 Computed Speed and Position Data

Given three-axis linear acceleration data of a conveyor sampled at 500 Hz, its linear speed can be computed with a numerical integration method. The computed speed and position does not reliably reflect the accurate stop and start scenarios. Any small linear trend or nonzero mean in acceleration data will accumulate over time and it will lead to significant error in computed speed and position. Hence the following pre-processing and filtering is necessary to compute realistic speed and position data from the acceleration data:

- First, the acceleration data is filtered at 5 Hz,
- Second, the linear trend of the linear acceleration is removed,
- Third, the mean is set to zero because of zero net acceleration,
- Finally, the speed and position are computed using the first order linear integration with 0.002 s interval.

Figure 12.4 left is a graph of three computed linear speeds for a smooth conveyor and right is for a jerky conveyor. Please note the scale difference between two graphs. Figure 12.5 left is a graph of computed 3D position of a smooth conveyor and right is for a jerky conveyor.

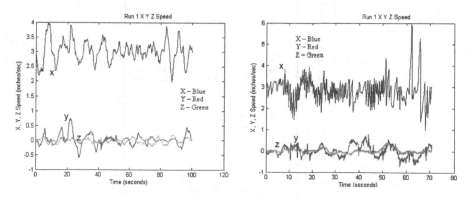

Fig. 12.4 Computed speed of a smooth conveyor (*left*) and a jerky conveyor (*right*)

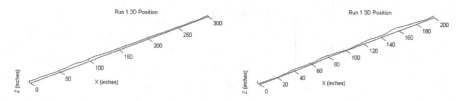

Fig. 12.5 Computed 3D position of a smooth conveyor (*left*) and a jerky conveyor (*right*)

Table 12.4 Statistics of three linear computed speeds

Speed	Mean (X) in./s	Mean (Y) in./s	Mean (Z) in./s	StdDev(X) in./s	StdDev(Y) in./s	StdDev(Z) in./s	Max(X) in./s	Max(Y) in./s	Max(Z) in./s	Min(X) in./s	Min(Y) in./s	Min(Z) in./s
Conveyor 1 Sample	3.0	0.0	0.0	0.4	0.2	0.1	4.0	0.7	0.2	2.0	−0.6	−0.2
Conveyor 2 Sample	2.9	0.0	−0.1	0.7	0.1	0.1	5.5	0.3	0.2	0.5	−0.3	−0.4
Conveyor 3 Sample	3.0	0.0	0.0	0.7	0.2	0.3	5.9	0.7	2.1	−2.3	−1.3	−1.6
Conveyor 4 Sample	2.9	0.0	0.0	0.9	0.1	0.1	6.2	0.5	0.2	−0.4	−0.4	−0.2
Conveyor 5 Sample	2.0	0.3	0.0	0.8	0.3	0.1	4.9	2.4	0.3	−0.8	−1.2	−0.3
Conveyor 6 Sample	2.0	0.0	0.0	1.1	0.6	0.1	5.1	2.1	0.7	−1.3	−1.8	−0.8

Basic statistical analysis of three linear speeds has been performed and the results are summarized in Table 12.4.

The range of speed distribution is measured by one standard deviation. Hence the conveyor motion stability can be similarly evaluated by the value of standard derivation. Although Conveyor 6 moves at lower average speed, it is most unstable conveyor with the widest range of speed and highest magnitude of acceleration (change rate of speed). Conveyor 1 is significantly stable as shown by their standard deviation as well as maximum and minimum values.

12.2.6 Conclusion and Summary

This section presented a method of collecting conveyor motion data. Representative raw linear acceleration data samples was presented. Standard statistical analysis has been performed for both linear acceleration and computed linear speed to evaluate the conveyor motion stability. The Fast Fourier Transformation (FFT) has been performed on linear acceleration data to identify the frequency domain pattern of the conveyer acceleration. The FFT analysis has shown that the jerky motion along the conveyor main travel direction is of random nature because the frequencies are not at fixed values and their magnitudes vary among different samples of the same conveyor type.

Based on the actual conveyor motion data collected from the assembly plants, the dynamic motion profile library has been built for each conveyor type that is commonly used in automotive general assembly. These motion profiles can be used in laboratory experiments for developing new line tracking solutions and validating vendor supplied black box solutions that are targeted for plant floor deployment.

12.3 Calibration of a System of a Gray Value Camera and MDSI Range Camera

We call a calibration procedure for a system of cameras a cross-calibration procedure if the calibration parameters of each camera in the system are obtained iteratively from the calibration parameters of the others and from the spatial relation between the cameras in the system. In this section we present a cross-calibration procedure for a system composed of range cameras based on the Multiple Double Short-Time Integration (MDSI) approach and a regular gray-value camera.

A range camera produces a grid of range pixels. Each range pixel provides the distance between the observed object and the pixel surface measured along a viewing ray. A feasible way to realize a range camera based on the time-of-flight principle has been presented by Mengel et al. [13]. According to the MDSI approach each range pixel integrates the radiation emitted by a near infrared laser and reflected by the observed objects during two time periods of different length. Two intensity acquisitions, a short one s and a long one l are made. The distance to the observed object is linear in the quotient *s/l* (see e.g., [13]). In [14] we have shown that this is also true for the z-coordinate of the observed point. However, to obtain full 3D information of the observed point the complete mapping of the range camera has to be determined. Since the range pixels are located in an image-like grid behind a common lens, a calibration of the MDSI range camera includes computing the linear mapping for the z-coordinate for each pixel.

In this session we present a simple and flexible range camera calibration method, which utilizes a calibrated gray-value camera. We apply the cross-calibration approach by using corresponding data to calibrate the whole camera system. From the calibrated system a gray-value camera independent range camera calibration can easily be obtained.

12.3.1 Cross-Calibration Procedure

The estimation of the intrinsic parameters of a gray-value camera is a well studied problem. All intrinsic camera parameters can be obtained by at least three observations of a planar calibration target in general position (see e. g. [15]). Therefore, a regular grid of black circles on white background forms a suitable pattern for a calibration of the intrinsic parameters of the gray-value camera. Hence, we assume the gray-value camera to be calibrated.

The coordinate system with respect to the gray-value camera is related to range camera's coordinate system by an isometric transformation $T{:}R^3 \Rightarrow R^3$, $T(p) = R(p - t)$ for each space point $p \in R^3$ with a rotation matrix $R \in R^{3 \times 3}$ and a translation vector $t \in R^3$.

We model the camera mapping of the range camera as a pinhole camera with respect to the gray-value camera: the range camera mapping is defined by $P_r R$ and t, where $P_r \in R^{3 \times 3}$ is an upper triangular matrix with positive entries such that for each space point $p \in R^3$ and its corresponding range pixel coordinate $p \in R^2$, there exists a $\lambda \in R$ with

$$\lambda \begin{pmatrix} p' \\ 1 \end{pmatrix} = P_r R (p - t). \tag{12.1}$$

This equation is commonly known as the pinhole equation (see e.g. [16]). Hence, the mapping of the range camera is determined by a regular matrix $A = P_r R \in R^{3 \times 3}$ and a vector $b = - P_r R t \in R^3$ with

$$\lambda \begin{pmatrix} p' \\ 1 \end{pmatrix} = Ap + b. \tag{12.2}$$

To obtain the relevant parameters A and b, let j be an image pair index within a sequence of images. Suppose we have obtained the centers of the black circles

$p_{1j}, \ldots, p_{mj} \in R^2$ in the gray-value image and the corresponding centers of the same circles $p'_{1j}, \ldots, p'_{mj} \in R^2$ in the short or long intensity image of the range camera.

Given a calibrated gray-value camera $K_g: R^3 \Rightarrow R^2$ we are able to reconstruct the centers of circles on a planar calibration pattern by its observation (see e.g. [17]). Let C_{ij} be the position of the center of the i-th circle observed in the j-th image pair. With this we obtain for each circle the relation

$$\lambda \begin{pmatrix} p'_{ij} \\ 1 \end{pmatrix} = AC_{ij} + b \qquad (12.3)$$

(for $\lambda \in R$) which is linear in A and b. The equations for all i and for different positions of the calibration pattern form a linear least-square problem. Notice that in contrast to the calibration procedure in [15], we do not need to impose additional constraints on A and b, such as orthogonality constraints. This approach to an estimate of A and b is known as the Direct Linear Transform (DLT) algorithm (see e.g., [16]). Since A is regular, we can determine P_r from A from an RQ-decomposition $A = P_r R$ into an upper triangular matrix P_r with positive Eigen values and an orthogonal matrix R. Such a decomposition can be obtained by a slightly modified QR-decomposition algorithm, which guarantees the positivity of the diagonal entries of P_r (see e.g., [18]). Observing that $b = -P_r Rt = -At$, we obtain the translation $t = -A^{-1}b$.

Since the coordinate systems are related by an isometric movement, the z-coordinate of every object observed by the range camera is also linear in the fraction of the short s and long intensity l of the range pixel. Thus, for every range pixel there exist $\alpha, \beta \in R$ with

$$z(s, l) = \alpha \frac{s}{l} + \beta. \qquad (12.4)$$

with respect to the coordinate system of the gray-value camera with intensity values s and l. In fact, z(s, l) is the unknown λ in Equation (12.2). So, a number of observations of the center of the circle in the range pixel to calibrate allow us to determine α and β by a linear least squares problem with two unknowns α, β. In order to obtain a valid solution for every range pixel, at least two observations of a circle are needed for every range pixel are required. To enhance the number of valid z-coordinates for each range pixel one can generate new data by extrapolation.

Once α and β are determined for each range pixel p, a back-projection results not in a viewing ray $\left\{ \lambda A^{-1} \begin{pmatrix} p' \\ 1 \end{pmatrix} - A^{-1}b : \lambda \in R_+ \right\}$, but - since λ is known in a 3D point with coordinates with respect to the gray-value camera. With α and β fixed we note $w_{i,j}(A, b)$ for the reconstructed 3D-point of the i-th target point pair p_{ij}, p'_{ij} in the j-th gray-value image along the viewing ray determined by A and b. We then determine the parameters A and b by minimizing the back-projectionerror

$$e(A,b) = \sum_{i=1}^{j} \sum_{i=1}^{m} (K_g(w_{ij}(A,b)) - p_{ij})^2 \qquad (12.5)$$

Note that e(A, b) is an error measured in the gray-value image.

A final optimization step involves all collected data to refine the parameters A and b and the parameters of the gray-value camera. Besides optimal surface reconstruction properties, we also need optimal back-projection properties into the image of the gray-value camera. Thus, a suitable error function is:

$$e(A,b,K_g) = \sum_{i=1}^{j} \sum_{i=1}^{m} (K_g(w_{ij}(A,b)) - p_{ij})^2 \qquad (12.6)$$

where K_g denotes the gray-value camera mapping. We solve the optimization problem induced by the error function in Equation (12.6) by the BFGS algorithm [19]. The error function in Equation (12.6) involves also the optimization of the gray-value camera parameters K_g which are improved by this final optimization step. This completes the cross-calibration procedure.

12.3.2 Experimental Results

We have applied the calibration procedure to a system comprised of a gray-value camera and a 64×8-pixel MDSI range camera developed within the European project UseRCams and a common gray-value camera. For our calibration we use a white plate with a set of black circles (see Fig. 12.6). We expose the board continuously to the camera system from different directions and distances.

After the final calibration a different scene of the same calibration plate at a distance of approximately 5 m was evaluated to cross-validate our calibration result. For each calibrated range pixel and for each plane reconstructed by the gray-value camera, we measure the absolute range reconstruction error with respect to our calibration plate. This error is the Euclidean distance between the calibration plate and the reconstructed surface point. To obtain comprehensive results for our experiment, we assume the range reconstruction noise of all range pixels to be identically distributed. Hence, we can aggregate sample data from all pixels together as being taken from a single population. Figure 12.7 shows the distribution of the absolute range reconstruction error for all calibrated pixels.

The distribution in Fig. 12.7 shows that errors with an absolute value of at most 16.5 cm occur in 75% of all cases and errors with an absolute value of at most 29.5 cm occur in 95% of all cases. The error expectation of 2.23×10^{-4} m validates the accuracy of our calibration procedure and the correctness of our model. The measured errors seem to be sensor noise.

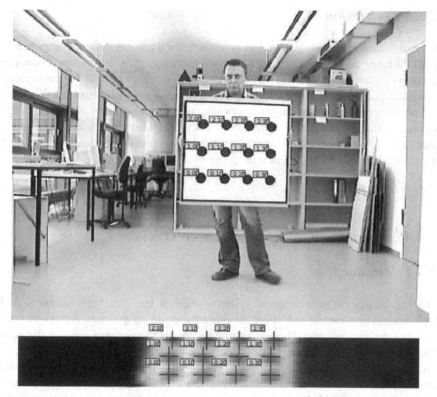

Fig. 12.6 Calibration plate observed in the image of the gray-value camera (*above*) and in the intensity image of the range camera (*below*) with marked correspondences

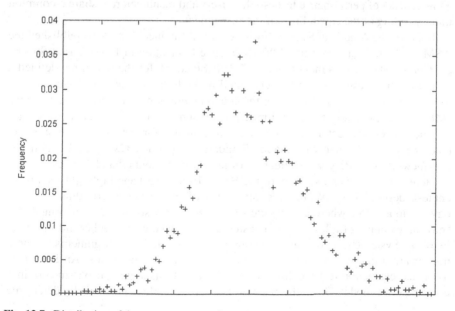

Fig. 12.7 Distribution of the range reconstruction error

12.3.3 Conclusion

We have presented a cross-calibration procedure for a camera system composed of an MDSI range camera and a regular gray-value camera. The calibration procedure also provides regular, gray-value camera independent parameters for the range camera with respect to the coordinate system defined by the gray-value camera. We have applied the procedure to the range camera system combined of a 64×8-pixel range camera developed within the European project UseRCams and a common gray-value camera. We are able to show that our calibration procedure yields an unbiased distance and surface reconstruction procedure.

12.4 Performance Evaluation of Laser Trackers

12.4.1 Introduction

Spherical coordinate measurement systems such as laser trackers, scanners and other devices are increasingly used in manufacturing shop floors for inspection, assembly, automation, etc. These instruments are also sometimes used in the calibration of other lower-accuracy instruments such as industrial robots and even certain Cartesian coordinate measuring machines (CMMs). The spherical coordinate systems provide high accuracy at relatively low cost (in comparison to more conventional Cartesian CMMs), and are portable and convenient to use. Because of the proliferation of such devices in recent years, there has been an increasing need for a uniform set of performance tests so that users and manufacturers share a common understanding of the capabilities of the instrument.

In 2007, the American Society for Mechanical Engineers (ASME) published the ASME B89.4.19 Standard titled "Performance Evaluation of Laser-Based Spherical Coordinate Measurement Systems," This Standard, for the first time, defined a common set of tests that can be performed by both the user and the manufacturer to establish if an instrument meets the manufacturer's performance specifications (MPE). This Standard, although limited to instruments that use a cooperative target such as retro-reflector, represents a significant step forward. It is the first and to date, the only performance evaluation Standard for spherical CMMs and establishes a framework for testing and evaluation of laser trackers and related devices.

Here, we present an overview of the B89.4.19 Standard and highlight the different tests described in it. We discuss capabilities of the large-scale coordinate metrology group at NIST where a complete set of B89.4.19 tests may be performed. We show an example of a laser tracker tested at NIST where the tracker did not meet its MPE. Systematic errors due to geometrical and optical misalignments within a tracker are a major source of uncertainty in tracker measurements. An ideal performance evaluation test has high sensitivity to all misalignment parameters in a tracker's error model. Given the error model, it is possible to numerically determine the sensitivity of each of the B89.4.19 tests to different misalignment parameters. We have performed such analysis and briefly discuss our method.

12.4.2 The ASME B89.4.19 Standard

The ASME B89.4.19 Standard describes three types of tests to be performed on trackers to evaluate their performance. These are the ranging tests, the length measurement system tests and two-face system tests.

Ranging tests assess the distance (or displacement) measurement capability of the instrument. The ranging system (an interferometer or an absolute distance measurement (ADM) system) establishes the unit of length and is therefore a critical component of the system. The tests as described in the Standard require the tracker to measure several calibrated lengths aligned along the line-of-sight of the tracker. The reference lengths employed may be calibrated artifacts, realized by free standing targets, or a laser-rail system.

The length measurement system tests are similar to volumetric length tests on Cartesian CMMs. A calibrated reference length is placed at different positions and orientations in the measurement volume and is measured by the tracker. The error in the measured length is compared against the MPE to determine conformance to specification. There are several sources of mechanical and optical misalignments within the construction of a tracker that produce systematic errors in the measured angle and range readings and therefore in measured lengths. The length measurement system tests are designed to be sensitive to these misalignments.

Some geometric misalignments are such that the errors in the measured angles of a fixed target change in sign when the same target is measured in the backsight of the instrument. Such frontsight-backsight measurements of a single target are called two-face tests. These tests are extremely useful because they capture a large number of geometric misalignments and they do not require a calibrated reference length. The Standard requires two-face tests to be performed at different positions within the work volume of the instrument. More details on the test positions may be found in [20, 21].

12.4.3 Large-Scale Metrology at NIST

The large-scale coordinate metrology group at NIST has the capability of performing the complete set of B89.4.19 tests. The ranging tests are performed in the 60 m long-length test facility that includes a laser-rail and carriage system. The carriage has two opposing retro-reflectors. One retro-reflector is used for the tracker under test while the other is used for the reference interferometer on the other end of the rail. The expanded uncertainty ($k = 2$) of reference length L is $U(L) = 5$ μm $+ 0.3 \times 10^{-6} L$.

The length measurement and two-face system tests are performed in the large-scale laboratory. Currently, the reference length for the length measurement tests is realized using the laser-rail and carriage system (LARCS) [22]. The LARCS (different from the 60 m laser-rail facility used for range calibration) employs a reference interferometer mounted on a rail (about 3 m long) that can be oriented in different ways (horizontal, vertical, inclined) to meet the B89.4.19 requirements. A carriage with two retro-reflectors rides on the rail. The tracker uses one retro-reflector while the reference interferometer utilizes the other. The expanded uncertainty ($k = 2$) of a nominal reference length L is $U(L) = 3.4 \ \mu m + 0.5 \times 10^{-6} \ L$ for the LARCS system. We are now evaluating different artifacts that may be used as the reference length instead of the LARCS system [23].

12.4.4 Tracker Calibration Examples

We have performed the B89.4.19 tests on different trackers at our facility at NIST. Some trackers that were tested met the manufacturer's specifications while others did not. We present the results from the length measurement and two-face system tests for a tracker in Fig.12.7.

The length measurement system test chart (Fig. 12.7 top) may be interpreted as follows. The 35 length tests are in the order in which they appear in the Standard. Test 1 is the horizontal length measurement at the near position (1 m away, azimuthal angle of 0°). Tests 2–5 are the horizontal length measurements at four orientations of the tracker (0°, 90°, 180° and 270°) at the 3 m distance.

Tests 6–9 are the horizontal length measurements at four orientations of the tracker at the 6 m distance. Tests 10–17 are the vertical length measurements. Tests 18–25 are the right diagonal length measurements and tests 26–33 are the left diagonal length measurements. Tests 34 and 35 are the user-defined positions.

The two-face chart (Fig. 12.7 bottom) may be interpreted as follows. The 36 two-face tests are in the order in which they appear in the Standard. Therefore, tests 1–4 are the two-face measurements at the near position (1 m) with the target on the floor for four orientations of the tracker (0°, 90°, 180° and 270°).

Tests 5–8 are the two-face measurements at the near position (1 m) with the target at tracker height for four orientations of the tracker. Tests 9–12 are the two-face measurements at the near position (1 m) with the target at twice the tracker height for four orientations of the tracker (0°, 90°, 180° and 270°).

Tests 13–24 are a repetition of tests 1–12 but with the tracker 3 m away from the target. Tests 25–36 are a repetition of tests 1–12 but with the tracker 6 m away from the target.

Figure 12.8 shows the results of length measurement and two-face system tests for a tracker tested at NIST. This tracker appears to have satisfactory length measurement performance, but demonstrates large two-face errors.

Notice that the two-face error demonstrates periodicity that is a function of the azimuth. In addition, the average two-face error (approximately 1.2 mm in Fig. 12.8)

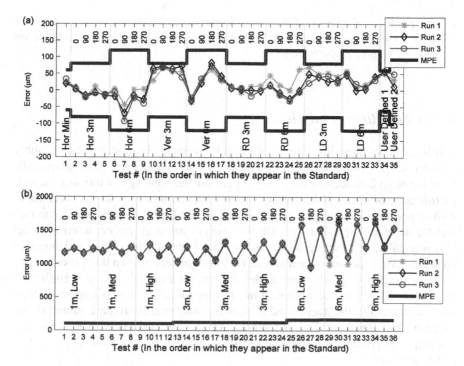

Fig. 12.8 Two-face system test results for a tracker

does not change with distance of the target from the tracker. The average two-face error with increasing distance may arise from an offset in the beam from its ideal position (The tracker does not have a beam steering mirror. Rather, the source is located directly in the head). Such an offset will result in decreasing error in the measured angle farther away from the tracker; consequently, the two-face error will be independent of range. The observed periodicity in the errors may potentially arise from an encoder eccentricity; in this particular case, it was observed that the vertical angle, rather than the horizontal angle, demonstrated this periodicity. There are no geometric error sources where such behavior may be modeled upon and therefore it is believed that the observed effects may be due to non-geometric effects such as stressing and relaxation of cables.

Several interesting points raised in this section are worth summarizing:

- Large length measurement or two-face system test errors typically suggest that geometric misalignments have not been properly compensated.
- Two-face errors as reported in the B89.4.19 Standard are the convolved errors in both the horizontal and vertical angles, and scaled by the range. Raw horizontal and vertical angle errors from a two-face test contain more diagnostic information.

- The purpose of extracting the magnitude of physical misalignments from B89.4.19 tests is to estimate the error in other length measurements made within the work volume of the tracker.

12.4.5 Sensitivity Analysis

A spherical coordinate instrument such as a laser tracker is a mechanical assembly of different components and therefore subject to misalignments such as offsets (offset in the beam from ideal position, offset between the standing and transit axes, etc), tilt (tilt in the beam, tilt in the transit axis, etc) and eccentricity (encoder eccentricity with respect to axis) during construction and assembly. It is general practice to correct for these misalignments by software compensation. A geometric error model [24] is required for this purpose that relates the corrected (true) range and angles to measured range and angles, and geometric misalignments within the tracker.

The corrected range (Rc) and angles (horizontal angle: Hc, vertical angle: Vc) of any coordinate in space are functions of several misalignment parameters within the construction of the tracker and also of the measured coordinate values at that location (Rm, Hm, Vm). The corrections ΔRm, ΔHm and ΔVm in Rm, Hm and Vm respectively may be expressed as (linear models may be sufficient as a first approximation),

$$Rc - Rm = \Delta Rm = \sum_{i=1}^{n} x_i u_i(Rm, Hm, Vm)$$
$$Hc - Hm = \Delta Hm = \sum_{i=1}^{n} x_i v_i(Rm, Hm, Vm) \qquad (12.7)$$
$$Vc - Vm = \Delta Vm = \sum_{i=1}^{n} x_i w_i(Rm, Hm, Vm)$$

where x is any misalignment parameter (eccentricity in encoder, beam offset, transit axis offset from standing axis, beam tilt, etc), and u, v and w are functions of measured range and angles.

Because different commercially available laser trackers have different mechanical constructions, an error model applicable to one tracker may not necessarily be applicable to another. At NIST, we have modeled three broad classes of trackers: (a) tracker with a beam steering mirror for which the Loser and Kyle [24] model is applicable, (b) tracker with the laser source in the rotating head and (c) scanner with source mounted on the transit axis with a rotating prism mirror that steers the beam to the target.

As an example, an error model for a tracker with the source located in the head is given by (see Fig. 12.9)

$$Rc = Rm + x_2.\sin(Vm) + x_8$$
$$Hc = Hm + \frac{x_{1t}}{Rm.\sin(Vm)} + \frac{x_{4t}}{\sin(Vm)} + \frac{x_5}{\tan(Vm)} + x_{6x}\cos(Hm)$$

Fig. 12.9 Schematic of beam offset in a tracker where the beam originates from the head (there is no beam steering mirror). Axes OT, ON and OM are fixed to the tracker's head and therefore rotate with the head about the Z axis

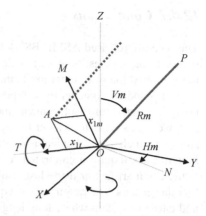

$$- x_{6y} \sin (Hm) + x_{9a} \sin (m.Hm) + x_{9b} \cos (m.Hm)$$

$$Vc = Vm - \frac{x_{1m}}{Rm} + \frac{x_2 \cos (Vm)}{Rm} + x_3 + x_{7n} \cos (Vm) \qquad (12.8)$$

$$- x_{7z} \sin (Vm) + x_{9c} \sin (m.Vm) + x_{9d} \cos (m.Vm)$$

where x_{1t} and x_{1m} are beam offsets along the transit axis and its normal, x_2 is the transit offset, x_3 is the vertical index offset, x_{4t} is the beam tilt, x_5 is the transit tilt, x_{6x} and x_{6y} are the horizontal angle encoder eccentricities, x_{7x} and x_{7y} are the vertical angle encoder eccentricities, x_8 is the bird-bath error, and x_{9a}, x_{9b}, x_{9c} and x_{9d} are the components of the mth order scale error in the encoder (1st order is not distinguishable from encoder eccentricity. Higher orders beyond 2nd order may be neglected).

This error model may now be used to numerically estimate the sensitivity of any given test to geometric misalignments that are included in the model.

12.4.6 Summary

The complete set of B89.4.19 tests may be numerically simulated and the sensitivity of each test to every parameter in the error model can be determined. The results, tabulated as a two-dimensional matrix with B89.4.19 tests in one axis and misalignment parameters in the other, is a "sensitivity matrix." We have created such sensitivity matrices for the three classes of trackers for which we have developed error models. Such matrices are useful in assessing the capabilities and limitations of any set of performance evaluation tests. Analysis of such sensitivity matrices further leads to the identification of optimal length positions that are sensitive to specific geometric misalignments.

12.4.7 Conclusions

The recently released ASME B89.4.19 Standard provides a common set of performance evaluation tests that may be performed by both the manufacturer and the user to evaluate if the instrument meets the manufacture's specifications.

The Standard contains three types of tests. The ranging tests assess the instrument's distance (or displacement) measuring capability. The length measurement and two-face system tests identify any systematic errors within the instrument's construction, such as mechanical and optical misalignments. The length measurement system tests require a calibrated reference length (typically 2.3 m long) realized either as an artifact or using laser-rails, or between free standing targets calibrated by other means. The ranging tests require a reference interferometer and a laser-rail and carriage system where long lengths may be calibrated or some other means to independently measure long lengths reliably. The two-face tests require no reference lengths. They are simple and easy to perform, and capture a large number of geometric misalignments.

The B89.4.19 test results provide valuable diagnostic information as well. Using geometric error models of the tracker, it may be possible to estimate magnitudes of misalignments in the construction of the tracker. Such information may then be used in determining errors in other length measurements made within the work volume of the tracker.

Geometric error models also serve a more general role. They may be used to determine the sensitivity of any given test to any geometric misalignment within the tracker. Such sensitivity analysis is useful in determining if a given set of performance evaluation tests effectively captures the misalignments, or if any modifications in the placement of reference lengths are necessary.

12.5 Performance of Super-Resolution Enhancement for LIDAR Camera Data

Laser detection and ranging (LIDAR) is a crucial component for navigation in autonomous or semi-autonomous robots. Current small robots generally employ a 2D scanning LIDAR that scans along a single line parallel to the ground and therefore cannot detect objects above or below the detection line [25, 26]. In indoor urban environments where the setting is highly cluttered with overhanging objects such as tabletops, the 2D scanning LIDAR systems may not be sufficient for navigation and obstacle avoidance [25]. A new generation of small and compact 3D LIDAR devices, named LIDAR camera, offers a promising solution to small robot navigation in urban environments where modern warfare is often conducted.

LIDAR camera devices are compact and lightweight sensors that acquire a 3D range image of the surrounding environment. The LIDAR camera device (Fig. 12.10) used in this study weighs 162 g and measures (5.0 × 6.7 × 4.23) cm [27]. LIDAR camera devices emit diffuse modulated near-infrared light and measure

Fig. 12.10 LIDAR camera

the subsequent phase shift between the original emitted light and the reflected light. The phase measurements are combined to calculate the range data based on the time-of-flight principle [27]. The detector utilized by LIDAR camera devices is a focal plane array (FPA), which is typically limited to a maximum size of 256×256 detectors. Consequently, these devices cannot achieve the resolution of scanning LIDAR systems. This disadvantage of LIDAR camera systems may be rectified by the application of super-resolution image reconstruction.

Super-resolution algorithms utilize a series of low-resolution frames containing sub-pixel shifts to generate a higher resolution image. These algorithms are typically composed of two major stages: registration stage and reconstruction stage. During the registration stage, the shift with respect to a reference frame (usually the first frame of the series) is computed to sub-pixel (i.e., decimal pixel) accuracy. The second stage utilizes this sub-pixel information to interpolate the low-resolution frames onto a higher resolution grid. A necessary condition for successful super-resolution enhancement is the presence of differing shifts between the frames in the series. The differing shifts of each frame provide additional information from which to reconstruct the super-resolved imagery. Previous work by Rosenbush et al. [28] applied a super-resolution algorithm [29] to LIDAR camera data, and observed improvement in image quality in terms of number of edges detected. In this work, the super-resolution algorithm of Young et al. [30] is applied to LIDAR camera imagery. This algorithm separates the registration stage into a gross shift (i.e., integer pixel shift) estimation stage and a sub-pixel shift (i.e., decimal pixel shift) estimation stage for improved registration accuracy. Both sub-stages use the correlation method in the frequency domain to estimate shifts between the frame series and the reference image. The reconstruction stage of Young et al.'s algorithm applies the error-energy reduction method with constraints in both spatial and frequency domains to generate a high-resolution image [30]. Because LIDAR camera imagery is inherently smoother than visible light imagery (LIDAR camera data does not capture the texture or color of the scene), this work develops a preprocessing stage for improved image registration. Specifically, a wavelet edge filtering method [31] and a Canny edge detection method [28] are investigated and compared against the accuracy achieved with no preprocessing. The wavelet edge filtering method provided more accurate shift estimation for LIDAR camera data.

To assess the improvement in super-resolution enhanced LIDAR camera data, the authors conducted perception experiments to obtain a human subjective measurement of quality. The triangle orientation discrimination (TOD) methodology [32, 33] was used to measure the improvement achieved with super-resolution. The TOD task is a four-alternative forced-choice perception experiment where the subject is asked to identify the orientation of a triangle (apex up, down, right, or left) [33]. Results show that the probability of target discrimination as well as the response time improves with super-resolution enhancement of the LIDAR camera data.

12.5.1 Methodology

12.5.1.1 Preprocessing Stage

The purpose of the preprocessing stage is to emphasize LIDAR camera image edges for improved frame registration. One investigated method was the use of multi-scale edge-wavelet transforms [34] to calculate the horizontal and vertical partial derivatives of the input image at the second wavelet scale for each frame of the series. The two derivatives were then combined using sum of squares to produce a wavelet edge enhanced frame series. Another investigated method was the use of Canny edge detection algorithm to generate binary edge frame series.

To assess the benefit of preprocessing, the following procedure was followed. A synthetic frame series was generated with known sub-pixel shifts. First, an over-sampled non-aliased scanning LIDAR reference image (204 × 204 pixels) was interpolated by an upsampling factor of eight (1,632 × 1,632 pixels) using a Fourier windowing method [34]. Then, the simulated high-resolution image was sub-sampled at different factors to produce several low-resolution frame series, each with a different degree of aliasing. Figure 12.10 shows the un-aliased spectrum of a discrete space signal (e.g., scanning LIDAR image) produced by oversampling a continuous space signal at a sampling frequency greater than the Nyquist frequency. If the sampling frequency is below Nyquist (simulated by sub-sampling the over-sampled image), the spectrum of the sampled signal is aliased with distorted higher frequency components as depicted in red in Fig. 12.11.

A synthetic frame series was generated by sub-sampling every m pixels in both dimensions of the simulated high-resolution image, where $m = 4, 8, 12, 16, 20, 28, 36, 48, 56$. Therefore the undersampling factors are $m/8$ (i.e., 0.5, 1, 1.5, 2,

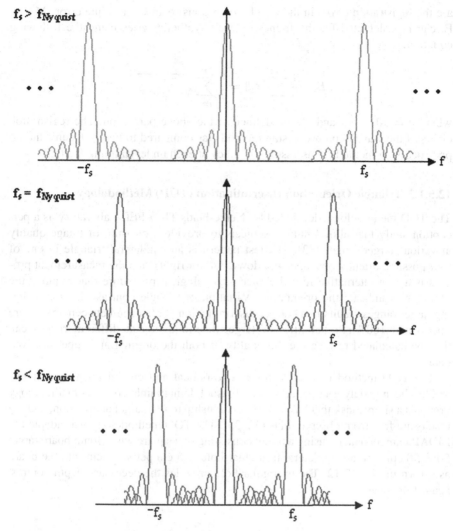

Fig. 12.11 (*Top*) Un-aliased spectrum of signal sampled above Nyquist frequency, (*mid*) at Nyquist, and (*bottom*) aliased at below Nyquist frequency

2.5, 3.5, 4.5, 6, 7), simulating different degrees of aliasing. For each undersampling factor, the sub-pixel shifts for each frame of the synthetic series were generated by varying the starting pixel position of sub-sampling according to a uniform random distribution (30 frames for each series). Then preprocessing using either the wavelet or Canny method was performed. Sub-pixel shift estimates from the preprocessed and no preprocessing series were compared to the known shifts. Let $\varepsilon_i = (\varepsilon_{xi}, \varepsilon_{yi})$ denote the registration error vector of the ith frame where ε_{xi} and ε_{yi}

are the registration errors in the x and y directions. A mean absolute error (MAE), E, can be calculated for the frames of each synthetic series using the following equation:

$$E = \frac{1}{n} \sum_{i=1}^{n} \|\varepsilon_i\| = \frac{1}{n} \sum_{i=1}^{n} \sqrt{\varepsilon_{xi}^2 + \varepsilon_{yi}^2}$$

where $n = 30$ ε_i, ε_{xi} and ε_{yi} are defined in the above paragraph. The registration errors of the wavelet preprocessing method was compared to that of Canny and no preprocessing methods to assess the accuracy at each undersampling factor.

12.5.1.2 Triangle Orientation Discrimination (TOD) Methodology

The TOD methodology, developed by Netherlands TNO-FEL Laboratory, is a perception study that allows human subjects to provide a measure of image quality at various target ranges [33]. The test pattern is an equilateral triangle in one of four possible orientations (apex up, down, left, or right), and the measurement process is a four-alternative forced-choice psychological procedure that requires the observer to indicate the orientation. Variation of triangle contrast/size by changing target ranges results in a correct discrimination percentage between 25% (pure guess) and 100%. Probabilities of target discrimination at different ranges can then be calculated to measure the quality of both the original and super-resolved data.

The TOD method is suitable for electro-optical and optical imaging systems, and has been widely used in thermal and visual domain imagers. This methodology provides a simple task that has a close relationship to real target acquisition, and the results are free from observer bias [32, 33]. The TOD methodology was adapted to LIDAR camera data by using a target consisting of a square white foam board target 50×50 cm with an equilateral triangular hole (7.5 cm per side) cut into the board as shown in Fig. 12.12. The triangular hole provides the necessary depth contrast against the board.

Fig. 12.12 TOD setup

12.5.2 LIDAR Camera

The device utilized in this study is the SwissRanger LIDAR camera. The camera emits diffuse 850 nm near-infrared light modulated at a default frequency of 20 MHz from a bank of 55 light emitting diodes. The non-ambiguity range achieved at this modulation frequency is 7.5 m. The LIDAR camera a pixel array of 176 × 144 with a field-of-view of 47.5° × 39.6°, and can capture images at a maximum rate of 50 frames/s (variable with respect to the integration time setting).

12.5.3 Data Collection

Data collection for the experiment was conducted at a laboratory in the National Institute of Standards and Technology. The LIDAR camera was placed 6.5 m from a beige wall as depicted in Fig. 12.11 . The target was positioned at (3, 3.5, 4, 4.5, 5, 5.5, and 6) m from the camera. The investigated ranges were limited to between 3 and 6 m because inaccurate behavior of LIDAR cameras was observed at very close and very far target distances [35]. At each range, the triangle was positioned at one of four possible orientations (apex up, down, left, right) with the center approximately 1 m high. For each orientation at each range, four trials were conducted. Each trial consisted of a sequence of 32 frames acquired by holding the camera. The natural motion of the hand while holding the camera provided the shifts required for the super-resolution algorithm. Motion is assumed to be limited to translations in the x (horizontal) and y (vertical) planes. Though slight rotation and translation in the z-plane (depth) might have occurred from holding the camera, these parameters were not considered in the current study.

12.5.4 Data Processing

For each series of 32 frames, the first 25 frames are utilized for super-resolution image reconstruction. The first frame was used as the reference frame from which pixel shifts were calculated for successive frames. The use of 25 frames resulted in a resolution improvement factor of five in each direction for the super-resolved image. To ensure that the monitor modulation transfer function (MTF) was not a limiting factor in the experiment, the super-resolved images 250 × 250 pixels were bilinearly interpolated by a factor of two to 500 × 500 pixels. The original imagery 50 × 50 pixels was bilinearly interpolated to 500 × 500 pixels for consistency between the baseline and super-resolved imagery.

12.5.5 Perception Experiment

The perception experiment was a four-alternative forced-choice procedure (up, down, left, right). The imagery in this experiment was organized in the image cells

Table 12.5 Image cell format and naming convention

Range m	A 3.0	B 3.5	C 4.0	D 4.5	E 5.0	F 5.5	G 6.0
Original image	AA	BA	CA	DA	EA	FA	GA
Super-resolved image	AB	BB	CB	DB	EB	FB	GB

and their naming convention is shown in Table 12.5. As shown in the row of original images in Table 12.5, the grayscale baseline range imagery was grouped into seven cells corresponding to the seven different target ranges. Each cell consisted of 16 original low-resolution LIDAR camera images (4 orientations × 4 trials). Similarly, the grayscale super-resolved range imagery was grouped into seven cells consisting of 16 images each as shown in the row of super-resolved images in Table 12.5. The experiment therefore consisted of 14 cells with a total of 224 images.

Nine subjects participated in the experiment in July 2008. The subjects were shown one image at a time with randomized presentation of cells and randomized presentation of images within each cell. The display was a 19 in. flat panel with an array size of 1,280 × 1,024 pixels.

12.5.6 Results and Discussion

12.5.6.1 Assessment of Registration Accuracy

Figure 12.13 shows the mean absolute error of registration at each undersampling factor for the generated synthetic experiments. The unit of error is fraction of a pixel. Wavelet preprocessing outperformed both the Canny method and no preprocessing method for undersampling factors of less than 6. Wavelet preprocessing was especially effective at low and moderate degrees of aliasing (undersampling

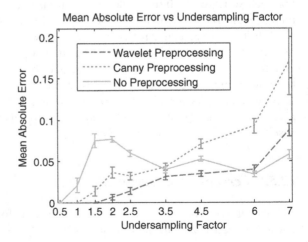

Fig. 12.13 Mean absolute registration error with standard deviation for each undersampling factor

factor of less than 3.5). If the imagery contained severe aliasing (undersampling factor greater than 6), then no preprocessing resulted in higher registration accuracy.
The observed trend is expected. LIDAR camera data is characteristically smooth
due to the lack of texture information, so edge filtering with the wavelet method
will improve registration. But if the data is severely undersampled that its middle to high frequency components are corrupted by aliasing, then wavelet edge
filtering (which uses these severely corrupted frequency components) will result
in poorer registration. The degree of aliasing in the imagery acquired with the
LIDAR camera is expected to be in the moderate range as super-resolved imagery
using wavelet preprocessing yields fewer artifacts than imagery produced without
preprocessing.

12.5.6.2 Triangle Orientation Discrimination Perception Experiment

Figure 12.14 shows grayscale and color images (color-coded to distance) of the
TOD target oriented up at a distance of 5 m from the camera. The orientation of the
equilateral triangular hole is difficult to discern in the original images at this distance
as the triangular hole appears like a blurred circle. By contrast, the orientation is
clear in the super-resolution enhanced imagery. For imagery with target distances
greater than 5 m, the orientation, as expected, was still more difficult to discern
using the original LIDAR camera imagery. But super-resolution at these greater
distances proved to be effective.

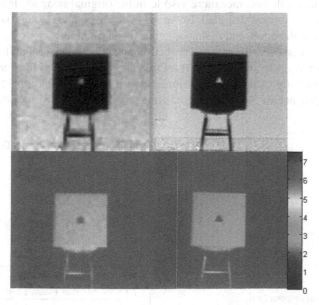

Fig. 12.14 Grayscale (*top*) and color-coded (*bottom*) LIDAR camera imagery for (*left*) original
image and (*right*) super-resolved image of TOD target at 5 m

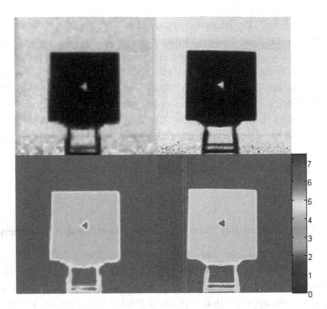

Fig. 12.15 Grayscale (*top*) and color-coded (*bottom*) LIDAR camera imagery for (*left*) original image and (*right*) super-resolved image of TOD target at 4 m

Figure 12.15 shows grayscale and color images of the TOD target oriented left at a distance of 4 m from the camera. As the target distance decreases, the orientation of the triangular hole becomes more visible in the original imagery, though the triangular hole still appears distorted. In the super-resolved image, the triangular hole does not appear distorted, and is shaped more like a triangle.

Figure 12.16 shows the group-averaged chance-corrected probability of target discrimination at each target range. At all ranges, super-resolved imagery produced a higher probability of target discrimination with smaller inter-subject variability. At a target distance of 3 m, the original imagery had a 77% of the probability of

Fig. 12.16 Chance-corrected probability of target discrimination at each target range

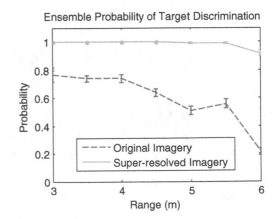

Fig. 12.17 Response times at each target range with standard error bars showing inter-subject variability

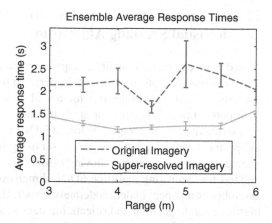

target discrimination, while the super-resolved imagery reached 100%. The target discrimination performance is increased by 30% using the super-resolution algorithm. As the target distance increased, subjects had more difficulty to discriminate the target orientation. At a target distance of 6 m, the original imagery had a 20% of the probability of target discrimination, while the super-resolved imagery reached 90%. That is a 350% improvement in target discrimination performance. In summary, the probability of target discrimination is increased by 30% to 350% for the target ranges from 3 to 6 m using the super-resolution algorithm.

Not only were subjects able to achieve higher accuracy at all ranges with super-resolved imagery, but the response times were also faster with less variability for super-resolved imagery at all ranges. Figure 12.17 shows the group-averaged response times at each range with standard error bars representing inter-subject variability. At a range of 6 m, subjects responded in an average time of 1.58 s using the super-resolved imagery, 23% faster than the response time using original imagery.

12.5.7 Conclusion

Super-resolution image reconstruction, complemented by a wavelet preprocessing stage for improved image registration, yields significant benefits for LIDAR camera imagery. In the triangle orientation discrimination experiment, subjects achieved higher accuracy at all investigated target ranges with faster response times and reduced inter-subject variability for super-resolved imagery. Complemented by super-resolution image reconstruction, the high frame rate, small size, and lightweight LIDAR camera sensors will be ideal for autonomous or semi-autonomous robot navigation in urban indoor environments. In semi-autonomous robot navigation, super-resolution enhancement will provide human operators with increased target discrimination. In fully autonomous mode, super-resolved imagery will allow guidance software to improve obstacle avoidance. The incorporation of super-resolution into the manufacturing robotic applications will improve small robot performance, contributing to safety of manufacturing environments.

12.6 Dynamic 6DOF Metrology for Evaluating a Visual Servoing Algorithm

In order to eliminate inflexible, expensive stationary stations on manufacturing assembly lines, robots must be able to perform their tasks on moving targets. However, manufacturers will not risk the use of dynamic visual servoing systems until their performance is better measured and understood. This section explores the use of a dynamic, six-degree-of-freedom (6DOF) laser tracker to empirically evaluate the performance of a real-time visual servoing implementation where a robot arm attempts to insert a peg in a moving object. The objective is to establish a general method for evaluating real-time 6DOF dimensional measurements of an object or assembly component under moderately constrained motion. Laser trackers produce highly accurate position and orientation data at a high data rate and can be considered ground truth, thereby enabling the evaluation of sensors for visual servoing.

This work primarily focuses on how to collect and analyze data taken simultaneously from a laser tracker and a visual servoing sensor under test. Issues include synchronizing the data streams, so individual data points are taken at the same time; external calibration of the two sensors, so individual data points can be compared in the same coordinate system; and comparison metrics, so individual data points can be compared between the two sensors to determine how close the system under test comes to the ground truth.

12.6.1 Purdue Line Tracking System

In this study, the NIST Manufacturing Engineering Lab collaborated with the Purdue Robot Vision Lab on initial experiments to evaluate the Purdue line tracking system. The Purdue line tracking system consists of a sensor system and robot arm that can track and mate with an object under random motion. The system addresses this goal using a subsumptive, hierarchical, and distributed vision-based architecture [36, 37, 39, 40]. The system consists of multiple real-time control modules running in parallel, where each module is controlled by a different tracking method with unique capabilities with respect to accuracy, computational efficiency, sensitivity to varying conditions, etc. By taking the most reliable input from all the modules, the system is able to achieve a high level of fault tolerance and robustness to noncooperative conditions such as severe occlusions and sudden illumination changes.

The Purdue system consists of three control modules, a system arbitrator, and a robot controller interface. The three control modules are: coarse control, model-based fine control, and stereo-based fine control. Each control module independently estimates the real-time position and pose of a part under view and the system arbitrator selects the most reliable for use by the robot controller to move the end effector towards the part.

The coarse control module resides at the lowest level in the system hierarchy, meaning that it will work only when the other two modules become unavailable or unreliable. It provides an initialization point for the controllers with higher hierarchy. That is, the coarse control is intended to track the target and command the robot to an approximate location in front of the target. This module only requires the level of accuracy that would place the end-effector in front of the target such that cameras for the other controllers can view the target. Coarse control visual-tracking uses a camera mounted on the ceiling with a view of the entire workspace.

The stereo-based fine control module resides at the highest hierarchy level in the system, meaning that while tracking the target it subsumes the other two modules. It uses a stereo pair of cameras located on the robot's end-effector. The visual-tracking loop in this module uses a blob analysis algorithm to detect, in both cameras, three prominent coplanar features on an engine cover. The engine cover is used as an example part from a real manufacturing application, but the specific part and application are not essential to this experiment—the goal is to dock the robot end effector with the part, not replicate an actual manufacturing operation. Then, using the calibration information the 3D coordinates of those features are reconstructed, and the 6DOF 3D pose of the target is estimated in the robot coordinate system [38]. Based on the estimated pose, the visual-servoing loop then performs the peg-and-hole motion using a Proportional-Integral-Derivative (PID) control law.

The model-based fine control provides redundancy to the stereo-based fine controller. It uses a monocular vision system and a known wire-frame model of the target [2]. The visual-tracking loop of this module first projects the model into the input scene with respect to the initial pose that is given by the coarse control module or the stereo-based control module. Then, it sequentially matches the straight lines of the wire-frame model to the detected edges in the scene for an updated calculation of the pose of the target. For robust pose estimation, it uses a backtracking scheme for correspondence search [2].

12.6.2 Experimental Set-up and Results

We present here preliminary results from a series of experiments conducted at Purdue in April, 2008 using an experimental set-up originally designed by the Purdue Robot Vision Lab to test real-time visual servoing (see Fig. 12.18) In the experimental set-up the visual servoing system performs a peg-and-hole task using an engine cover part as the target. As the engine part moves, the robot system attempts to insert a peg into a hollow wooden cylinder attached to the engine part. The engine part was either stationary or moved by an overhead linear rail at a constant velocity. The part was suspended from the linear rail to allow the experimenters to move the part back and forth by an attached string. The goal of this experiment is to command the robot to insert the peg into the cylinder on the engine cover while the engine cover is in motion, resembling the automation needed to perform tasks such as glass decking or wheel decking on-the-fly [2].

During each pass of the experiment, the position of the engine part was measured simultaneously by the Purdue visual tracking system and the NIST laser tracker. Data acquisition from each system was triggered by a common hardware signal, and the data time stamped and streamed to storage for off-line comparison.

Fig. 12.18 Peg and Hole experiment

The NIST laser tracker system used consisted of two major components, a base unit and an active target [40]. The active target is a SmartTRACK Sensor (STS) [40] capable of determining its orientation in 3D space. Weighing 1.4 kg, the STS has an angular resolution specification of ± 3 arc-s (or $\pm 0.000833°$). The complete manufacturer's specification can be found in [40]. The base unit is the Tracker3^{TM} Laser Tracking System (T3) which tracks the 3D position of the STS. The T3 system has ± 50 μm absolute accuracy at 5 m. Together, the T3 and STS provide an accurate but limited 6DOF pose estimation system. The STS sensor can be seen in Fig. 12.17 at the bottom of the engine part to the left.

Both the T3/STS system and the Purdue visual-tracking system allow an external signal to trigger their data acquisition. We use a single 30 Hz trigger signal shared by both systems. Both systems synchronize their computer clock with an NTP (Network Time Protocol) server every 10 s throughout the entire data collection. The data streams from both sensors were logged to files for later analysis.

The protocol was to run three sets of experiments, one with the target stationary, one with the target moving with a simple linear velocity, and one with the target moving with a linear velocity but randomly displaced manually by the experimenters. The results are given in the next section for the three sets.

12.6.2.1 Stationary Tests

The stationary tests allowed us to evaluate the basic performance of both systems and assure that the laser tracker was performing to specifications after shipping. The target was placed in four positions and data collected for 15–30 s for each. The results showed that both systems performed within specifications.

Table 12.6 lists the standard deviations of the stationary data set measured by the STS/T3 system. The results are consistent with the specifications [38]. The laser tracker stayed in a fixed position so the target distance varied.

Table 12.6   STS/T3 system: repeatability for stationary data

	Sample size	T3/STS mean distance (mm)	2 std (mm)
Position 1	466	3,550.054	0.006
Position 2	1,157	3,781.466	0.006
Position 3	1,050	3,882.787	0.005
Position 4	1,018	4,002.035	0.008

Table 12.7   Purdue system: repeatability for stationary data

	Sample size	Purdue system mean distance (mm)	2 std (mm)
Position 1	466	2,670.582	0.629
Position 2	1,157	2,625.820	0.582
Position 3	1,050	2,625.036	0.560
Position 4	1,018	2,636.701	0.561

Table 12.7 lists the standard deviations of the stationary data set measured by the Purdue system. The Purdue system moves the robot end-effector near the object, so the distance remains relatively constant. The results show a consistent value near 0.6 mm at a range of approximately 2.6 m.

For use as a reference system, a metrology technology should have an accuracy at least one order of magnitude greater than the system under test. In this case, the laser tracker is two orders of magnitude more accurate and precise than the Purdue vision system.

12.6.2.2   Linear Motion Tests

In the linear motion tests, the target was moved about 1.5 m left to right and tracked by both the laser tracker and the Purdue system. For each trial, the motion was repeated 30 times as the target moved and then was quickly moved back to the start position. The backward sweep was ignored in the data analysis as the Purdue system only tracked during the forward motion.

The differential motion as measured by both systems was used to determine the consistency between laser tracker and Purdue system. In effect, the comparison is being made again the measured speed of the target in each separate coordinate (X, Y, Z, roll, pitch, yaw).

The data below are from pass 6 of the first trial, and are typical of the linear runs. The sample size is 453, with 33 ms between data points. In the graphs below the horizontal axis is frame number and the vertical axis is the difference between the laser tracker motion change (velocity) in each coordinate and the Purdue system velocity. Since the laser tracker was accepted as ground truth, the difference is defined as the error in the Purdue system.

In Fig. 12.19 the errors can be seen to be consistent and relatively independent of position along the path, although that is yet to be evaluated statistically. The error

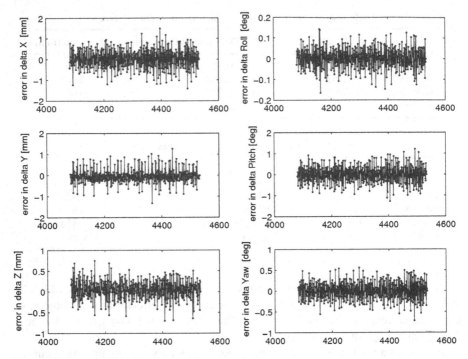

Fig. 12.19 Pass 6—Velocity errors vs. frame number

does vary by coordinate—different coordinates proved to be more sensitive to the left-to-right motion. In object coordinates the target is moving in the Y-Z plane, primarily in the Y direction, and the Z coordinate has slightly less error. Similarly, the roll angle is around the X axis and proved an order of magnitude reduced in error over the other rotations. Table 12.8 quantifies the error values while Fig. 12.20 gives a histogram of the error values to show roughly symmetric, zero centered error distributions. For coordinate Y there appears to be a secondary peak in positive error.

Table 12.8 Pass 6—Coordinate error statistics (n=453)

	Mean	Std	Max	Min
X in mm	0.01614	0.45376	1.49003	−1.41190
Y in mm	−0.03513	0.33355	1.27399	−1.32852
Z in mm	0.05206	0.21991	0.74652	−0.69851
distance	−0.04305	0.33273	1.21320	−1.50436
Roll in deg	0.00168	0.04915	0.14368	−0.16532
Pitch in deg	0.00088	0.40811	1.21672	−1.27938
Yaw in deg	0.00184	0.21589	0.56194	−0.71260
Total angle	−0.04798	0.05195	0.03383	−0.28635

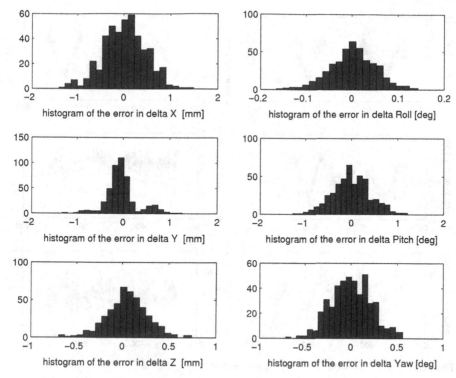

Fig. 12.20 Pass 6—Velocity errors (histogram)

12.6.2.3 Shaking Motion Tests

In the shaking motion tests, the basic target motion and repetitions were identical to those in the moving target test, but the experimenters could pull a string to swing the target with an impulse motion. This was done through a series of different motions—first with no extra motion to match the linear case, then with the impulse motion varying by amplitude and frequency.

The data in Fig. 12.21 are from pass 3 of the first shaking motion trial. The sample size is 455, with 33 ms between data points. In the graphs below, the horizontal axis is frame number, and the vertical axis the difference between the laser tracker velocity in each coordinate and the Purdue system velocity.

The graphs show four impulse motions as the target was pulled back four times, relatively smoothly and consistently. The error first goes positive as the speed of the target slows down and the Purdue system undershoots the speed and then goes negative as the Purdue system overshoots the speed. The graph scales have changed from the linear motion case as the error range has approximately doubled. The Z axis remains the one with lowest error while the roll angle error is greater compared to the linear motion tests as the impulse motion rotated the target around the X-axis.

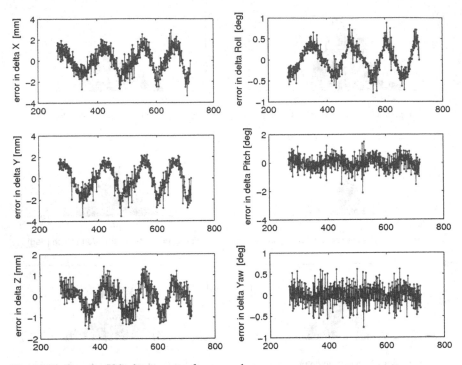

Fig. 12.21 Pass 3—Velocity errors vs. frame number

The Purdue experimental set-up only allowed either a smooth linear motion controlled by the linear rail, or the linear rail motion with shaking motions superimposed by a user pulling a string. The latter motions are not precisely repeatable, and therefore problematic in analysis. In the future we intend to address this in two ways. First, by characterizing the user motions as generally consisting of certain amplitudes or frequencies. In real world applications, motions will consist of randomized, non-repeating motions, and we need statistical models to handle them. Second, by working to generate repeatable motion at multiple frequency and amplitude scales by using a robot arm and other mechanisms.

12.6.3 Conclusions

In this chapter we have demonstrated the use of a precision laser tracker to evaluate a state-of-the-art visual servoing perception system, with the objective of establishing general techniques for evaluating 6DOF sensors. The demonstration involved a synchronized data collection system that used a hardware trigger to collect at a 30 Hz rate, while the laser tracker has the ability to collect at up to 150 Hz. The laser tracker was verified accurate enough for the approximate 1 mm range of error that needed to be measured.

Future work will involve a more detailed analysis of the data and the establishment of better calibration techniques and metrics to insure consistent comparisons between laser tracker and sensor data streams.

12.7 Summary

This Chapter presented five separate research results, from five research groups, which fit within general framework for the evaluation of perception algorithms. Each result stands alone as a distinct research contribution, but taken together, they illustrate what is required for a complete system for sensor and algorithm evaluation.

The individual contributions interrelate in a number of ways. The first paper on the "Preliminary Analysis of Conveyor Dynamic Motion," gave statistics on the motion of a specific conveyer used in automotive manufacturing. The maximum acceleration recorded during the experiments was on the order of 100 cm/s^2. This maximum is within the specifications of the API laser tracker used during the experiments in the final paper "Dynamic 6DOF Metrology for evaluating a visual servoing system," which is 3 g. For laser tracker technology to be suitable for use as a standard metrology technique in the evaluation of motions sensors on the manufacturing line, the technology should be capable of handling the full range of motions. For this case it does. Further research should be directed at understanding other conveyer mechanisms and their full range of standard and exceptional motions, and developing a suitable statistical model to determine if the range of motions of a specific conveyer can be properly sensed by a perception system.

The second paper, "Calibration of a System of a Gray Value Camera and an MDSI Range Camera," illustrates the cross-calibration of two very different sensors: a grey value camera and relatively low-resolution range sensor. For the purposes of sensor evaluation, it is important to be able to cross-calibrate various types of sensors, particularly 3D sensors, and to characterize the accuracy of the calibration. In the final paper, "Dynamic 6DOF Metrology," difficulties in cross-sensor calibration influenced the available methods of analysis. Further research here needs to be directed at understanding how to choose, apply and evaluate calibration algorithms for different pairings of sensors, including standard reference sensors, so calibration can be accounted for when comparing reference and test sensors.

The third paper, "Performance evaluation of laser trackers," presents current standards for static evaluation of laser trackers. The well-established, highly accurate techniques used for static dimensional metrology give a foundation for extending the techniques to dynamic metrology suitable for perception system evaluation. The laser tracker used in the final paper was evaluated by these standards. Further research should be directed at developing versions of static evaluation metrics that are appropriate for dynamic evaluation.

The fourth paper, "Performance of Super-Resolution Enhancement for LADAR Camera Data," gives a methodology for evaluating the practical resolution of 3D range sensors. This methodology, Triangle Orientation Discrimination (TOD), is an innovative approach to give a quantitative metric based on human perception. The paper also gives a method for data enhancement from a standard range camera. For the evaluation of 3D and 6D sensors, it is important to have solid metrics, and to best understand the nature of the sensors so as to know if their base algorithms and data are the best possible. Further research should be directed here at a full suite of metrics for 3D range data evaluation.

The last paper, "Dynamic 6DOF Metrology for evaluating a visual servoing system," brought all elements together in one framework for preliminary experiments to evaluate a 6DOF sensor system. The framework, as given in the introduction, includes three steps to model task requirements, design and calibrate a standard measurement system, and establish metrics for system evaluation. To extend these preliminary experiments to rigorous, general standards will require further research both in each individual element, and in their integration. The objective of the experiments, developing a reference method for tracking an object's pose and position as it moves in an environment with uncontrolled lighting and background, is a central problem in manufacturing perception and a solution to this general problem would have applications for other manufacturing perception tasks.

References

1. Hutchinson, S.A., Hager, G.D., and Corke, P.I., A tutorial on visual servo control. IEEE Transactions on Robotics and Automation, 12(5):651–670, October 1996.
2. Yoon, Y., DeSouza, G.N., and Kak, A.C., Real-time tracking and pose estimation for industrial objects using geometric features. Proceedings of the Int. Conference in Robotics and Automation, Taiwan, 2003.
3. Yoon, Y., Park, J., and Kak, A.C., A heterogeneous distributed visual servoing system for real-time robotic assembly applications. Proceedings of the International Conference on Robotics and Automation, Orlando, Florida, 2006.
4. Montemerlo, M., Thrun, S., Dahlkamp, H., Stavens, D., and Strohband, S., Winning the DARPA grand challenge with an AI robot. Proceedings of the AAAI National Conference on Artificial Intelligence, Boston, MA, 2006.
5. Darrell, T., Gordon, G., Harville, M., and Woodfill, J., Integrated person tracking using stereo, color, and pattern detection. International Journal of Computing Vision, 37(2):175–185, June 2000.
6. Scharstein, D., Szeliski, R., and Zabih, R., A taxonomy and evaluation of dense two-frame stereo correspondence algorithms. Workshop on Stereo and Multi-Baseline Vision (in conjunction with IEEE CVPR 2001), pp. 131–140, Kauai, Hawaii, December 2001.
7. Phillips, P.J., Flynn, P.J., Scruggs, T., Bowyer, K.W., Jin Chang, Hoffman, K., Marques, J., Jaesik Min, Worek, W., Overview of the face recognition grand challenge. Computer Vision and Pattern Recognition, 2005. CVPR 2005. IEEE Computer Society Conference on, 1:947–954, 20–25 June 2005.
8. Shi, J., Preliminary Analysis of Conveyor Dynamic Motion, PerMIS, August, 2008.
9. Hanning, T., and Lasaruk, A., Calibration of a System of a GrayValue Camera and an MDSI Range Camera, PerMIS, August, 2008.
10. Muralikrishnan, B., Sawyer, D., Blackburn, C., Phillips, S., Borchardt, B., and Estler, T.W., Performance Evaluation of Laser Trackers, PerMIS, August, 2008.
11. Hu, S., Young, S.S., and Hong, T., Performance of Super-Resolution Enhancement for LADAR Camera Data, PerMIS, 2008.

12. Chang, T., Hong, T., Shneier, M., Holguin, G., Park, J., and Eastman, R., Dynamic 6DOF Metrology for Evaluating a Visual Servoing Algorithm, PerMIS, August, 2008.
13. Mengel, P., Listl, L., Koenig, B., Toepfer, C., Pellkofer, M., Brockherde, W., Hosticka, B., Elkahili, O., Schrey, O., and Ulfig, W., Three-dimensional CMOS image sensor for pedestrian protection and collision mitigation. Advanced Microsystems for Automotive Applications 2006, Berlin, Springer, pp. 23–39, 2006.
14. Hanning, T., Lasaruk, A., and Wertheimer, R., MDSI range camera calibration. Advanced Microsystems for Automotive Applications, Berlin, Springer, 2008.
15. Zhang, Z., A Flexible new technique for camera calibration. Technical report, Microsoft Research Technical Report MSR-TR-98-71, 1998.
16. Hartley, R., and Zisserman, A., Epipolar Geometry and the Fundamental Matrix. In: Multiple View Geometry in Computer Vision. Cambridge, MA, Cambridge University Press, 2000.
17. Hanning, T., Schone, R., and Graf, S., A closed form solution for monocular re-projective 3d pose estimation of regular planar patterns. International Conference of Image Processing, Atlanta, Georgia, pp. 2197–2200, 2006.
18. Golub, G.H., and van Loan, C.F., Matrix Computations. 3 ed. Baltimore, John Hopkins University Press, 1996.
19. Press, W.H., Flannery, B.P., Teukolsky, S.A., and Vetterling, W.T., Numerical Recipes in C, 2 ed. Cambridge, MA, Cambridge University Press, 1992.
20. ASME B89.4.19-2006 Standard – Performance Evaluation of Laser-Based Spherical Coordinate Measurement Systems, www.asme.org, 2006.
21. Estler, W.T., Sawyer, D.S., Borchardt, B., and Phillips, S.D., Laser-scale metrology instrument performance evaluations at NIST. The Journal of the CMSC 1(2):27–32, 2006.
22. Sawyer, D.S., Borchardt, B., Phillips, S.D., Fronczek, C., Estler, W.T., Woo, W., and Nickey, R.W., A laser tracker calibration system. Proceedings of the Measurement Science Conference, Anaheim, CA, 2002.
23. Sawyer, D.S., NIST progress in the development of a deployable high-accuracy artifact for laser tracker performance evaluation per ASME B89.4.19. Proceedings of the CMSC conference, Charlotte, NC, July 21–25, 2008.
24. Loser, R., and Kyle, S., Alignment and field check procedures for the Leica Laser Tracker LTD 500. Boeing Large Scale Optical Metrology Seminar, 1999.
25. Ng, T.C., SIMTech Technical Reports, 6(1):13–18, 2005.
26. Committee on Army Unmanned Ground Vehicle Technology, Technology development for Army unmanned ground vehicles, Sandia Report, 2002.
27. MESA Imaging, 2006. SwissRanger SR-3000 Manual. http://www.mesa-imaging.ch/.
28. Rosenbush, G., Hong, T.H., and Eastman, R.D., Super-resolution enhancement of flash LADAR range data. Proceedings of the SPIE, 6736:67314, 2007.
29. Vandewalle P., Susstrunk S., and Vetterli, M., A frequency domain approach to registration of aliased images with application to super-resolution. EURASIP Journal on Applied Signal Processing, 2006, Article ID 71459, 2005.
30. Young, S.S., and Driggers, R.G., Super-resolution image reconstruction from a sequence of aliased imagery. Applied Optics, 45:5073–5085, 2006.
31. Devitt, N., Moyer, S., and Young, S.S., Effect of image enhancement on the search and detection task in the urban terrain. In Proceedings of SPIE, Vol. 6207, 62070D 1-13, 2006.
32. Driggers, R.G., Krapels, K., Murrill, S., Young, S., Thielke, M., and Schuler, J., Super-resolution performance for undersampled imagers. Optical Engineering, 44(1):14002, 2005.
33. Bijl, P., and Valeton, J.M., Triangle orientation discrimination: the alternative to MRTD and MRC. Optical Engineering, 37(7):1976–1983, 1998.
34. Young, S.S., Driggers, R.G., and Jacobs, E.L., Signal Processing and Performance Analysis for Imaging Systems. Norwood, MA, Artech House, 2008.
35. Anderson, D., Herman H., and Kelly A., Experimental characterization of commercial flash LADAR devices. In Proceedings of the International Conference of Sensing and Technology, 2005.

36. DeSouza, G.N., and Kak, A.C., A Subsumptive, Hierarchical, and Distributed Vision-Based Architecture for Smart Robotics, IEEE Transactions on Systems, Man, and Cybernetics – Part B: Cybernetics, 34(5):1988–2002, 2004.
37. Hirsh, R., DeSouza, G.N., and Kak, A.C., An Iterative Approach to the Hand-Eye and Base-World Calibration Problem, Proceedings of the IEEE International Conference on Robotics and Automation, Seoul, May 2001.
38. Hutchinson, S.A., Hager, G.D., and Corke, P.I, A tutorial on visual servo control. IEEE Transactions on Robotics and Automation, 12(5):651–670, October 1996.
39. Yoon, Y., Park, J., and Kak, A.C., A Heterogeneous Distributed Visual Servoing System for Real-time Robotic Assembly Applications, in Proceedings of the International Conference on Robotics and Automation, Orlando, Florida, 2006.
40. SMARTTRACK SENSOR, Automated Precision Inc., http://www.apisensor.com/PDF/Smart TrackeuDE.pdf, 2007.

Chapter 13
Quantification of Line Tracking Solutions for Automotive Applications

Jane Shi, Rick F. Rourke, Dave Groll, and Peter W. Tavora

Abstract Unlike line tracking in automotive painting applications, line tracking for automotive general assembly applications requires position tracking in order to perform assembly operations to a required assembly tolerance. Line tracking quantification experiments have been designed and conducted for a total of 16 test cases for two line tracking scenarios with three types of line tracking solutions: encoder based tracking, encoder plus static vision based tracking, and the analog sensor-based tracking for general assembly robotic automation. This chapter presents the quantification results, identifies key performance drivers, and illustrates their implications for automotive assembly applications.

13.1 Introduction

In automotive general assembly (GA), vehicle bodies are being carried on a continuous moving assembly line through hundreds of workstations where a variety of parts are being assembled together as shown in Fig. 13.1.

In order to robotically assembly parts onto a continuous moving vehicle body, the performance of current available line tracking solutions for automotive general assembly applications have to be well understood. For automotive general assembly applications, the vehicle body position at the point of assembly operations on a moving conveyor needs to be measured by encoders or other sensors for the duration of the part assembly so that robot can track the moving vehicle body for assembly operations. This requires that a specific position is to be tracked for a fixed time period with a required position tolerance.

In the past several months, a collaborative study has been conducted to thoroughly quantify the performance of three types of robotic line tracking solutions,

J. Shi (✉)
General Motors Company, Warren, MI, USA
e-mail: jane.shi@gm.com

R. Madhavan et al. (eds.), *Performance Evaluation and Benchmarking of Intelligent Systems*, DOI 10.1007/978-1-4419-0492-8_13,
© Springer Science+Business Media, LLC 2009

Fig. 13.1 Automotive general assembly conveyors transport vehicle bodies for general assembly operations

encoder-based, encoder plus static vision, analog laser sensor-based. This Chapter describes the quantification methodology, illustrates the experimental setup, summarizes the quantification results, and identifies two key performance drivers of line tracking for automotive assembly applications.

13.2 Quantifying Line Tracking Solutions

The operating principal of the robotic line tracking is for a robot to use a sensor's outputs to track a moving part as shown in Fig. 13.2.

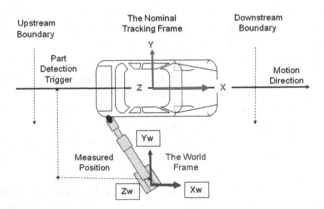

Fig. 13.2 Robotic line tracking

A world frame is specified for the robot location. A nominal tracking frame is set to be rigidly attached to the moving part. Once part detection is triggered, a sensor, such as an encoder, measures the part position and robot uses the sensor's reading to track the moving part.

In typical automotive robotic welding and material handling applications, robots move to fixed locations without any external sensor inputs. In the line tracking assembly applications, robots adjust their positions based on sensor's inputs to achieve the goal of tracking a moving vehicle body to their best capabilities. This is one rudimental form of robotic intelligent system. Often, the performance of such a robotic line tracking system is not given by solution vendors. Occasionally, the resolution of the sensor outputs is given as the system tracking accuracy. The system dynamic response characteristics and the application environment affect how well robots can track a moving part. Thus they should been considered for the system performance evaluation. When quantifying the performance of a line tracking solution. The following key technical issues need to be addressed:

1. How to measure the part position with respect to time?
2. How to measure the robot position with respect to time?
3. How to synchronize above two position measurements?
4. Which case of part movement should be evaluated?
5. How many of part movement cases should be evaluated?

The relative position between the robot and the moving part is critical to the success of the line tracking application when robot is in a tracking mode. It is valid to combine the issue (1) and (2) to measure the relative position between the robot and the moving part. This means that only one relative position needs to be evaluated with respect to time. In the actual evaluation experiments described in the next section, the relative position is measured with Dynalog's CompuGauge. At the same time, both part position and the robot position in the tracking mode are recorded and then relative position is computed with synchronized trigger. It has been determined that the computed robot position data is accurate enough for the evaluation purpose.

Application environment plays a critical role in deciding which and how many evaluation test cases for the part movement. Several key relationships we would like to determine:

- Does the tracked position change for the same constant moving speed reached by different acceleration?
- Does the tracked position change for different constant moving speeds?
- How does the tracked position change for an unstable moving speed?

A total of sixteen evaluation test cases are designed to cover three classes of motion profile including a variety of accelerations, an emergency stop case, and an unstable conveyor bounce motion as detailed in a later section.

13.3 Experimental Setup and Performance Data Collection Method

In the experimental setup for the line tracking evaluation, the moving vehicle body is emulated by a robot on one dimensional rail as illustrated by Fig. 13.3.

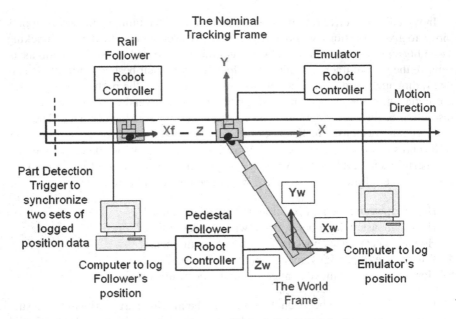

Fig. 13.3 Quantification Experimental Setup and Data Collection Method

Three main components of the experimental setup are as followings:

- *Emulator*: it is emulating a moving vehicle body on the assembly plant floor. A variety of motion profiles have been implemented by the Emulator as summarized in next section.
- *Rail Follower*: it is a robot transport unit (RTU) that utilizes one linear rail to follow the Emulator's motion.
- *Pedestal Follower*: it is a stationary six axes robot that utilizes its arm to follow the Emulator's motion.

Two types of tracking solutions have been evaluated: the rail tracking and the arm tracking:

- The rail tracking: the Rail Follower is tracking the Emulator with one linear rail motion.
- The arm tracking: the Pedestal Follower is tracking the Emulator with its 6DOF arm motion.

Both Emulator and Follower's positional data are collected using a position recording software provided by the robot controller. The collected data are then calculated to derive the relative position between the Follower and the Emulator. The computed relative position was validated with Dynalog's CompuGauge at the start of project. It has been determined that the computed relative position data is accurate to be within 5%. For all evaluation tests and their associated data graph, the recorded position data by the Follower and the Emulator are analyzed.

13.4 Quantification Test Cases

Three classes of vehicle motion profiles have been emulated by the Emulator for the quantification tests:

- Stable motion with a variety of accelerations
- E-Stop
- Actual measured conveyor bounce motion [1]

A total of 16 test cases are designed and conducted for both the rail tracking and the arm tracking scenarios. For high volume assembly plants, three line rates are used as representative cases: 40, 60, and 80 in Jobs Per Hour (JPH) which translates to conveyor speed in mm per second (mm/s) The test cases are organized for three stable constant speeds (60,120, and 160 mm/s) with four different acceleration profile (rising edge of the speed chart) plus an emergency stop cases (acceleration with very high negative number). In addition, one special test case is included that is based on the actual measured conveyor bounce motion.

Table 13.1 summarizes all test cases with a speed profile graph that illustrates the constant top speed and four accelerations and one emergency stop deceleration.

13.5 Quantification Results of the Encoder Based Line Tracking Solution

For the encoder based line tracking, the encoder is connected to the robot controller that drives the Follower as illustrated in Fig. 13.4.

A set of gear boxes drive the encoder while the Emulator is moving per the specification of the speed profile in each test case. With a properly selected gear ratio, the resolution for the encoder is set to be at least 500 counts per mm. This represents minimum resolution of 0.002 mm per count. The encoder digital counts are applied to the Follower in both rail and arm tracking cases every 0.012 s.

A digital I/O handshaking signal between the Emulator and the Follower simulates the part detection trigger to start the line tracking. In the data collection process, the trigger signal is emulated by a digital handshaking between the Emulator and Follower to start recording robot position data at the same time. Thus the relative position between the Emulator and the Follower can be computed during the data analysis.

13.5.1 Rail Tracking Results

All test cases are conducted for the encoder-based rail tracking and repeated at least three times to validate the repeatability. The position lag errors are plotted in Figs. 13.5–13.7 for the constant conveyor motion cases.

Table 13.1 Evaluation test cases

Line rate (JPH) conveyor constant speed (mm/s)	Acceleration (mm/s²)	Speed profile (mm/s vs. second)
40 JPH 60 mm/s	150 320 650 745 −11,190	
60 JPH 120 mm/s	150 320 650 745 −11,190	
80 JPH 160 mm/s	150 320 650 745 −11,190	
40 JPH 60 mm/s	Bounce motion with various magnitudes	

For the constant motion cases, several observations can be made of encoder based rail tracking results:

- Lag error is proportional, although not exactly linearly, to the stable top speed.
- Higher acceleration does not increase the total lag error: it seems to have reduced the total lag error slightly (about 1 mm) for the rail tracking.
- And all test cases are repeatable within 1 mm.

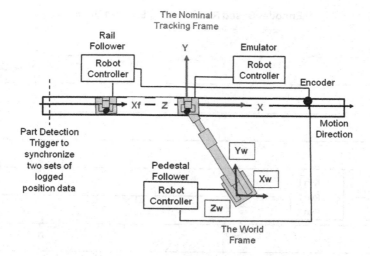

Fig. 13.4   Encoder connects to the robot controller of the rail or pedestal follower

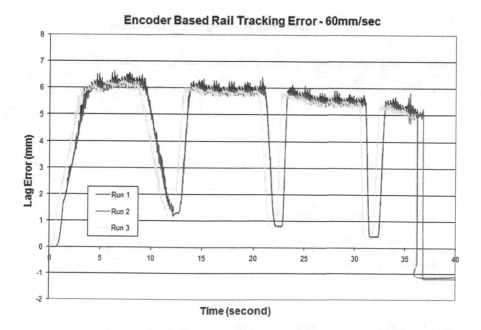

Fig. 13.5   Position lag error for the encoder based rail tracking: 60 mm/s case

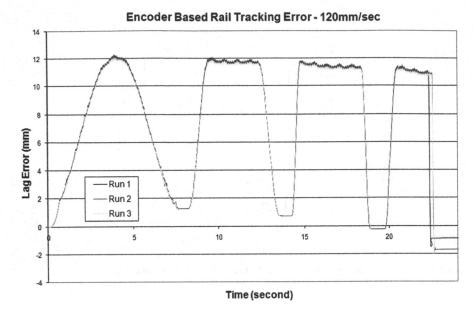

Fig. 13.6 Position lag error for the encoder based rail tracking: 120 mm/s case

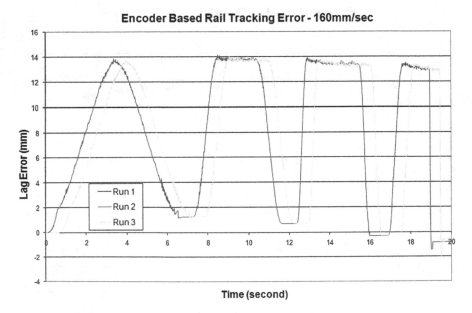

Fig. 13.7 Position lag error for the encoder based rail tracking: 160 mm/s case

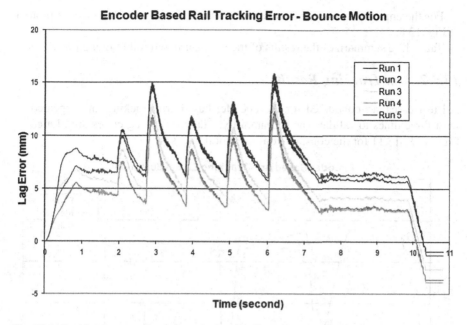

Fig. 13.8 Position lag error for the encoder based rail tracking: emulated conveyor bounce case

Table 13.2 Encoder (0.002 mm resolution) based rail tracking performance

Line rate (JPH) conveyor constant speed (mm/s)	Acceleration (mm/s²)	Range of position tracking error (mm)	Repeatability (mm)
40 JPH 60 mm/s	150 320 650 745	< 6.8 mm	< 0.5 mm
	−11,190 (emergency stop)	Overshoot −1.2 mm	< 0.5 mm
60 JPH 120 mm/s	150 320 650 745	< 12.5 mm	< 0.5 mm
	−11,190 (emergency stop)	Overshoot −2.1 mm	< 0.5 mm
80 JPH 160 mm/s	150 320 650 745	< 14.5 mm	< 0.5 mm
	−11,190 (emergency stop)	Overshoot −2.1 mm	< 0.5 mm
40 JPH 60 mm/s	Bounce motion with various magnitudes	−3.8 mm ∼ 16.0 mm	< 4.0 mm

For the emulated conveyor bounce motion case, the position lag error is plotted in Fig. 13.8.

Table 13.2 summarizes the results of the encoder based rail tracking test cases.

13.5.2 Arm Tracking Results

All test cases are conducted for the encoder-based arm tracking and repeated at least three times to validate the repeatability. The position lag errors are plotted in Figs. 13.9–13.11 for the constant conveyor motion cases.

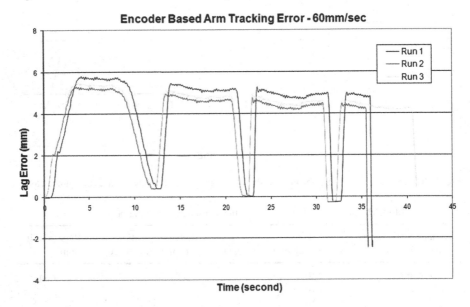

Fig. 13.9 Position lag error for the encoder based arm tracking: 60 mm/s case

For the constant motion cases, similar observations can be made of encoder based arm tracking results:

- Lag error is proportional, although not exactly linearly, to the stable speed.
- Higher acceleration does not increase the total lag error: it seems to have reduced the lag error slightly (about 1 mm) for the arm tracking.
- And all test cases are repeatable within 2 mm.

For the emulated conveyor bounce motion case, the position lag error is plotted in Fig. 13.12. The arm tracking error is within +16 and −3 mm with a repeatability of about 5 mm. Table 13.3 summarizes the results of the encoder based arm tracking test cases.

Compared with the encoder based rail tracking, the arm tracking has

- similar position lag error,
- slightly worse repeatability, and
- worse emergency stop performance.

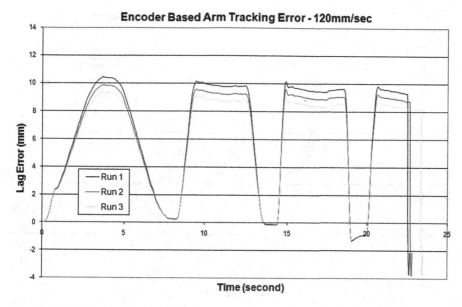

Fig. 13.10 Position lag error for the encoder based arm tracking: 120 mm/s case

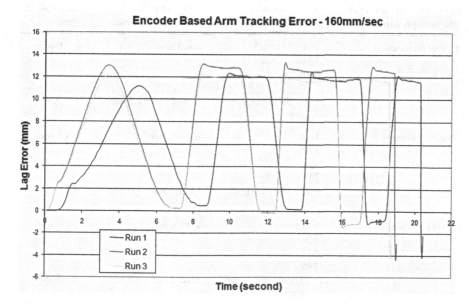

Fig. 13.11 Position lag error for the encoder based arm tracking: 160 mm/s case

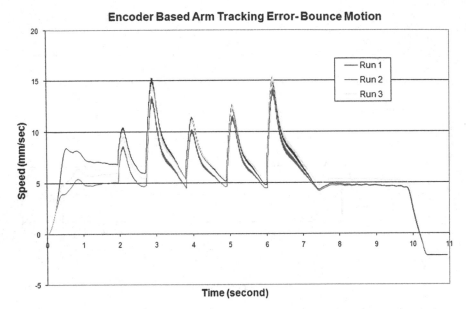

Fig. 13.12 Position lag error for the encoder based arm tracking: emulated conveyor bounce case

Table 13.3 Encoder (0.002 mm resolution) based arm tracking performance

Line rate (JPH) conveyor constant speed (mm/s)	Acceleration (mm/s^2)	Range of position tracking error (mm)	Repeatability (mm)
40 JPH 60 mm/s	150	< 6.0 mm	< 1.0 mm
	320		
	650		
	745		
	−11,190 (emergency stop)	Overshoot −2.5 mm	< 1.0 mm
60 JPH 120 mm/s	150	< 11.5 mm	< 1.5 mm
	320		
	650		
	745		
	−11,190 (emergency stop)	Overshoot −4.0 mm	< 1.0 mm
80 JPH 160 mm/s	150	< 13.5 mm	< 2.0 mm
	320		
	650		
	745		
	−11,190 (emergency stop)	Overshoot −4.5 mm	< 1.0 mm
40 JPH 60 mm/s	Bounce motion with various magnitudes	−3.0 mm ∼ 16.0 mm	< 5.0 mm

13.6 Quantification Results of the Encoder Plus Static Vision Line Tracking Solution

When a static vision system is used for the line tracking, the Follower uses the detected object location to correct its position relative to the object. This one time correction of robot's position based on the vision results is "static" compared with the use of vision results at a fixed time period to track an object's position [2]. The encoder counts are the sensor inputs that drive the Follower to track the Emulator.

Figure 13.13 illustrates the setup of the vision system used in the line tracking evaluation. The vision camera is attached to the Pedestal Follower and the vision grid object is attached to the Emulator so that there is no relative motion when the Follower is tracking the Emulator at the constant speed.

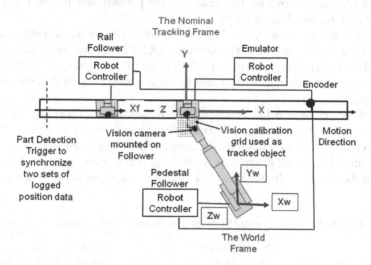

Fig. 13.13 Setup of vision system for line tracking evaluation

Table 13.4 lists the vision system setup used in the evaluation testing.

The following procedures are implemented to quantify the static vision use in the encoder based line tracking:

(1) *Before* the Emulator starts to move, the Arm Follower is centered with the vision calibration grid so that the vision can use it to "set reference". This means that the vision snap and find operation will return a zero offset.

Table 13.4 Vision system parameters

Camera lens (mm)	Camera standoff distance (mm)	The grid object dimension (mm × mm)	Pixel resolution (mm/pixel)	Object resolution (mm)
8	640	320 by 320	0.599	0.15

(2) Enter a static 2D offset, (x1, y1, R1), to the Emulator robot so that the robot arm moves the specified offset *before* the Emulator starts moving.

(3) *After* the Emulator reaches its *constant* speed, the camera mounted on the Pedestal Follower captures the image of the vision calibration grid. The second vision snap and find operation determines the vision offset (x2, y2, R2).

(4) Apply the vision offset (x2,y2,R2) from step 3 to the Pedestal Follower so that the Pedestal Follower compensates the offset value.

(5) Verify the applied vision offset: the camera mounted on the Follower captures the image of the vision calibration grid again. The third vision snap and find operation determines the offset (x3, y3, R3) that, in theory, should be zero if compensated correctly and perfectly.

Table 13.5 summarizes the test cases and the performance for the static vision with the encoder based line tracking.

The first three columns (X1,Y1,R1) are the static offset of the grid object applied by the Emulator in step (2). The columns (X2,Y2,R2) are the object position detected by the vision system in step (3). The last three columns (X3,Y3,R3) are the object position detected by the vision systems in step (5) after the correction has been applied in step (4).

The shaded column in Table 13.5, X2–X1, is the difference between the detected X offset and the applied X offset. The values in this column are the relative position lag detected by the vision when the Emulator is traveling at the steady state speed of 114 mm/s. This is this steady state position lag that is proportional to the constant speed of the Emulator. In this case (114 mm/s), it is averaged at 8.35 mm with one

Table 13.5 Static vision (0.15 mm resolution) arm tracking evaluation results

Constant speed (mm/s)	Offset applied			Offset detected by vision at constant Speed				Vision offset after compensating detected offset		
	X1 (mm)	Y1 (mm)	R1 (deg)	X2 (mm)	X2+X1 (mm)	Y2 (mm)	R2 (deg)	X3 (mm)	Y3 (mm)	R3 (deg)
114	0	0	0	−8.4	−8.4	−0.9	0	1.3	0	0
	0	0	0	−8.4	−8.4	−0.8	0	−1.2	0	0
	0	0	0	−8.4	−8.4	−0.8	0	−1.2	0	0
	−5	0	0	−3.3	−8.3	−0.9	0	−0.5	0.1	0
	5	0	0	−13.3	−8.3	0	0	−1.9	0	0
	−15	0	0	6.7	−8.3	−0.8	0	0.7	−0.1	0
	15	0	0	−23.3	−8.3	−1.1	0	−3.1	−0.1	0
	10	−15	3	−18.6	−8.6	13.9	3.1	−2.2	2.2	0
	10	−15	3.1	−18.6	−8.6	13.9	3.0	−2.2	2.2	0
	−10	15	−2	1.9	−8.1	−15.8	−2	−0.2	−2.4	0
	−10	15	−2	1.8	−8.2	−15.8	−2	−0.1	−2.3	0

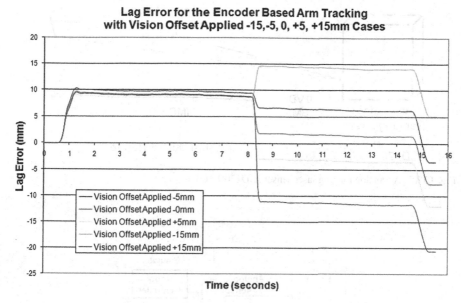

Fig. 13.14   Corrected position is initial object offset plus static lag

standard derivation of 0.15 mm. Figure 13.14 plots the position lag between the Emulator and the Pedestal Follower for the first seven rows in Table 13.5.

Detailed analysis of the collected robot position data reveals that the corrected position is the combination of the static position lag (caused by the motion) and the initial object position offset. How accurately the robot can compensate the position is a different evaluation question that could be a separate investigation study. The rough indicator is the last three column (X3,Y3,R3) in Table 13.5.

13.7 Quantification Results of the Analog Laser Based Line Tracking Solution

An analog ranging laser can be used in the place of an encoder where mounting an encoder is not feasible in the line tracking applications. The analog laser ranging sensor, ILD1700-100 [3], uses the optical triangulation principle to measure the distance between the laser sensor and the smooth surface target as shown in Fig. 13.15. The measurement range is between 70 mm the Start of Measuring Range (SMR) and 170 mm the End of Measuring Range (EMR). The midrange (MMR) is used for the line tracking application. This means that the analog laser signal drives the Rail or Pedestal Follower to maintain a constant distance with the Emulator. Customized software has been developed to use the analog laser sensor for the line tracking function.

Figure 13.16 illustrates the setup for the analog laser based line tracking. While the Emulator is moving, the analog laser measures the distance between the Emulator and the Follower. This analog laser sensor based line tracking method is

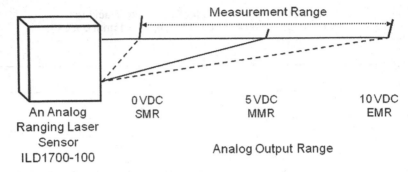

Fig. 13.15 An analog ranging laser sensor ILD1700-100

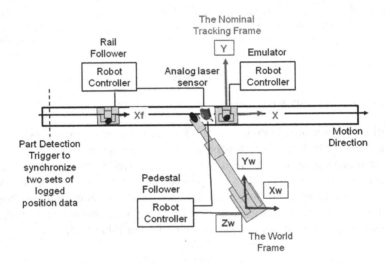

Fig. 13.16 Analog laser based line tracking solution

only concerned with one dimension: the conveyor main travel direction. It does not require the sensor to recognize any object geometry feature as in a traditional tracking method. Any flat reflecting surface can be used for the laser signal to return to its receiver. The goal is to keep the laser reading constant by driving the Follower to track the Emulator's position.

In order to use the analog laser sensor to track the Emulator, additional system components have to be set up. The analog signal is first digitized by an analog to digital (A/D) converter. Then the digital signal goes through a DeviceNet adaptor to interface with the DeviceNet module inside the robot controller as shown in Fig. 13.17.

A scan rate can be set for the device net interface module at 12 ms. This means that the digitized analog sensor can be read by the robot controller every 12 ms. The line tracking software then utilizes the analog sensor inputs every scan period. There

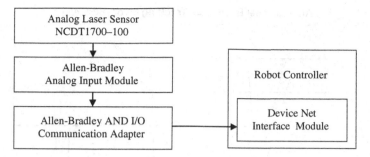

Fig. 13.17 Analog laser ranging sensor connection to a robot controller

is an unknown system time delay between the sampled sensor reading and the use of such reading by the robot controller to track the Emulator.

The analog laser sensor has measurement resolution of 0.0006 mm. Experiments have shown that the stable reading of the analog laser is around 0.0012 mm when the analog sensor and the measured surface are both stationary.

13.7.1 Analog Sensor Based Rail Tracking Results

All test cases are conducted for the analog sensor based rail tracking and repeated at least three times for validating the repeatability. The position lag errors are plotted in Figs. 13.18–13.20 for the constant conveyor motion cases.

For the constant motion cases, several observations can be made of analog sensor based rail tracking results:

- Lag error is proportional to the top stable speed.
- Higher acceleration causes significant overshoot before the Rail Follower can stabilize its speed to match the Emulator speed as illustrated in Fig. 13.21.

For the emulated conveyor bounce motion case, position lag error is plotted in Fig. 13.22.

Compared with the encoder based rail tracking result shown in Fig. 13.8, The analog laser based rail tracking exhibits significant overshoot and oscillation.. Further examination of the Rail Follower speed identifies the root cause of the large position lag error as shown by Fig. 13.23. With the analog sensor's inputs, the Rail Follower cannot closely follow the Emulator speed.

Table 13.6 summarizes the results of the analog laser based rail tracking test cases.

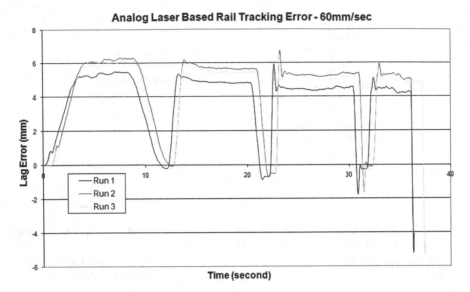

Fig. 13.18 Position lag error for analog sensor based rail tracking: 60 mm/s case

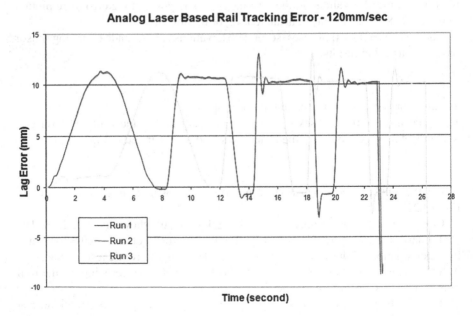

Fig. 13.19 Position lag error for analog sensor based rail tracking: 120 mm/s case

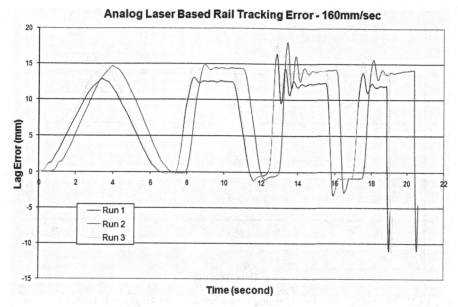

Fig. 13.20 Position lag error for analog sensor based rail tracking: 160 mm/s case

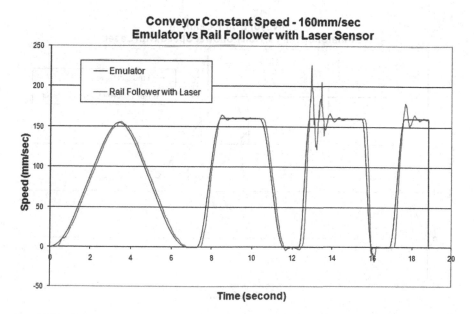

Fig. 13.21 The rail follower overshoots before it stabilizes in high acceleration case

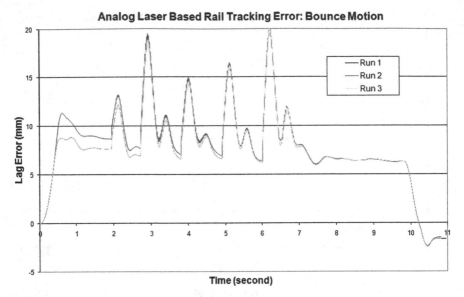

Fig. 13.22 Position lag error for the analog sensor based rail tracking: emulated conveyor bound motion case

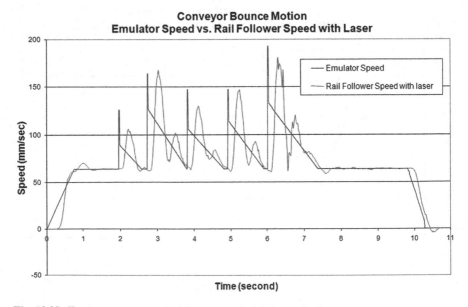

Fig. 13.23 Emulator speed vs. rail follower speed with the analog laser

Table 13.6 Analog laser (0.0012 mm resolution) based rail racking performance

Line rate (JPH) conveyor constant speed (mm/s)	Acceleration (mm/s^2)	Range of position tracking error (mm)	Repeatability (mm)
40 JPH 60 mm/s	150	< 7.0 mm	< 2.5 mm
	320		
	650		
	745		
	−11,190 (emergency stop)	Overshoot −5.5 mm	< 1.0 mm
60 JPH 120 mm/s	150	< 14.0 mm	< 2.0 mm
	320		
	650		
	745		
	−11,190 (emergency stop)	Overshoot −8.0 mm	<1.0 mm
80 JPH 160 mm/s	150	< 18.5 mm	< 4.0 mm
	320		
	650		
	745		
	−11,190 (emergency stop)	Overshoot −12.0 mm	< 2.0 mm
40 JPH 60 mm/s	Bounce with various magnitude	−3.0 mm ~ 21.0 mm	< 5.0 mm

Compared with the encoder based rail tracking of Table 13.2, the rail tracking with the analog laser has

- larger position lag error for constant motion speeds,
- worse repeatability,
- significantly worse emergency stop performance, and
- significantly larger position lag error for the dynamic motion speed.

13.7.2 Analog Sensor Based Arm Tracking Results

All test cases are conducted for the analog sensor based arm tracking and repeated at least three times for validating the repeatability. The position lag errors are plotted in Figs. 13.24–13.26 for the constant conveyor motion cases.

For the constant motion cases, similar observations can be made of encoder based arm tracking results:

- Lag error is proportional to the top stable speed.
- Higher acceleration causes significant overshoot before the Arm Follower can stabilize its speed to match the Emulator speed as illustrated in Fig. 13.27.

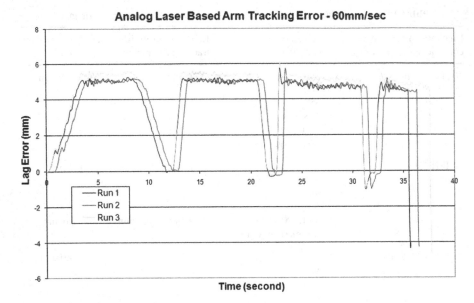

Fig. 13.24 Position lag error for the analog sensor based arm tracking: 60 mm/s case

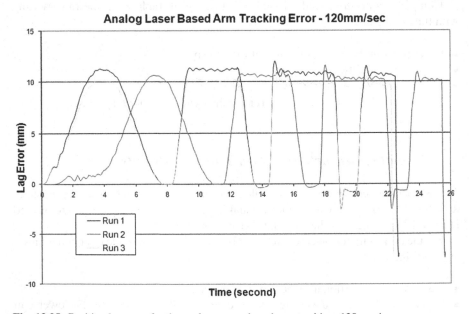

Fig. 13.25 Position lag error for the analog sensor based arm tracking: 120 mm/s case

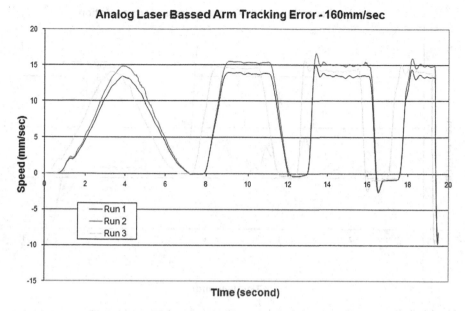

Fig. 13.26 Position lag error for the analog sensor based arm tracking: 160 mm/s case

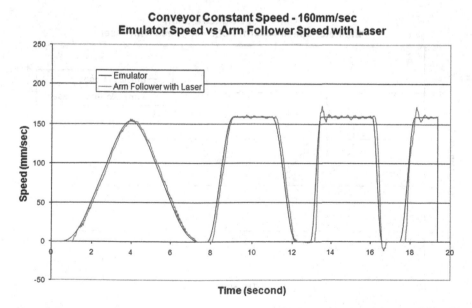

Fig. 13.27 Emulated speed vs. arm follower speed with the analog laser sensor

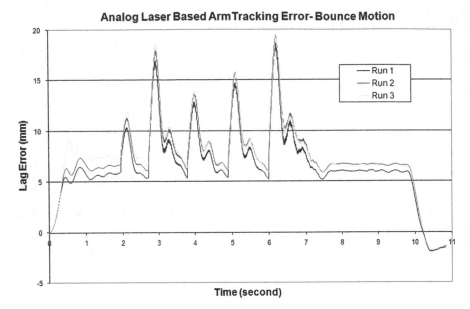

Fig. 13.28 Position lag error for the analog laser sensor based arm tracking: emulated conveyor bound motion case

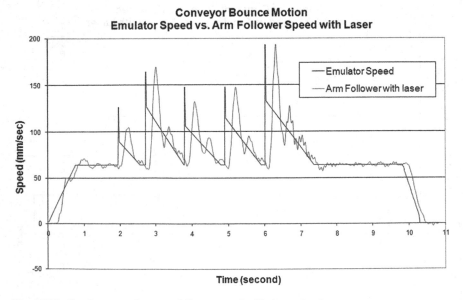

Fig. 13.29 Emulator speed vs. arm follower speed with the analog laser sensor

Table 13.7 Analog laser (0.0012 mm resolution) based arm tracking performance

Line rate (JPH) conveyor constant speed (mm/s)	Acceleration (mm/s²)	Range of position tracking error (mm)	Repeatability (mm)
40 JPH 60 mm/s	150	< 6.5 mm	< 1.0 mm
	320		
	650		
	745		
	−11,190 (emergency stop)	Overshoot −5.0 mm	< 1.0 mm
60 JPH 120 mm/s	150	< 12.0 mm	< 2.0 mm
	320		
	650		
	745		
	−11,190 (emergency stop)	Overshoot −7.5 mm	< 1.0 mm
80 JPH 160 mm/s	150	< 17.0 mm	< 2.0 mm
	320		
	650		
	745		
	−11,190 (emergency stop)	Overshoot −10.0 mm	< 1.0 mm
40 JPH 60 mm/s	Bounce with various magnitude	−2.0 mm ~ 19.0 mm	< 6.0 mm

For the emulated conveyor bounce motion case, the position lag error is plotted in Fig. 13.28. Compared with the encoder based arm tracking results shown in Fig. 13.12, the analog laser based arm tracking exhibits significant overshoot and oscillation. Further examination of the Arm Follower speed identifies the root cause of the large position lag error as shown by Fig. 13.29. With the analog sensor's inputs, the Arm Follower can not closely follow the Emulator speed.

Table 13.7 summarizes the results of the analog laser sensor based arm tracking test cases.

Compared with the encoder based arm tracking of Table 13.3, the arm tracking with the analog laser has

- larger position lag error for constant motion speeds,
- worse repeatability,
- significantly worse emergency stop performance, and
- significantly larger position lag error for the dynamic motion speed.

13.8 Conclusion and Automotive Assembly Applications

Unlike the line tracking for automotive paint applications where the speed match between the robot and the vehicle body plays critical role for the paint quality, the line tracking for automotive general assembly applications requires position tracking of the system to perform assembly operations within a required assembly tolerance for a period of time. Two simple line tracking performance metrics are used here for evaluating the robotic line tracking solutions (1) the range of relative positional tracking error between the robot and the vehicle body in the main direction of travel, and (2) the repeatability of the relative positional tracking error as measured in three standard deviations. These metrics are two key measures that are indicators of how well a robot can track a moving vehicle body for a given application environment. Ideally, comprehensive metrics of the robotic line tracking performance are the relative position tracking error between the robot and the moving vehicle body in a three dimensional space as measured by six degree of freedom (6DOF) and its repeatability.

The line tracking capabilities, as quantified by the results highlighted in Section 13.5–13.7, varies in a wide range. A highly-accurate tracking (<3 mm) can be achieved for a smooth and steady conveyor using encoder and vision to compensate the steady state offset. A very low performance tracking (about 21 mm) occurs for a dynamic and rough conveyor motion with an analog laser. By comparing and contrasting experimental results, key performance drivers for the robotic line tracking can be concluded as follows:

(1) Sensor resolution: it limits how tight the robot system can track the conveyor line motion.
(2) Actual time delay of applying the sensor data: it limits how dynamically responsive the robot system can respond to the change in the conveyor line motion.

Many robotic general assembly applications can be implemented on a moving assembly line. Examples are seat load, wheel and tire (W&T) install, body chassis marriage (BCM), glass install, fastening, weather strip (WS) install, body side (BS) molding install, emblem install, instrument panel (IP) load, as well as door assembly. Some of these applications only install one part while others are more complex that involves sub-assembly components. Even for the same application, the assembly environment can be different for different vehicle programs or even within the same vehicle program in different assembly plants. First, the assembly tolerance may be different on different vehicle platforms. Second, different conveyance methods and speeds are used at different assembly plants. Third, vehicle carrier that sits on the top of conveyor and carries the vehicle body can be different at different assembly plants. Finally the workstation layout is different at different assembly plant. For example, for the seat load application, one assembly plant requires robots to track vehicle body tightly so that the seats can be placed onto specific stud locations within the tolerance of 0.5 in. while vehicle body is transported by the most rough conveyance system. At another assembly plant, the seat load requires a loose assembly tolerance (<1.0 in.) while the vehicle body is being transported on a conveyor with stable and smooth motion. For this reason, the quantification results can provide the first cut of assessment of possibility of achieving required assembly tolerance for a chosen line tracking solution given the assembly

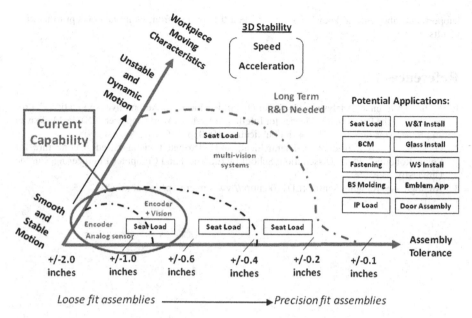

Fig. 13.30 Current line tracking capabilities and application environment

environment. Also, the results can assist in selecting a proper tracking solution for a given assembly requirement when a solution is feasible. Figure 13.30 is a simple illustration of current line tracking capabilities in terms of required assembly tolerance and conveyor motion characteristics. The horizontal axis illustrates gradually tighter assembly tolerance as it goes from the left to the right while the vertical axis illustrates gradually unstable and more dynamic motion characteristics as it goes from the bottom to the top. It is obvious that current line tracking capabilities only cover the bottom left portion of application environment where loose assembly tolerance and a smooth and stable conveyor motion occurs. Apparently there is a capability void where long term research and development (R&D) is needed.

When considering a line tracking solution for general assembly, all relevant factors, such as assembly tolerance and the conveyor motion characteristics need to be thoroughly examined with the current line tracking capabilities. By understanding current capabilities of line tracking solutions, appropriate robotic automation systems can be developed and designed for a variety of GA applications. Key fundamental system drivers can be specified for robots to track the moving conveyor accurately enough for the specific assembly tolerance plus the environment uncertainty with adequate dynamic response at the appropriate location for the general assembly automation.

Acknowledgments The authors would like to acknowledge Bob Scheuerman, Marty Linn, Marjorie Winston, Thomas McGraw, Ke Zhang-Miske, Brooks A. Curtis, Ronald Reardon, Loring Dohm, Michael Poma, Neil McKay, Charles W. Wampler, and Roland Menassa. Their inputs, support, contribution, and knowledge have shaped this project from its initial concept to its final results.

References

1. J. Shi, "Preliminary Analysis of Conveyor Dynamic Motion for Automotive Applications", Proceedings of the Performance Metrics for Intelligent Systems Workshop (PerMIS), R. Madhavan and E. Messina (eds.), NIST Special Publication 1090, p. 156–161, August 2008.
2. G. DeSouza, "A Subsumptive, Hierarchical, and Distributed Vision-based Architecture for Smart Robotics", Ph.D. Dissertation, School of Electrical and Computer Engineering, Purdue University, May 2002.
3. Analog Laser Ranging Sensor ILD1700 http://www.micro-epsilon.com.